BIOLOGY:

A LABORATORY GUIDE TO THE NATURAL WORLD

Second Edition

Dennis J. Richardson and Kristen E. Richardson
Quinnipiac University

PEARSON

Benjamin
Cummings

San Francisco Boston New York
Cape Town Hong Kong London Madrid Mexico City
Montreal Munich Paris Singapore Sydney Tokyo Toronto

Editor-in-Chief, Science: John Challice
Executive Editor: Gary Carlson
Project Manager: Crissy Dudonis
Vice President of Production & Manufacturing: David W. Riccardi
Executive Managing Editor: Kathleen Schiaparelli
Assistant Managing Editor: Becca Richter
Production Editor: Elizabeth Klug
Supplement Cover Manager: Paul Gourhan
Supplement Cover Designer: Joanne Alexandris
Manufacturing Buyer: Ilene Kahn
Electronic Composition and Formatting: Laserwords
Cover Photo Credit: Bird-of-Paradise (Strelitzia Reginae) Molokai, Hawaii, close up @Getty Images/Image Bank

Printed in the United States of America

10 9 8 7 6 5 4

ISBN: 978-0-13-144935-0
       0-13-144935-4

Pearson Education Ltd., London
Pearson Education Australia Pty., Limited, Sydney
Pearson Education Singapore, Pte. Ltd.
Pearson Education North Asia Ltd., Hong Kong
Pearson Education Canada, Ltd., Toronto
Pearson Educación de Mexico, S.A. de C.V.
Pearson Education—Japan, Tokyo
Pearson Education Malaysia, Pte. Ltd.

To Kate, Maggie, and Emma
Our favorite little varmints

and

To Our Parents, Bill & Rita, Margie & Roger

Contents

Preface

We often hear the argument that nonmajor students shouldn't have to learn much about biology because it seems unimportant in the real world. Yet, we humans are animals– linked to all other life on this planet, governed by the same natural laws as everything else. A biology laboratory course gives you a better understanding of yourself and reveals your relationship to all other living organisms. As we face increasing disease outbreaks, energy crises, and environmental mismanagement, an understanding of biology has never been more important.

And so, with nonmajors in mind, we present *Biology: A Laboratory Guide to the Natural World.*

The exercises in this guide accompany David Krogh's *Biology: A Guide to the Natural World*, 3rd edition. This manual is built upon the solid foundation of a traditional approach, but it also provides new avenues for you to explore the natural world, through web-based exercises and a touch of inquiry-based investigation. These exercises encompass many levels of biology with a touch of aesthetic and philosophical aspects of natural history.

We hope you will walk away with a new understanding of the natural world around you, a world of wonder and beauty.

D.J.R. and K.E.R.

REVIEWERS

Mike Barger, *Peru State College*
David Byres, *Florida Community College at Jacksonville, South Campus*
Patricia Cox, *University of Tennessee*
Isaure deBuron, *College of Charleston*
Deborah Dodson, *Vincennes University*
Christine M. Foreman, *University of Toledo*
Carl S. Frankel, *Pennsylvania State University, Hazelton*
Carolyn Glaubensklee, *University of Southern Indiana*
Catherine Hurlbut, *Florida Community College at Jacksonville*
Julie Schroer, *Bismarck State College*
George Sideris, *Long Island University*
Kelly Sullivan, *Albuquerque Tech Vocational Institute*
Janet L. Vigna, *Southwest State University*
M. Eva Weicker, *Pennsylvania State University*
Jamie Welling, *South Suburban College*
Catherine J. Wong, *Quinnipiac University*

ACKNOWLEDGMENTS

We wish to extend our sincere thanks to everyone who helped us complete this book: Elizabeth Klug and Crissy Dudonis at Prentice Hall for their professional advice; Dan Holiday for helping prepare many of the digital images; Rich Clopton and several reviewers for contributing ideas for exercises; Susan Jessup and Barbara Nitchke, who provided line drawings; Nick Aliano, Dan Benesh, Delia Buerstetta, Avery J. Dennis, and Kyle McWhirter at the University of Nebraska-Lincoln, who helped compile specimens for Appendix 3; and Harvey Levine, whose invertebrate collection at Quinnipiac University was used for many of the images in Appendix 3. Thanks also to our mentors Wilbur Owen, Brent Nickol, John Janovy, Jr., the late Mike Mathis, William Glider, and Neal Buffaloe, who provided us with an insightful view of life while teaching us how to teach.

We are especially grateful to our children, Kate, Maggie, and Emma. They have patiently tolerated us while growing up with this project, even through the second edition. We owe them for providing much-needed laughter and encouragement.

1

Science as a Way of Learning

We start off confused and end up confused on a higher level.
A.E. Chalmers, 1976

Here you are, in a biology course that you've put off as long as possible. Does the word *science* give you an uneasy feeling? Science can be overwhelmingly complex, but learning about it doesn't have to be. A famous scientist once said that science is "nothing but trained and organized common sense" *(Huxley, 1854)*. Take it one step at a time, and you will see how the pieces fit together.

Start by observing something about the living world around you. You may notice a vine-like morning glory climbing on a fence, for example (Figure 1.1). Then ask a question. Why do the flowers of a morning glory bloom (open) only for a few hours each day? By asking a question, you are already on your way to understanding biology. You may not realize how many steps you take just to ask a question: you look at the plant on several occasions, and you observe how the flowers are open in the morning, but closed later in the day.

Why do they bloom only for a few hours? The steps you take to answer the question will flow in a logical manner. You think of some factors that may affect blooming. Do the flowers open only on sunny days? Does blooming depend on the temperature? Does humidity play a part in blooming? You decide to answer your question by investigating one of the possibilities. If blooming is temperature-dependent, then the flowers should

FIGURE 1.1   Morning Glory.

1

open as the temperature climbs and close as higher temperature approaches.

After observing the flowers for a few days, you may see a pattern that is consistent with the temperature-dependent approach. Or you may see a pattern that isn't. In either case, you could start the investigation over to learn additional information. If the flowers bloom in a temperature-dependent manner, you can investigate why they might do so. Or if temperature doesn't seem to have an effect on blooming, you can eliminate that possibility and investigate another.

In this manner, we do indeed start off confused (with a question) and end up confused (with another question) on a higher level! Much of biology is discovered with this approach, called the **scientific method [Krogh section 1.2]**. As you can see, it is only organized common sense, but it can lead to fascinating knowledge.

THE SCIENTIFIC METHOD

How do we know what we know? The study of the acquisition of knowledge is called **epistemology**—the very root of our methods of learning. Four major intellectual endeavors have been pursued by human cultures throughout history—religion, philosophy, art, and science. Let's examine how science differs from these other three intellectual pursuits.

Religion is defined as the service and worship of God or the *supernatural*; it is a commitment or a devotion to *religion or faith*.

Philosophy is the pursuit of wisdom; a search for a general understanding of values and reality by chiefly *speculative* rather than observational means; an analysis of the concepts expressing fundamental *beliefs*.

Art is the conscious use of skill and *creative imagination*, particularly regarding aesthetic objects. Aesthetics is a branch of philosophy dealing with the nature of beauty, the quality in something that gives pleasure to the senses or pleasurably exalts the mind or spirit.

Science is knowledge covering natural phenomena or the operation of general laws *as obtained and tested through the scientific method*.

Science differs from the other three intellectual pursuits in that it is grounded in **empiricism** (study of phenomena that can be seen or measured), whereas the other three tend to be more abstract or subjective. This fundamental difference does not make science better than religion, art, or philosophy. On the contrary, it could be argued that this difference places extreme limitations upon science, because none of the others have empirically defined boundaries. Yet, empiricism gives great strength to science because it enables hypotheses to be tested objectively.

The morning glory example presented in the introduction fits into the steps of the scientific method.

1   **Observation** You observe that *morning glory blooms only during certain hours of the day*. In many experiments, this step includes personal observation as well as reading what is already published. Such pieces of information are sometimes referred to as *facts*.
    In science, a *fact* refers to a piece of information; this is unlike the vernacular (everyday) use, where *fact* refers to a statement of truth, or something that is undeniable.

2   **Question** The question that follows is, *Why do the blooms close at midday?*

3   **Hypothesis** A hypothesis is a tentative, general statement based on specific observations. You have several hypotheses: *Sunlight affects blooming. Temperature affects blooming. Humidity affects blooming.*

4   **Prediction** Predictions are logical conclusions to be expected in view of the hypothesis. They may be phrased in the form of an if-then statement: If the hypothesis is correct, then a certain outcome is expected. Predictions provide opportunities to find independent pieces of evidence supporting a hypothesis, as well as opportunities to reject a faulty hypothesis. From the temperature hypothesis, this prediction follows: *If blooming is temperature-dependent, then the flowers should open as the temperature climbs and close as higher temperature approaches.*

5   **Experiments** Predictions are tested by observation and/or experimentation. If you conduct your experiment over a few days, you may find that *blooming is temperature-dependent*. Or you may not.

6   **Conclusion** New data are the product of observation and experimentation. *The new data will either support or fail to support the hypothesis.* (*Datum* is singular, and *data* is plural).

A hypothesis can never be proven. To claim that you have proven a hypothesis is to claim that you have a hold on absolute truth. Within the conceptual boundaries of science, absolute truth is unattainable. This is not to say that absolute truth does not exist, just that it can never be known in regard to science. It is most appropriate for a scientist to use terms like *data* rather than *facts* and *principles* rather than *truths*. It is better for a scientist to say that data *suggest* than to say that data *prove* something.

Why is a scientific discovery not "absolute truth?" There are several reasons. One is because hypotheses are based on observations and data from a particular point in time. We are continuously learning more about the world around us, and, sometimes, new information pushes old conclusions aside. You may say "Eureka! I have proven that blooming in morning glories is temperature-dependent!" But you'll be scratching your head if someone else finds that the flowers bloom even in colder (yet very sunny) weather next week. And so, you'll have to show your objectivity, step aside, and consider their findings.

Once a hypothesis has amassed a great amount of supporting evidence, it may be afforded the status of theory. A theory is a general set of principles that explain some aspect of nature. Note that a theory is *not* a hypothesis that has been proven. It is a hypothesis with a tremendous amount of supporting evidence.

In scientific use, a *theory* is a widely accepted paradigm, or model, that is supported by a tremendous amount of evidence. In vernacular use, *theory* refers to an idea, often with no more validity than a guess.

Exercise 1.1   EXPERIMENTAL DESIGN

In designing an experiment, several variables are defined:

The **independent variable** is the variable that is manipulated. *You manipulate the temperature of the plant's atmosphere.*

The **dependent variable** is the variable that will be measured. *You measure plant blooming, noticing the temperature at which the plant blooms.*

Controlled variables are kept constant throughout the experiment. *The soil type,*

amount of water applied, and amount of light are kept constant.

The **experimental group** contains plants that are subjected to manipulation. *Morning glory plants in this group will be subjected to rising temperature.*

The **control group** contains plants held under standard conditions, in which the independent variable is not manipulated. The control group usually approximates "normal" conditions. However, these plants should be grown in the lab to eliminate effects of humidity and plant growth, variables we do not want to test in this experiment. *Morning glory plants in this group will be exposed to constant temperature.*

Read this very carefully: When comparing the experimental group to the control group, the effects of the independent variable can be determined by measuring the dependent variable in both groups. *You compare blooming in the experimental group to blooming in the control group. Does temperature appear to have an effect on blooming? If yes, then the data support the hypothesis. If no, then the hypothesis is rejected.*

When scientists compare experimental groups to control groups, they do so using statistical analyses, which are mathematical tests to determine whether observed differences in the two groups are "real" or are a result of "pure chance." When a scientist claims that something is significantly different from something else, she is implying that the differences are real according to statistical analysis.

After reading the *Big Red Fertilizer* example, identify the components of its experimental design in the following questions:

Farmer Hen read an article in *Sodbuster* magazine that some researchers "down at the U" had developed a new fertilizer called *Big Red Fertilizer*. The product was supposed to be nothing short of an agricultural miracle. It produced corn plants that had more ears per plant. The article even claimed that kids who ate the corn grown with *Big Red Fertilizer* would turn into star athletes.

On the basis of what he read and what he had heard, Farmer Hen generated the hypothesis that *Big Red Fertilizer enhances corn production better than regular fertilizer*. From this hypothesis, Farmer Hen predicted that *if Big Red Fertilizer enhances corn production better than regular fertilizer, then corn plants raised on Big Red Fertilizer will have more ears of corn per plant than corn raised on regular fertilizer.* He designed an experiment to test the prediction as follows: Corn was planted at the same time in two adjacent fields. One field was fertilized with *Big Red Fertilizer* and the other received regular fertilizer. The fields were plowed on the same days and were irrigated, receiving equal amounts of water on the same days.

At the end of the season, Farmer Hen observed the corn plants and found that plants maintained on *Big Red Fertilizer* indeed had more ears than those that had received regular fertilizer. Thus his prediction was found to be correct, and the hypothesis that *Big Red Fertilizer* enhances corn production better than regular fertilizer was supported. Note that although the hypothesis was supported, it was not *proven* to be correct!

Questions

1. **What is the hypothesis?**

2. **What is the prediction?**

3. **Which plants are included in the experimental group?**

4. **Which plants are included in the control group?**

5. **Identify the independent variable.**

6. **Identify the dependent variable.**

7. **Identify the controlled variables.**

8. **What conclusion would be drawn if the corn plants in the control field were found to have more ears of corn per plant than plants from the experimental field?**

9. **What conclusion would be drawn if the corn plants in the control field were found to be taller than plants from the experimental field?**

10. **What other prediction(s) might be generated from the hypothesis in the preceding example?**

Exercise 1.2   A LIBRARY EXERCISE IN THE SCIENTIFIC METHOD

Another step in the scientific method is the publication of the findings. Findings are of no value unless they are communicated throughout the scientific community! Results of scientific investigation are sometimes reported in journals maintained by professional scientific societies. Examples of these include *Journal of Protozoology, American Journal of Botany*, and *American Midland Naturalist*. Table 1.1 gives the common parts of a scientific paper and describes the purpose for each section.

The vast majority of good experiments follow the scientific method, even if the hypothesis and prediction(s) are not explicitly stated. You will be assigned a scientific paper from a professional journal that reports the results of an experiment. Summarize the publication and answer the following questions:

1. State the hypothesis.
2. State the prediction(s) upon which the experiment is based.
3. Describe the experiment.
4. Identify the independent variable(s).
5. Identify the dependent variable(s).

TABLE 1.1	CONTENTS OF A SCIENTIFIC PAPER
Abstract	A brief summary of the entire paper.
Introduction	Provides background from the scientific literature that leads the researcher to the question at hand. Provides conceptual framework for experiment.
Materials and methods	Provides a description of the experiments with enough detail that the experiment could be replicated by other scientists.
Results	Data from the experiment are presented.
Discussion	Data from the experiment are interpreted. Incorporates results into previous findings and links results back to the objectives and purpose of the study.

6. Identify controlled variables.
7. Summarize the results of the experiment.
8. Do the results support the hypothesis? Defend your answer.

Exercise 1.3   EFFECTS OF NUTRIENTS ON PLANT GROWTH

Extended five-week exercise

Plants are organisms that produce their own food. During photosynthesis, plants use sunlight, carbon dioxide from the air, and water to make food. It is tempting to assume that plants have all they need with sunlight, water, and carbon dioxide. However, in addition to these substances, we submit the hypothesis that *plants also require other essential nutrients to develop and grow properly*. Those needed in the greatest amounts are called **macronutrients**. Those needed in smaller amounts are called **micronutrients**. See Table 1.2 for a list of essential nutrients.

TABLE 1.2 ESSENTIAL NUTRIENTS FOR PLANT GROWTH	
Macronutrient	**Function**
Calcium	
Magnesium	
Nitrogen	
Phosphorus	
Potassium	
Sulfur	
Micronutrient	
Iron	

Now investigate the effect(s) of a nutrient deficiency on the growth and survival of plant seedlings. You will use plant seedlings that have been germinated and maintained without nutrients. You will maintain the seedlings using *hydroponics*, a method that bathes the plant roots in nutrients, without using soil, in order to eliminate variables associated with soils.

Week 1

1. Divide into groups, and use your library and Web resources to investigate the effects of micronutrients and macronutrients on plant growth. Use this information to complete Table 1.2. Answer Question 1.

2. Four of your plants will be grown in a solution that has no mineral deficiency. The other four will be placed in a solution that is deficient in one of the essential nutrients listed in Table 1.2. Select which essential nutrient to use as your independent variable.

3. Figure 1.2 shows you how to set up your plants for maintenance in the hydroponic system. Leaving a 2-cm margin at the top of each test tube, fill the four test tubes of your experimental group with the proper solution. Fill the four test tubes of your control group with the proper solution, also leaving a 2-cm margin at the top of each.

4. Roll eight, 4 cm × 2 cm strips of cotton to wrap around your seedlings. Choose seedlings of uniform size with root tips at least 3 cm long. Using forceps, pick up a seedling by its thick seed coat, taking great care not to injure the root or shoot tips. Place the seedling in the cotton and gently wrap the cotton around the seed coat. The root and shoot tips should be visible on either side of the cotton. If the root or shoot tips are covered with cotton, they will not grow properly!

5. Use the forceps to insert the wrapped seedling into a test tube. Make sure that the root is bathed in the solution, but take care not to saturate the cotton. Push the cotton aside with forceps and use a clean pipette to drop additional solution into the test tubes, if necessary. Neatly wrap each test tube in foil so that the plant roots are protected from light.

6. Use a permanent marker to label the foil of each test tube with your group's initials, type of solution,

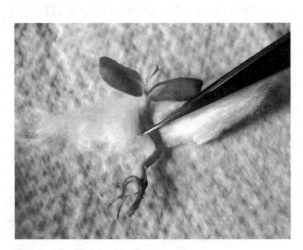 

FIGURE 1.2   Hydroponics Setup.

and plant number. Place the rack under a grow light. You will need to replenish your nutrient supply every other day. Work out a schedule with your lab partners, dividing this responsibility. Ask your instructor when you will have access to the lab.

7. On the basis of the hypothesis that plants require essential nutrients to develop and grow properly, state your hypothesis and prediction(s) in Questions 2 and 3. Feel free to use library materials to formulate and refine your hypothesis and predictions.

8. Discuss with your lab partners how you will make your observations. For example, you may want to measure root length, shoot length, and leaf color. Answer Questions 4 to 8. In the bottom section of Table 1.3, write additional dependent variables that you want to measure.

TABLE 1.3 GROWTH OF EXPERIMENTAL AND CONTROL GROUP PLANTS							
Experimental group				**Control group**			
Nutrient deficiency:							
Root length				**Root length**			
Week	**Plant 1**	**Plant 2**	**Plant 3**	**Plant 4**	**Plant 1**	**Plant 2**	**Plant 3**	**Plant 4**
2								
3								
4								
5								
Shoot length				**Shoot length**			
Week	**Plant 1**	**Plant 2**	**Plant 3**	**Plant 4**	**Plant 1**	**Plant 2**	**Plant 3**	**Plant 4**
2								
3								
4								
5								
Leaf color				**Leaf color**			
Week	**Plant 1**	**Plant 2**	**Plant 3**	**Plant 4**	**Plant 1**	**Plant 2**	**Plant 3**	**Plant 4**
2								
3								
4								
5								
Week	**Plant 1**	**Plant 2**	**Plant 3**	**Plant 4**	**Plant 1**	**Plant 2**	**Plant 3**	**Plant 4**
2								
3								
4								
5								

Weeks 2 to 5

9. You will measure your plants once a week for four weeks. Be sure to keep track of the growth of each individual plant, and record these measurements in Table 1.3. Also, record any general observations that you think might be important to know at the end of the experiment.

10. At the end of the experiment, discuss how to communicate your results to the rest of your class. Construct charts to record your data. The class may prepare a table, a bulletin board, or a Web page summarizing the effects of various deficiencies on the survival and the growth of plant seedlings.

Questions

1. **What is the effect of your nutrient on plant growth and overall health, according to your library research?**

2. **Hypothesis:**

3. **Prediction(s):**

4. **Independent variable:**

5. **Dependent variable:**

6. **Controlled variables:**

7. **Experimental group:**

8. **Control group:**

9. **Did your data support your hypothesis?**

10. **If your results were different than you expected, why do you think this is so?**

11. **Which nutrients seem to be the most important in regard to plant growth and survival?**

12. **Why is it an advantage to use four different seedlings?**

13. **Did all of your plants grow in the same manner? What differences did you note in the rate of growth? What can you attribute these differences to?**

Exercise 1.4 SCIENCE AND TECHNOLOGY

Web exercise

You may hear the terms *science* and *technology* used together and often interchangeably; however, science and technology are not the same thing! **Technology** is derived through applied science and may be defined as *the tools we use to carry out science and to provide objects necessary for human sustenance and comfort.* Thus, *science* is an intellectual process with the scientific method as its foundation; *technology* provides useful tools needed to carry out that process. It is important to bear in mind that although powerful new technology provides us with tools used to increase our understanding of concepts, technology should be viewed as a means to an end, rather than an end in itself.

Find out what you want first of all, and then if a machine will help you, use it. (*W.H. Auden, 1933*)

Science and technology offer the potential for doing good as well as for doing evil. Today, we have more knowledge at our disposal than at any other time in human history. We also have amazingly powerful technology. How will we use these fruits of science? What will our legacy be? What do we want for ourselves? What do we want for our children and grandchildren? Through science, we have the knowledge and power to realize our greatest dreams or our worst nightmares.

1. Use your library's online resources and other search engines such as Google (*http://www.google.com*) and Altavista (*http://www.altavista.com*) to gather information about one technological tool used in scientific discovery. Please note that the "tool" does not have to be a machine. It can be a process, product, or another innovation that facilitates scientific discovery. For this particular Web exercise, it is recommended that you start with some keywords from this list: *science, technology, invention, innovation, ethics, implications.*

2. Once you choose a topic, investigate how the tool was discovered or developed and how its use has changed over time. Use a minimum of three reliable web sites to complete the assignment. See Appendix 1 for proper citation format, and type these citations alongside the questions as you answer them. Retype the questions listed below, and double-space your answers.

   Here are some web sites that may be helpful; this is not a complete list, by any means! You should conduct your own Internet search in addition to visiting these Web sites.

   http://web.mit.edu/invent/i-archive.html

   http://www.kcls.org/hh/invention.cfm

   http://www.si.edu/lemelson/dig/links.html

   http://www.nasm.si.edu/galleries/gal111/universe/

   http://historywired.si.edu/index.html

http://www.howstuffworks.com

This website lists topics to use, in case you are have trouble choosing one:

http://www.nationalhistoryday.org/03_educators /teach99/technology/topics.htm

Questions

1. Which search engine(s) and keyword(s) did you use?

2. What is the technological tool you are investigating?

3. When was it invented, and by whom?

4. How is the tool used, and in what way(s) is it beneficial?

5. What scientific discoveries led to the development of the tool?

6. What further discoveries were made after the tool was put into use?

7. As a whole, how will the use of that tool advance science?

8. Discuss positive and negative implications the use of this tool may introduce.

9. Do you think the discoverer(s) of the tool knew about possible harmful effects? Explain.

10. Be sure to include your citations within your answers to the above questions.

2

Natural History

*The field of biology can lead the dedicated student to much beauty,
and it can also teach respect.*

N.D. Buffaloe, 1962

The term **natural history** is often used in reference to descriptive or observational biology. It traditionally referred to all "natural knowledge" *(Hobbes, 1651)*, including geology, climatology, and astronomy, in addition to the study of living organisms. The term **biology** was first introduced by Jean-Baptiste de Lamarck **[Krogh section 16.3]** to designate the scientific study of plants and animals, separate from other aspects of natural history *(Huxley, 1876)*.

Biology does not have to involve experimentation to adhere to the scientific method; observation can be used as well. For instance, surveys may be conducted to determine what types of organisms are present in a habitat; no experiments are conducted, but hypotheses and predictions are still generated and data are collected. Descriptive (observational) biology has the same validity as any experimental study conducted in a laboratory. Observation is one of the most important skills a scientist must develop, and it is a key component of the scientific method. Careful observation of nature and natural phenomena leads to the generation of hypotheses.

Currently, naturalists are frantically working to document the tremendous, disappearing biodiversity that exists on this planet. Natural history promotes a better understanding of the natural world through observational and descriptive studies. This work will provide us with a better understanding of nature and our place in it. As you complete this course, we hope you become a naturalist, too.

Exercise 2.1 OBSERVING NATURE

Extended exercise

A pioneering environmentalist, Rachel Carson pointed out that we often take the tremendous gift of beauty that nature has to offer for granted; because we can see the beauty of nature anytime, perhaps we never get around to seeing it at all *(Carson, 1956)*. The great nineteenth-century American zoologist, Louis Agassiz, once said, "Study nature, not books." This statement can be extended to say, "Study nature, not books, computers, and television sets." There is no better way to become intimately familiar with the object of your study than through firsthand observation. Do not, however, misunderstand this statement to undermine the importance of scholastic study of biology through lecture, books, film, and modern computer technology. Observation of and interaction with nature are deeply enriched by diligent study.

11

FIGURE 2.1   *Nature Study* by Goleman, 1860.

1. Go outside into "nature." Choose a spot where you can comfortably sit, and observe what is around you. Use your senses. Then document your experience in a journal. What do you see, hear, feel, and smell? What are the weather conditions? What is "happening" in nature around you? What questions do you have regarding your spot? How do you think your spot will change over time? Your instructor may give you certain tasks to complete during each visit, such as collecting and pressing plants, or collecting various acorns, insects, leaves, and so forth.

2. Your first entry should describe your spot in detail. Try to photograph your spot. Take a field guide with you and identify as many organisms as you can, including insects, spiders, trees, and flowers.

3. Compose sketches of objects or organisms near your spot.

4. Include the biological significance of your observations. Relate what you see to what you learn in class.

5. Write any thoughts, ideas, memories, or emotions your spot evokes in you.

6. Construct a hypothesis based on your initial observations. Generate predictions based on the hypothesis; then plan further observations to test your predictions.

7. After several visits to the spot, summarize your observations in your journal. In your conclusion, state whether or not you may accept or reject your hypothesis. Don't forget to do steps 1–6 on each visit.

Exercise 2.2 NATURAL HISTORY IN ART

(Written with Laura M. Duclos)

Science, as an intellectual endeavor, is unique from philosophy, religion, and art; but each of these has also profoundly influenced the others in the cultural evolution of humanity. For instance, art has played a strong role in the tradition of natural history *(Janovy, 1985)*. Our perception of the full beauty of the world is enriched through the study of natural history. Humankind's fascination with the terror, solitude, and beauty of nature has played a prominent role in the evolution of society, art, literature, and religion by influencing the formulation of cultural views. Edward O. Wilson (1994) said, "Nature is part of us, as we are part of Nature." This is the essence of *biophilia*, the concept that human beings are born with an intense craving for the natural world and other forms of life. Biophilia directs us and influences our thoughts and actions.

Charles Darwin (see Figure 12.1) was one of the truly great natural historians. It is ironic that his classical approach in observing nature formulated the evolutionary paradigm that gave rise to modern biology. Indeed, Darwin gave biology the ultimate hypothesis, paving the way for the decline of the era of traditional natural history, with the concomitant rise of "modern experimental biology."

The Victorian Period refers to the years 1837 to 1901, during which Queen Victoria (Figure 2.2) reigned over the British Empire.

Within Darwin's lifetime, under the influence of Thomas Henry Huxley (Figure 2.3), biology began a rapid transition from a Victorian age of pleasure and exploration to an era marked by technological advancement, reductionism, and laboratory sterility.

Reductionism refers to the study of "parts" rather than the "whole."

During this time of scientific and technological revolution, science moved from the drawing

FIGURE 2.2 Queen Victoria, 1819–1901.

room into the schoolroom and from the parlor into the Parliament. Huxley began a cultural movement in which science was whisked to the forefront of societal progress, advocating a utilitarian approach. Huxley satirized the German philosopher and poet, Novalis, saying, "Philosophy can bake no bread." *(Desmond, 1994)*.

This movement transformed traditional natural history into modern experimental biology, and ultimately transformed European and American culture. Despite the many gains by humanity during this transition, much of the appreciation for raw beauty died in the field of biology. Undoubtedly, the illustrations of natural history texts of the Victorian period convey an artistic quality that is rare in texts of contemporary biology (Figure 2.4) Such illustration has become the exception as opposed to the rule.

Contemporary society's fast-paced trends and obsession with cutting-edge technology leaves little time for biologists to enjoy and contemplate the beauty and serenity of nature, as was intrinsic to the era of natural history. Today the lives of most scientists are spent in structural occupations, watching the clock and lab budgets.

FIGURE 2.3   Thomas Henry Huxley, 1825-1905. He was a noted English biologist and education reformer.

Perhaps the source of much of the beauty and "artistic flair" of Victorian works of natural history sprung from the interweaving of natural history with natural theology.

The term **natural theology**, coined by William Paley in 1802, referred to a philosophy that the perfections, beauty, and "wonder" of nature's design reflect the Creator's glory.

The goal of Victorian works of natural history was to encourage their readers "to see evidence of God's existence and attributes in the natural organisms around them, by means of natural theology" *(Barber, 1980)*. Up through the middle of the nineteenth century, natural history and natural theology were tightly interwoven. Instead of being concerned with biological accuracy, most Victorian natural history publications were primarily concerned with capturing and conveying divine beauty in the design of nature.

The love of nature goes back to prehistoric humans, and is first documented with the cave paintings and symbolic iconography of ancient cultures (Figure 2.5). An aptitude for art is present even in our nonhuman relatives *(Morris, 1962)* (Figure 2.6). Reverence for, understanding of, and harmony with nature were central themes in ancient cultures, whose early art

shows anatomical accuracy and anthropomorphic personification of animals. This may be directly attributed to a need for close observation of wildlife during the struggle for survival.

Anthropomorphism refers to the act of giving animals human characteristics. Walt Disney's Mickey Mouse is a grand example of this.

In many ancient cultures, animal and plant images possessed symbolic meaning and addressed the role that these organisms played in the everyday lives of people. For them, an understanding of nature was essential for survival. Communication of biological concepts emerged even before languages.

Internet links to sites containing the work of most artists mentioned in this chapter may be found at *http://www.arts.com.*

Art is a medium capable of transcending cultural boundaries. It explains complex theories and messages in a universal, visual manner. Of the great American painters, Winslow Homer is perhaps the most renowned for his depiction of nature. He was highly influenced by Japanese art and sought to capture not only the accuracy of the natural world but also the "moods" of the landscape and the "emotions" of the Earth. Originally focused on human subjects, Homer's later work began to shift from human images to that of the reverence and power of the sea. He painted northern and tropical seascapes. He found that oil paints, which are traditionally heavier and darker pigments, are appropriate for conveying the atmosphere in northern climates; and watercolor, a more delicate and transparent paint, depicts the bright and carefree landscape of the tropics. The vivid, clear watercolors were perfect for painting turquoise seas and neon red hibiscus in the Bahamas. In most of Homer's later works, humans tend to lie in the fringes, with nature monopolizing the canvas. Occasionally, he even painted his scenes from the nonhuman animal's point of view.

A group of American painters, the Hudson River School, is renowned for its members' masterpieces depicting the profound, beautiful, and grotesque faces of nature. These artists played a prominent role in American thought through the middle of the nineteenth century. In their paintings, like Japanese works,

FIGURE 2.4   Illustrations from Natural History Texts of the Victorian Era. a. Blue Jays from *Illustrated Natural History of the Animal Kingdom* by S. G. Goodrich, 1859. b. Skunk from Goodrich, 1859. c. *Ornithorhynchus* from Goodrich, 1859. d. Nightingale from *Popular Zoology* by J. D. Steele and J. W. P. Jenks, 1887. e. Moluccan Beetles from *The Malay Archipelago* by Alfred Russel Wallace, 1869. f. Wild boar from Steele and Jenks, 1887.

FIGURE 2.5   Cro-Magnon cave paintings in a cavern of font-de-Gaume, France.

FIGURE 2.6   a. A chimpanzee named Congo, hard at work.
b. Brush-painting by Congo showing remarkable organization and patterning.

nonhuman nature is the subject. The canvases epitomize biodiversity. Imagine walking into a gallery filled with American landscapes: The dim museum lighting softly spotlights each canvas; one can almost hear the sounds of native birds and the rush of falling water emanating from stoic canvases and yellowing varnish. In an eerie trance, the paintings ignite the soul.

Because of the Hudson School's romantic renderings of upstate New York, the area's population and tourism industry exploded. As more and more city folk wanted firsthand experience of the beautiful scenery and animals seen in art galleries, rural communities blossomed into towns and cities. It was the desire to fulfill primordial yearnings for grand views and lush forests that drove the American affinity for landscape paintings. Once again, biophilia resurfaced.

Some painters in the Hudson River School opted to use art as a means of sending powerful messages to the general public. Thomas Cole created a set of enormous paintings titled *The Course of Empire*. All the paintings focus on one setting and are paintings of the past, the present, and the future. Cole's message pertains to the passage of time and to the natural biological processes governing our world. *Past* shows the land before human intervention—rich in biodiversity, rich in primordial plant and animal life. *Present* depicts the same piece of land, but now the land has witnessed the birth of a medieval empire bustling with life and full of human activity. No longer is there great biodiversity; rather, the human species has overthrown the kingdom of nature. *Future* shows the same human empire in decay, with wild things returning to overtake those constructed by human hands. The action of natural forces, along with recolonization by plants and animals, aid in bringing the human empire to a close. *The Course of Empire* could serve as a commentary on contemporary environmental issues, particularly the loss of biodiversity.

The impressionist painters of the late nineteenth and early twentieth century were similarly influenced by biology, emphasizing nature in an emotional rather than realistic aspect. They were concerned with the effects of lighting upon everyday objects, with the changing seasons and moods of the land, and with the ways plants and animals interacted with their surroundings. These aspects conveyed a certain "feel" or "emotion" to the composite scene. This style of painting is characterized by the use of many subtle hues of the same color to evoke contrasting emotions, by the sloppy stroke of a brush, and by seemingly haphazard placement of paint to create the desired image.

The most famous of the French impressionists was Claude Monet (1840–1926). Monet truly embraced the concept of biophilia and let nature become an integral part of his life. At his home in Giverny, he designed a large flower and water garden in an attempt to capture the essence of the world he loved. His series *Water Lilies* would not be the same had he not taken time to contemplate how nature operated. The graceful lines of the plants, the intricate relationships between the shrubs and water's edge, and the placid reflection of the water superimposed upon the depth of the pond are achieved through patient observation. Only after acquiring a sense of familiarity for his subject matter was Monet able to blur the image into an emotional impression, while still retaining the identity of the plants themselves.

Animals (including *Homo sapiens*) are commonly painted as either portraits or integral components within a landscape. Correct anatomical placement of body parts and underlying bone and muscle structure is necessary for lifelike renditions. The best portraits (including those of the animalists James Audubon, Rosa Bonheur, George Stubbs, and Leonardo da Vinci) were completed by those artists possessing a complete knowledge of animal anatomy. It was Degas who first portrayed horses galloping in accordance with their anatomy. Prior to Degas' scrutiny of equine anatomy, horses were depicted with their hind legs outstretched like a hobbyhorse. Anatomically, horses are incapable of assuming such a position.

It was James Audubon who left a legacy of bird paintings, which are still used today by ornithologists and naturalists, not only for scientific reference but also for inspiration as works of aesthetic beauty. The power of a portrait and a measure of its quality are defined by the painting's ability to capture and convey personality and life. Once again, biology is central to achieving such a goal. See some of Audubon's works at *http://www.historical-museum.org/collect/audubon/al.htm*.

The role of nature and natural history, although diminished in modern art, remains an important influence. Modern artists will often choose to paint certain images based solely upon the emotional rush they experience when they see a particular landscape. For example, Jane McNichol painted *Heartland* in 1995, and it was inspired by a trip to the west coast of Ireland. As our culture becomes more and more focused on technology, McNichol says that we will need more paintings of the natural world to "keep our spirit rooted to the land." Once again, biophilia.

Vertigo is aptly named, painted by the American artist Katherine Bowling in 1988. When one looks up a tree trunk, there is not only a physiological effect of vertigo but also an emotional one. Nature's awe delivers a head rush equal to the physiological effect. Bowling said these were fleeting images, like those you might see when you close your eyes at night before going to sleep.

In photography, Ansel Adams is perhaps one of the most well known names. He photographed national parks such as Yosemite and Yellowstone, for he was fascinated with the interplay of light and shadow on the various textures of trees, rocks, soil, and water.

The microscopic world has also had a profound influence upon art. Inspired by Darwin's *Origin of Species*, the nineteenth century French artist Odilon Redon found microscopic organisms as an artistic inspiration. In his work *Origins*, a series of lithographs, he portrayed the emergence of complex life forms from the microfauna of "primordial soup." Microorganisms have even served as inspiration for architectural creation; Frank Lloyd Wright used microscopic cells as an inspiration for some of his work *(Gamwell, 2003)*.

In Sickness and in Health, by the American painter Ross Bleckner, was inspired by the microscopic world. This painting features the artist's rendition of blood cells and expresses "...the contemporary fear that the lovely pastel-colored cells of one's partner, whom one is committed to love in sickness and in health, may be invaded by the lethal HIV virus." *(Gamwell, 2003)*.

Huxley (1882) pointed out that "the province of art overlays and embraces the province of the intellect" and that many human endeavors, including natural history, "derive much of their quality from simultaneous and even unconscious excitement of the intellect." Thus, through study and interaction with nature, an aesthetic quality emerges. This, after all, is what art is all about. Art is the common language of humanity. It speaks volumes about natural history, and natural history provides a wellspring of inspiration to the artist.

Clearly, natural history has strongly influenced art. But what impact does art have on natural history? What biological knowledge does a person extract from artwork? In the spirit of Janovy (1985), we are asking you to complete a mental exercise that combines seemingly unrelated ideas. This may be an unusual assignment for a biology course, but enlighten your instructor with what you know. Your assignment is to evaluate the biological content of a work of art. The best way to do this is to visit a local or campus art museum.

Consider the following points in your discussion: What was your first impression of the artwork? What did it make you think of? Did it elicit certain memories or emotions? Does this work represent some aspect of modern society? (Does modern society mimic what it portrays?) What impact did the work of art have on the general public when it was introduced? In studying this work of art, what have you learned about biology? Here are the criteria for the paper you will be writing, based on the assigned work of art:

1. Artist and title of artwork
2. Artist's biography
3. Time period and location of completion
4. Biological significance of artwork
5. Its monetary or educational value
6. At least two pages long
7. Two references in bibliography from books or journals

8. If possible, provide a photocopy or computer printout of the art.

Absolutely No Clippings From Books Or Magazines Are Allowed!

Exercise 2.3   NATURAL HISTORY IN LITERATURE

An immense body of literature is also influenced by natural history. The profound influence of nature on the work of Lord Byron is evident in these lines from *Solitude* in *Childe Harold*. Note the reverence for nature and the biophilia evoked in the first stanza, and the attempt to define man's "place" in nature in the second stanza.

There is a pleasure in the pathless woods,
There is a rapture on the lonely shore,
There is society where none intrudes,
By the deep Sea, and music in its roar:
I love not Man the less, but Nature more,
From these our interviews, in which I steal
From all I may be, or have been before,
To mingle with the Universe, and feel
What I can ne'er express, yet cannot conceal.

Roll on, thou deep and dark-blue Ocean—roll!
Ten thousand fleets sweep over thee in vain;
Man marks the earth with ruin—his control
Stops with the shore; upon the watery plain
The wrecks are all thy deed, nor doth remain
A shadow of man's ravage, save his own,
When, for a moment, like a drop of rain,
He sinks into thy depths with bubbling groan,
Without a grave, unknell'd, uncoffin'd, and unknown.

Often, an attempt to understand nature gives us an enriched understanding of ourselves. But as Wilson pointed out, we are part of nature. In the following excerpt from *The Shell* in *Maud*, Alfred Lord Tennyson betrays the tremendous influence of natural theology on Victorian culture and addresses man's place in the larger scheme of things.

I
See what a lovely shell,
Small and pure as a pearl,
Lying close to my foot,
Frail, but a work divine,
Made so fairly well
With delicate spire and whorl,
How exquisitely minute,
A miracle of design!

II
What is it? A learned man
Could give it a clumsy name.
Let him name it who can,
The beauty would be the same.

We have given a couple of examples of literature by great poets that were profoundly influenced by natural history. The task of trying to pick out such representative material is overwhelmingly difficult. So now, we ask you to get lost—in your public, college, or university library, that is. Ask for help when you get there. A librarian will be able to point you in the right direction.

If poetry is your fare, we encourage you to delve into Charles Baudelaire, Lord Byron, Percy Shelley, John Keats, Alfred Lord Tennyson, John Burroughs, Emily Dickinson, T. S. Eliot, W. H. Auden, and Robert Frost, to name a few. You might begin by sampling from a good collection of poems. If you are primarily interested in nineteenth-century poetry, begin by sampling selections from *Songs of Nature*, edited by John Burroughs (1901). Although this book is out of print, copies should be available in your library or from interlibrary loan. If you are interested in narrative works but do not know where to start, go to the author screen of the computer in your library index. Type in one or more of the following: Ralph Waldo Emerson, Hugh Miller, Henry David Thoreau, Lewis Carroll, Herman Melville, Upton Sinclair, John Steinbeck, Rachel Carson, Konrad Lorenz, Loren Eiseley, John Fowles, Joseph Conrad, Aldo Leopold, John Janovy, Paul Johnsgard, Stephen J. Gould, or Edward O. Wilson. The preceding names are just a few from a very long list. Take a look at some other authors who may not be in the brightest spotlight. Oh, by the way, you might try your professor's bookshelf for some more eccentric fare.

Your assignment is to find a favorite book, article, or poem influenced by some aspect of natural history. (It doesn't have to be about natural history, just related to or influenced by natural history in some way. Be creative!) Then, when you have chosen a piece of literature, choose from the options listed below.

1. Option A: Write a parallel piece of work, mimicking the original, but choose your own organism or aspect of natural history to write about.
   Option B: Write an essay telling why this piece of literature moved you, and why it is worthy of study.
   Option C: Taking into account your experience in this course so far, and the ideas given by the literature you have read, write a poem or essay that integrates what you have learned about biology, or your own life.

2. Provide information about the author whose work you have chosen to study. If your instructor wishes, provide a photocopy of the poem or article, for your instructor's reference while reading your assignment.

3. Provide a "references" section at the end of your assignment.

4. Double space, include the date, and add a title. No title page is necessary.

UNIT 1
Essential Parts

Biological Molecules

The fundamental unit of life is the cell, which has a membrane and contains an aqueous solution of chemicals. The reactions taking place among these chemicals facilitate every physiological process that occurs in a living organism, including muscle contraction, nerve impulses, and blood filtering. Thus, an understanding of basic chemistry is important to understand living organisms. The exercises in this chapter provide you with a review of the basic groups of organic compounds and some of their fundamental properties **[Krogh Chapter 2, Chapter 3]**.

You will be conducting various biochemical tests designed to detect specific types of organic compounds. A **colorimetric assay** is a biochemical test in which a color change results from a chemical reaction, and the intensity of the color change is directly proportional to the concentration of the substance being tested. Such procedures are utilized in many "everyday" applications. For example, when crime laboratories conduct forensic investigations, they commonly employ biochemical tests in the analysis of semen and blood stains. Independent and government laboratories often conduct an array of biochemical assays to determine or confirm the dietary content of food items.

Urinalysis is frequently employed in clinical laboratories to determine the sugar and protein composition of urine, which provides important information about the health of the patient.

Exercise 3.1 THE NATURE OF ORGANIC COMPOUNDS

Caution! The crucible gets very hot. Do not touch it with bare hands!

Living organisms are made up of organic compounds. The most common elements comprising organisms are carbon, hydrogen, nitrogen, and oxygen. These constituent elements of organic compounds make up about 95% of your body weight. By definition, **organic compounds** are those that always possess carbon and hydrogen **[Krogh section 3.3]**. The ability of carbon to form up to four covalent bonds with other atoms enables carbon chains to serve as the backbone for large, energy-rich molecules **[Krogh essay *Notating Chemistry,* page 26]**. This exercise compares how much energy is stored in sodium chloride (NaCl) and sucrose ($C_{12}H_{22}O_{11}$).

The human body is made up of 55-70% water

1. Look at the chemical formulas for table salt and sugar. Answer Question 1.

2. Place about 2 g ($\frac{1}{4}$ teaspoon) of sodium chloride (table salt) in a crucible. Using tongs, hold it over the flame of a Bunsen burner or alcohol lamp.

   Record what happens over a period of 3 to 5 minutes.

3. Place about 2 g ($\frac{1}{4}$ teaspoon) of sucrose (table sugar) in a crucible. Using tongs, hold it over the flame of a Bunsen burner or alcohol lamp.

   Record what happens over a period of 3 to 5 minutes.

Questions

1. **Based on the chemical formulas given, which compound do you think stores more energy?**

2. **Which compound is organic and which is inorganic? How do you know?**

3. **What is the source of the black smoke?**

Exercise 3.2   BENEDICT'S TEST FOR SIMPLE SUGARS

(Adapted from Mathis, 1996)

Wear gloves and goggles! Use caution when heating substances. Direct test tubes away from your face.

Carbohydrates are composed primarily of carbon, hydrogen, and oxygen in the form of sugar molecules. Carbohydrates are classified as either simple sugars or complex carbohydrates. Glucose and fructose are **simple sugars** (monosaccharides) and are commonly found in fruit. Table sugar (sucrose) is a disaccharide, a combination of a glucose molecule bound to a fructose molecule.

Examples of **complex carbohydrates** (polysaccharides) are glycogen and starch, which are polymers (chains) of glucose molecules. Animals store energy in the form of glycogen, and plants store energy in the form of starch. Cellulose, the primary component of the cell walls of plants, is another example of a complex carbohydrate. Although carbohydrates such as monosaccharides and starches are composed of the same elements, they react differently with other chemical compounds **[Krogh Figure 3.12]**.

Benedict's test is for the detection of simple sugars. Benedict's solution reacts with monosaccharides such as glucose and fructose to cause a color change. In the presence of a small amount of monosaccharide, Benedict's solution will react weakly and turn from blue to green. In the presence of moderate or large amounts of monosaccharide, Benedict's solution will react strongly and turn from blue to yellow-orange to brick red.

In many of the following exercises, you will use substances that provide a **positive control** or **negative control**. For example, distilled water is used as a negative control in this exercise. You already know that distilled water has no sugar in it. The controls provide references to compare with the color of other tested substances, so that you know what color to look for.

1. Boil about two inches of water in a water bath. While you wait, add 1 ml of the solutions listed in Table 3.1 to properly labeled, clean test tubes. If a diet soda is used, a clear soda is better than a caramel-colored one.

2. Add 1 ml of Benedict's solution to each tube and mix well by gently tapping the side of the tubes with your index finger.

3. Place the tubes in the boiling-water bath.

4. Check the tubes for color changes after 3 minutes and record your results in Table 3.1. Record no reaction, or a weak, moderate, or strong reaction according to the color exhibited by each solution.

Exercise 3.3   IODINE TEST FOR STARCH

(Adapted from Mathis, 1996)

Wear gloves and goggles!

Starch is an example of a complex carbohydrate. A solution of iodine turns dark blue or black in the presence of starch, but not in the presence of simple sugars. The intensity of the reaction indicates the concentration of starch.

1. Add 1 ml of the solutions listed in Table 3.2 to five properly labeled, clean test tubes.

2. Add 1 drop of iodine solution to each test tube and mix the contents by tapping the tube with your index finger. Record the results in Table 3.2. Record no reaction, or a weak, moderate, or strong reaction according to the color exhibited by each solution.

Questions

1. **Did any of the substances in the sugar or starch tests yield results that surprised you? Which ones?**

2. **What did you expect, and why do you think the substance contains sugar, and not starch, or vice-versa?**

Exercise 3.4   ENZYMATIC DIGESTION OF STARCH

(Adapted from Mathis, 1996)

Wear gloves and goggles while transferring iodine and Benedict's solution.

When you eat a meal containing starch, before any of the energy stored in the starch molecules can be

TABLE 3.1 BENEDICT'S TEST (SUGAR)	
Solution	**Result**
Distilled water (negative control)	
10% Glucose (positive control)	
Potato extract	
Apple juice	
Diet soda	
regular soda	

TABLE 3.2 IODINE TEST (STARCH)		
Solution	**Result**	
Distilled water (negative control)	~~positive~~	neg
1% Starch (positive control)	~~negative~~	pos
Potato extract	~~negative~~	pos
Apple juice	~~positive~~	neg
Diet soda	~~positive~~	neg
regular soda	~~positive~~	neg

used, the molecules must be digested, or broken down into their simplest units by the action of enzymes **[Krogh section 6.5]**. **Amylases** are enzymes that break complex carbohydrates down into simple sugars. Enzymes in our bodies work at specific temperatures and pH levels. The process of starch digestion begins in the mouth, with an enzyme called salivary amylase. In this experiment, you will see how salivary amylase catalyzes the digestion of starch molecules.

Read through all directions before starting, and note that some solution transfers have to be done in a timely manner. Use separate pipettes for each of the solutions!

1. Refer to Figure 3.1 as you proceed. Obtain eight clean test tubes. Label them as follows: E (experimental solution), C (control solution), T_0C, T_0E, T_5C, T_5E, $T_{10}C$, and $T_{10}E$ (designating the E and C solutions tested at 0, 5, and 10 minutes. T_0C refers to "Time zero" for the control solution, for example.)

2. Place 5 ml of 1% starch solution into the E and C tubes. Answer Question 1.

3. Test for the presence of starch (obviously, this is a positive control): Place a drop of E solution into the well of a spot plate. Do the same for the C solution, placing it in a well in the next row on the spot plate. Add a drop of iodine to each well. These are your "reference wells," which give you an idea of what color to look for while carrying out the next steps. Record the color change in Table 3.3. This is the "0" elapsed time reading for the starch tubes, because the enzyme has not yet been applied to the starch. (Enzymatic digestion has not yet begun.) Indicate whether there is a weak, moderate, or strong reaction, or no reaction at all.

4. Test for the presence of simple sugars: Pipet 0.5 ml from the C tube into the tube labeled T_0C and 0.5 ml of the E tube into the tube labeled T_0E. Add 0.5 ml of Benedict's solution to each tube and gently mix them by tapping the tubes with your finger. Place the tubes in the boiling water bath, and note any color change after 3 minutes in Table 3.3. Again, record this color change in the "0" elapsed time row in Table 3.3.

5. Now, "collect" approximately 2 ml of your own saliva in a small beaker.

6. Add 1 ml of saliva to the E tube. Using a clean pipette, add 1 ml of water to the control tube. Mix the contents of each tube thoroughly by tapping them with your index finger?

7. Place the experimental and control tubes into the 37°C water bath and note the time.

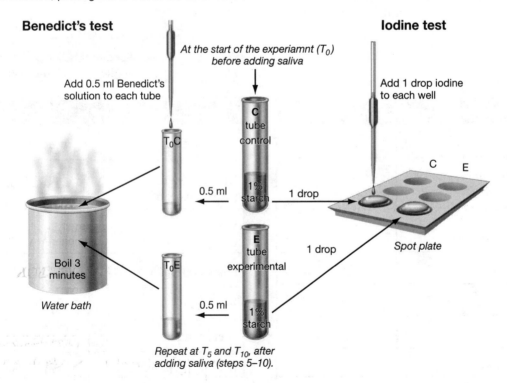

FIGURE 3.1   Steps 1 through 4 of Exercise 3.4 (Enzymatic Digestion of Starch).

TABLE 3.3 DIGESTION OF STARCH INTO SIMPLE SUGARS				
	Experimental (with enzyme)		Control (without enzyme)	
Time elapsed (minutes)	Iodine test (starch)	Benedict's test (sugar)	Iodine test (starch)	Benedict's test (sugar)
0				
5				
10				

Work together with your lab partners to carry out steps 8 and 9 simultaneously!

8. **After 5 minutes**, test each solution for the presence of simple sugars by pipetting 0.5 ml from the C tube into the tube labeled T_5C and 0.5 ml from the E tube into the tube labeled T_5E Add 0.5 ml of Benedict's solution to each tube and mix. Place the tubes in the boiling-water bath, and note any color change after 3 minutes in Table 3.3. (This is the 5-minute elapsed time reading.)

9. **After 5 minutes**, place a drop of solution from the E tube and from the C tube into clean wells of the spot plate. Test for the presence of starch by adding a drop of iodine to each well. Record the color of each well in Table 3.3. (This is the 5-minute elapsed time reading.)

10. Repeat steps 8 and 9 after 10 minutes have elapsed, using the $T_{10}C$ and $T_{10}E$ tubes for the sugar test and two more wells in the spot plate for the iodine test. Record your results in Table 3.3.

Questions

1. **How quickly do you think salivary amylase will act? Hint: How fast must it work in its natural location?**

2. **What happened to the starch concentration over time? Why?**

3. **What happened to the sugar concentration over time? Why?**

4. **Why do you think the tubes were kept at 37°C?**

Exercise 3.5   SUDAN TEST FOR LIPIDS

(Adapted from Mathis, 1996)

Wear gloves and goggles while transferring filter paper to and from Sudan III. Do not discard Sudan III in the sink—use the container provided by your instructor.

Lipids (fats) are made up of exactly the same elements as carbohydrates, but fats contain more hydrogen and less oxygen than the carbohydrates. Lipids are important as sources of stored energy, insulation, and components for the cell membrane **[Krogh section 3.6]**. They are also an efficient source of energy, yielding over twice as many calories per gram as either proteins or carbohydrates. (See the MediaLab in Krogh Chapter 3 for further information concerning fat substitutes.) Sudan III stain combines with lipids to form a vivid orange color.

1. Using a pencil, initial a piece of filter paper and draw several 1-cm circles on it, at equal distances apart, on the perimeter of the paper. If you brought your own substances to test, draw circles for them also. Label the circles W (distilled water), SM (skim milk), WM (whole milk), and so on, according to the substances listed in Table 3.4. For the butter substitute, use a product such as I Can't Believe It's Not Butter™ or Benecol™.

2. Put 1 drop of each substance on the appropriate circle and allow the paper to dry. Blot off any excess, if necessary.

3. Place the paper into the bowl containing Sudan III stain for 5 minutes, immersing the entire circle of paper into the solution.

4. Using forceps, remove your paper from the stain and gently wash it in a pan of water for approximately 1 minute.

5. Examine each spot for coloration. Record the results in Table 3.4, indicating the intensity of color as weak, moderate, or strong.

Questions

1. **According to your Sudan test, does the fat content for any of the products contradict what is listed on their nutrition label? Which ones?**

2. **Why do you think some nutrition labels might indicate no grams of fat per serving, while a Sudan test might indicate otherwise?**

TABLE 3.4 DETECTION OF LIPIDS	
Solution	**Sudan III test (fats)**
Distilled water (negative control)	
Butter (positive control)	
Butter substitute	
Whole milk	
Egg white	
Egg yolk	
Egg Beaters™	
Skim milk	
1% milk	

Exercise 3.6   BIURET TEST FOR PROTEINS

(Adapted from Mathis, 1996)

Wear gloves and goggles when transferring Biuret solution!

As you will see in Chapter 6, proteins are very important molecules in living organisms **[Krogh section 3.7]**. In the Biuret test, sodium hydroxide and copper sulfate react with proteins to form a purple color. The intensity of the color indicates the concentration of protein.

1. Add 1 ml of each of the substances listed in Table 3.5 to properly labeled, clean test tubes.

2. Add 1 ml of Biuret solution to each tube and mix thoroughly by tapping the tube with your index finger.

3. Record any color changes in Table 3.5, indicating the intensity of the color change as weak, moderate, or strong.

Questions

1. **Does the protein content of the substances listed in Table 3.5 surprise you? Explain.**

TABLE 3.5 DETECTION OF PROTEINS	
Solution	**Biuret test (proteins)**
1% protein (positive control)	Strong
Distilled water (negative control)	Weak
~~Skim milk~~ Milk ~~Whole milk~~	Moderate
Egg white	Strong
Egg yolk	Weak
~~Egg Beaters™~~ Cheese	Moderate

2. **Are your results consistent with the nutrition labels for each substance?**

Exercise 3.7   COLORIMETRIC ASSAY

Wear gloves and goggles when transferring Biuret solution!

The Biuret test is an assay that detects the concentration of proteins in a solution. The concentration of an unknown substance can be determined by comparing its color change to the color of samples for which the concentration is known. Using reference samples of known concentration, you can create what is known as a **standard curve**. In this exercise, you will determine the protein concentration of an unknown sample by comparing it to a standard curve.

1. Use the 1% protein stock solution, which contains 10 mg protein per ml of water. (The conversion is as follows: 1% = 1 g/100 ml = 0.01 g/ml = 10 mg/ml.) Prepare dilutions of 0% (water), 0.2%, 0.4%, 0.6%, and 0.8% protein concentration. As seen in Table 3.6, to make a 0.8% protein concentration, you add 8 ml of 1% solution to 2 ml distilled water in a test tube. Complete Table 3.6 and prepare the rest of the solutions using a similar method.

2. Obtain a solution of unknown protein concentration from your instructor.

3. Conduct a Biuret assay: add 1 ml of Biuret solution to each of the six tubes containing 0, 0.2, 0.4, 0.6, 0.8, and 1% protein. This will give you a standard curve.

4. Then conduct a Biuret assay on your unknown solution by adding 1 ml Biuret solution to 10 ml of your unknown solution in another test tube.

5. Compare the color of your unknown solution to the standard curve by holding your unknown near the standard curve samples. Which color does it most closely resemble?

TABLE 3.6 DILUTIONS OF PROTEIN SOLUTION		
Solution, mg/ml Stock (1%) = 10 mg/ml	ml of stock solution	ml of distilled water
1.0% = 10 mg/ml	10	0
0.8% = 8 mg/ml	8	2
0.6% = 6 mg/ml		
0.4% = 4 mg/ml		
0.2% = 2 mg/ml		
0 mg/ml		

Such colorimetric assays are common procedures in biochemistry and in cell and molecular biology laboratories. However, instead of estimating the concentration of an unknown solution as we have done, a spectrophotometer is used to obtain a much more precise estimate of the concentration of the unknown.

A standard curve is determined by measuring the absorbance of light waves by a series of samples with known concentrations. A beam of light of a known intensity is passed through the sample in the spectrophotometer. The amount of light coming out of the sample is recorded. The difference between the two indicates the amount of light absorbed by the sample. Absorbance is measured in "optical density units," and the units from each of the wells are compared. In this instance, the greater the protein concentration, the darker the sample, and the greater the absorbance. Again, the protein concentration of the unknown sample is determined by comparison to the standard curve.

Questions

1. **What is the protein concentration of your unknown protein solution?**

2. **Name some practical applications of a colorimetric assay.**

Unknown or food item	Benedict's (sugar)	Iodine (starch)	Sudan (lipids)	Biuret (protein)

TABLE 3.7   COMPOSITION OF AN UNKNOWN OR A FOOD ITEM

Exercise 3.8   COMPOSITION OF AN UNKNOWN SUBSTANCE

Continuous with previous exercises

You may be given a solution of unknown composition or you may bring a food item from home. Using the assays and materials available to you from today's laboratory, try to determine the composition of the item. Keep good records of each test in Table 3.7. You will be asked to report on the composition of your item and to provide evidence to support your claim.

Questions

1. In regard to the food item that you tested, which result was most surprising? Explain why.

2. In regard to the "unknown" sample you tested, what does the substance contain?

4

Microscopy and Basic Cell Structure

... by the help of Microscopes, there is nothing so small, as to escape our inquiry; hence there is a new visible World discovered to the understanding.

Robert Hooke, 1665

The compound microscope is one of the most useful instruments available to students of biology. It is to the biologist what the hammer is to the carpenter or the violin is to the musician. The microscope has an interesting history, and has evolved from hand-held contraptions to more complex, freestanding models.

Hans and Zacharias Janssen, Dutch spectacle makers, are credited with making the first crude microscope around 1590. Their microscope was 18 inches long and only 1 inch in diameter. It is not known whether the Janssens made any profound discoveries or observations with their microscope. Galileo, however, is said to have copied their idea for his own magnifying instruments, including his famous invention, the telescope.

These first crude microscopes had low magnifying power, yet the images they imparted of common objects were stunning to unsuspecting viewers. Galileo himself noted that his microscope "made flies look as big as lambs" *(Magner, 1979)*.

Because fleas were readily available to early microscopists and were often the object of study, the name *vitrum pulicare*, or flea glass, was given to the microscope. But, as it is with many new scientific ideas and inventions, the microscope was not readily accepted throughout the educated community. Magner gives this amusing account:

> The very life of George Stiernhielm, a Swedish poet who amused himself with scientific experiments, was placed in jeopardy by the use of magnifying lenses. A Lutheran clergyman who had been persuaded to look at a flea through the magnifying glass was so frightened by the unnatural dimensions of the creature that he denounced Stiernhielm as a sorcerer and an atheist. Without the intervention of Queen Christina, the poet might have been burnt as a witch.

Hopefully, your experience with the microscope will be much less perilous.

The scientist credited for being the greatest microscopist of the seventeenth century is Anton van Leeuwenhoek (pronounced "lay-ven-hook") **[Krogh essay *How Did We Learn? First Sightings: Anton van Leeuwenhoek*, page 92]**. Leeuwenhoek's most famous works were the discovery of red blood cells, the study of insects, and his observation of sperm and ideas about reproduction. His work with parasites and bacteria were so advanced that they were not put to good use until 200 years later.

33

USE AND CARE OF THE MICROSCOPE

Treat the microscope respectfully—it is an expensive instrument. Use both hands while carrying the microscope, and always set it down gently. Never touch the lenses. Occasionally the lenses need to be cleaned, and this should be done only with lens paper, which is available from your instructor. Do not use paper towels or your handkerchief, because this may scratch the lenses. Before returning the scope to the cabinet, remove the specimen, clean the microscope if necessary, and gently wrap its cord around the base and replace its cover.

You will be using two different types of microscopes. One is the **compound microscope** (also called the light microscope), which is used for small, thin specimens. Thicker, larger specimens will be studied with a **dissecting microscope** (sometimes called a stereoscopic microscope). Most of your work today will involve the compound microscope.

There are three basic concepts associated with microscopy *(Perry and Morton, 1995)*:

1. **Magnification**—the enlargement of an image
2. **Resolution**—the ability to distinguish between separate points of an image. As magnification increases, resolution does also, up to about 1000×.
3. **Contrast**—the degree to which an image stands out against its background

Exercise 4.1   PARTS OF THE MICROSCOPE

Refer to Figure 4.1, and fill in the microscope parts as they are introduced. The **base** of the microscope contains the **light source** and the **light switch**. The **stage** and **stage clips** hold a specimen in place. Under the stage are a built-in **condenser** and **iris diaphragm**. The condenser collects light beams and directs them upward. Notice that the light

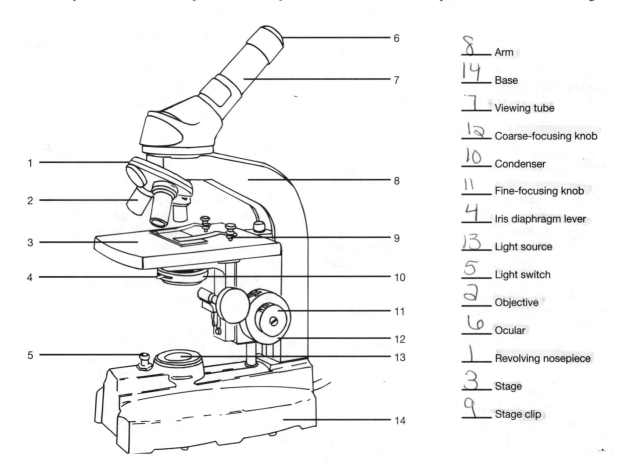

8 Arm
14 Base
7 Viewing tube
12 Coarse-focusing knob
10 Condenser
11 Fine-focusing knob
4 Iris diaphragm lever
13 Light source
5 Light switch
2 Objective
6 Ocular
1 Revolving nosepiece
3 Stage
9 Stage clip

FIGURE 4.1   Compound Microscope.

shines through an opening in the stage called the **aperture**. The iris diaphragm controls how much light comes through the aperture, and it opens and closes with the movement of a lever, which is located in front of the stage. You might have to adjust the amount of light coming through the iris diaphragm, depending on the thickness of the specimen to be viewed. Thicker specimens need more light and an open iris diaphragm; thinner specimens need less light, requiring the iris diaphragm to be closed a bit. Too much or too little light will result in a poor image of the specimen, so take time to adjust the iris diaphragm correctly. As you use the microscope, adjust the diaphragm until you see the best possible image.

The **viewing tube** contains the **ocular lens**, also called the eyepiece, which magnifies an image 10×. As you look through the ocular, you will see a circle of light called the **field of view**. There are three objective lenses on the revolving **nosepiece**, which quietly clicks when one of the objective lenses is moved into place. See Figure 4.2. Note the distance between the end of each objective lens and the stage. This is called the **working distance**. The **scanning objective**—the shortest objective—imparts a magnification of 4×, the **low-power objective** a magnification of 10×, and the longest objective—the **high-power objective**—a magnification of 40× (or 43×). Some microscopes have more than three objectives and can magnify images up to 1000×. The total magnification imparted to an object being viewed through the microscope is the product of the magnifying powers of the ocular and the objective lens being used. Complete Table 4.1.

The **arm** connects the base to the body of the microscope. The focus knobs are located near the base of the microscope. The larger knob is the

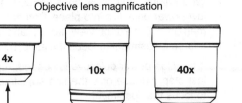
Objective lens magnification

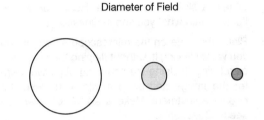
Diameter of Field

FIGURE 4.2   Relationship of objective lenses to working distance and field of view.

coarse adjustment, and the smaller knob is the **fine adjustment**. With the scanning objective in place, turn each knob while looking at the side of the microscope. Notice how the coarse adjustment moves the nosepiece and objective a great deal, while the fine adjustment moves them so little that it is hardly noticeable. With this concept, you must remember this very important rule:

NEVER use the coarse adjustment while the high-power objective is in place.

Doing so may drive the objective into the slide, crushing the slide and possibly damaging the objective lens.

BE CAREFUL!

Here are important guidelines to follow:

1. Always start with the scanning objective when viewing a specimen.

2. Place the slide on the stage so that the specimen is centered over the aperture. Locate the specimen in the field of view by looking through the microscope. Center the specimen in the field of view.

3. Use the coarse adjustment knob to focus on the specimen. Adjust the iris diaphragm to provide optimal light intensity.

TABLE 4.1 COMPOUND MICROSCOPE MAGNIFICATION			
	Magnification		
Name of Objective	**Objective**	**Ocular**	**Total**
Scanning	4	10	4 × 10 = 40
Low power		10	
High power		10	

4. When higher magnification is needed, move the low-power objective into place. Center the specimen in the field of view again, and use the coarse adjustment to focus the specimen.

5. If the high-power objective is needed, repeat the last step, but use only the fine adjustment knob to focus the specimen!

Exercise 4.2   THE LETTER E

Now you will look at your first specimen with the compound microscope.

1. Obtain a slide of the letter **e**. Hold it up to the light. Sketch what you see in Question 1.

2. Place the slide on the microscope stage just as you were viewing it, without flipping the slide over or rotating it. Using the scanning objective, center the image and bring it into focus with the coarse adjustment. Make a sketch of what you see in Question 1.

3. Move the slide to your right while looking at the image through the microscope. Note which way the image moves in the field of view. Answer Questions 2 and 3.

4. Make sure that the letter **e** is in the center of your field of view and clearly in focus. Put the low-power objective in place. Focus until a clear image is visible.

5. Make sure the image is clearly in focus. Note the working distance. Again, refer to Figure 4.2.

6. Put the high-power objective in place. Bring the image into clear focus using the fine adjustment. Again, note the working distance.

USE ONLY THE FINE ADJUSTMENT!

Questions

1. **Sketch the letter e.**

    As seen with the unaided eye:

As seen through the microscope:

2. **How does the image appear, relative to your initial sketch?**

3. **What direction does the image move in the field of view when you push the slide to the right?**

4. **What happens to the image as you increase magnification?**

5. **What happens to the size of the field of view?**

6. **Why is it important that the object is centered in the field of view before switching to a higher power objective?**

7. **What happens to the working distance as you switch to a higher power objective?**

8. **Why is it crucial to use only the fine adjustment when the high-power objective is in place?**

Exercise 4.3 CROSSED THREADS

The **depth of field** is the thickness of the specimen that you can see in focus *(Morgan and Carter, 1996).* The depth of field is short with a compound microscope and decreases with increasing magnification; thus, it is necessary to focus up and down in order to view all levels of the specimen. A good microscopist will continually use the fine adjustment. It is important to remember that the objects being viewed are three-dimensional. To illustrate this point and the concept of depth of field, you will view a prepared slide of three crossed, colored cotton threads.

1. Obtain a slide of crossed cotton threads and place it on the microscope stage.

2. Using the scanning objective, bring the specimen into focus. How many threads are in focus?

3. Switch to the low-power objective. How many threads are in focus now?

4. Switch to the high-power objective. How many threads are in focus now?

5. Remember that specimens are three-dimensional, meaning that they have depth. With the high-power objective in place, determine which thread is on top, which is in the middle, and which is on the bottom.

Questions

1. **What is the sequence of the colored threads on the slide? How do you know?**

red - in the middle
blue - on top
yellow - on bottom

Exercise 4.4 THE DISSECTING MICROSCOPE

The dissecting (stereoscopic) microscope has lower magnification (7× to 30×), and is used for examining larger specimens that cannot be viewed with a compound microscope. See Figure 4.3. Using a stereoscopic microscope, observe a variety of objects such as coins, rings, and fingers, and note how this scope differs from the compound microscope.

1. Using the focus knob, bring your specimen into focus. Note that the depth of field is much greater than with the compound microscope, so you can see the contours and outer characteristics of the specimen.

2. Note that the image seen through the microscope is not reversed or inverted.

3. Your microscope may allow you to adjust the amount and direction of the light hitting your specimen. Manipulate this by moving the external light source (unless your microscope has an internal light source), switching the knob on the light-source box, and by adjusting the mirror below the stage, if there is one.

4. The working distance of a stereoscopic microscope is much greater than that of a compound microscope. This greater working distance enables the microscopist to actually manipulate or dissect the specimen, thus the name, **dissecting microscope.** Using forceps or dissecting tools, move your specimen around while looking through the oculars.

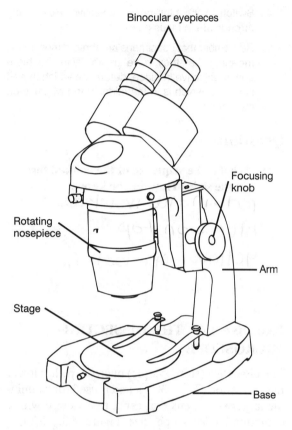

FIGURE 4.3   Stereoscopic Microscope.

Questions

1. **Sketch 2 different objects as they appear through the dissecting microscope. Label each sketch with the magnification used on the microscope.**

Exercise 4.5   PLANT CELLS

A **cell** is the basic unit of life. A cell is a highly organized structure that consists of a cell membrane with cytoplasm and DNA inside. The term *cell* was coined by the English scientist Robert Hooke

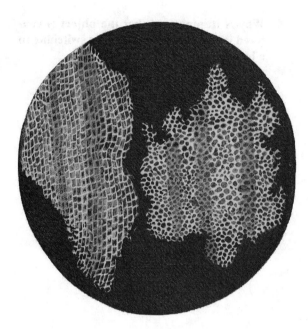

FIGURE 4.4   Robert Hooke's drawings of cells observed with a microscope.

about 300 years ago when he observed box-like structures in slices of cork and leaves (Figure 4.4). The **modern cell theory** states that all living organisms are composed of cells, and that all cells are derived from preexisting cells.

Most types of organisms have nuclei in their cells; therefore, they are called **eukaryotic** cells. The plant and animal cells you will examine are eukaryotic. Bacteria, on the other hand, have no nucleus, and are called **prokaryotic** (see Chapter 13). For additional traits of eukaryotic and prokaryotic cells, see Figure 4.5 **[Krogh Chapter 4]**.

The slides used in previous exercises were "prepared" slides, made by biological supply companies. For the next exercises, you will be preparing your own slides, called **temporary wet mounts**. See Figure 4.6.

Always deposit used microscope slides and cover slips into a designated "sharps" or "glass" container, not into the trash. Do not discard the prepared slides made by a biological supply company!

---

1. Obtain a clean glass microscope slide and place a drop of water on the center of it.

2. Get a small piece of sliced onion. Snap the piece of onion in half and peel the thin, transparent surface off the inner lining of the onion.

know this chart

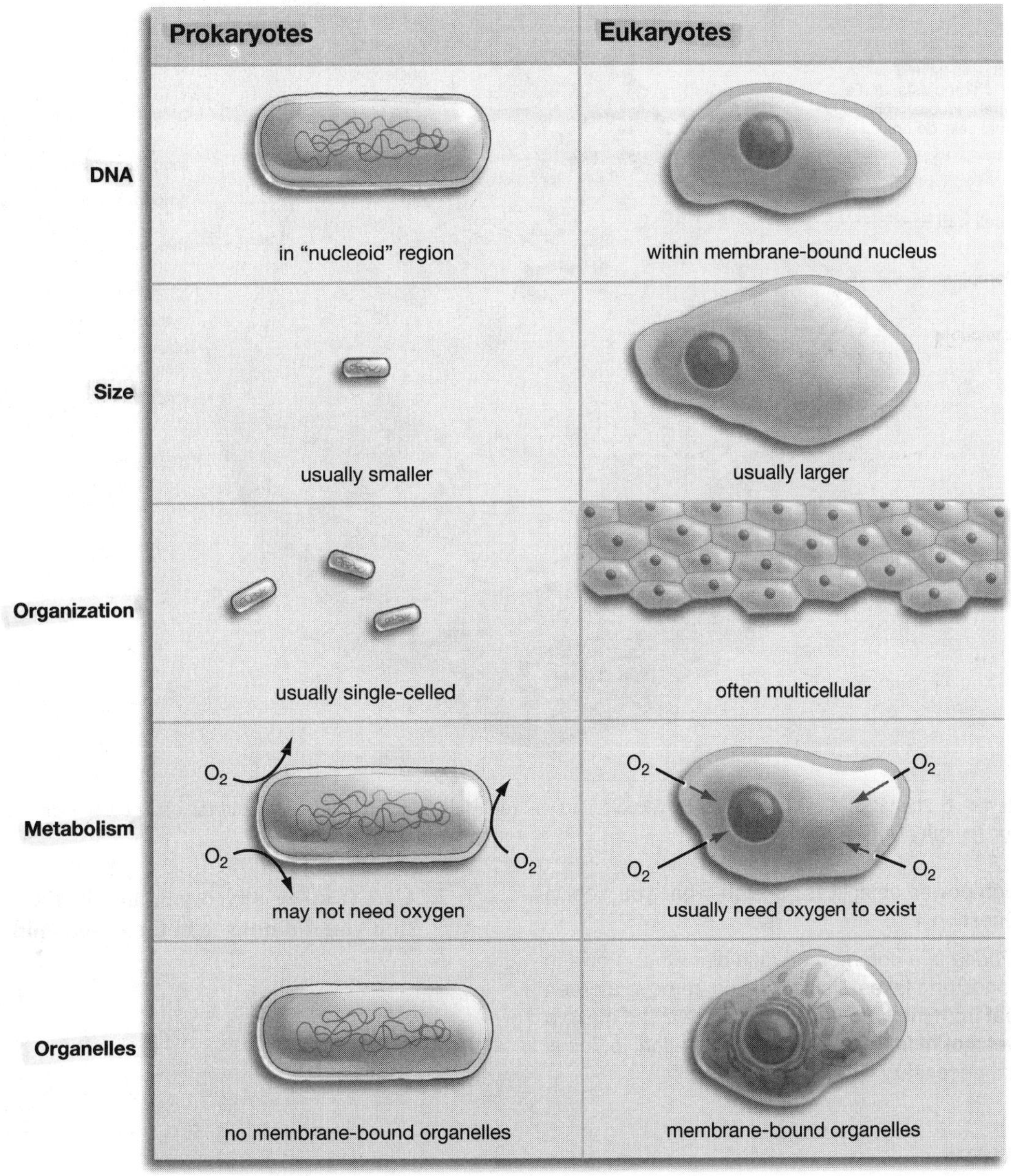

Prokaryotes	Eukaryotes
DNA	
in "nucleoid" region	within membrane-bound nucleus
Size	
usually smaller	usually larger
Organization	
usually single-celled	often multicellular
Metabolism	
may not need oxygen	usually need oxygen to exist
Organelles	
no membrane-bound organelles	membrane-bound organelles

FIGURE 4.5   a. Prokaryotic and eukaryotic cells compared. A prokaryote cell is a self-contained organism, like the bacteria. eukaryotic organisms are either single-celled or multicellular, and include protists, fungi, plants, and animals.

This tissue layer is known as an epithelial lining. See Figures 4.6a and 4.6b.

3. Place the onion epithelial tissue flat into the drop of water on the slide. If it rolls up or folds over, use dissecting pins to unfold it.

4. See Figure 4.6c. Place one edge of the cover slip on the slide, and carefully lower it with a dissecting needle at an angle to reduce the chance of trapping an air bubble. Examine the specimen with your compound microscope.

5. Now take the cover slip off your onion specimen, and stain the tissue by adding a drop of iodine. Replace the cover slip and look at the specimen again with your microscope, using your low- and

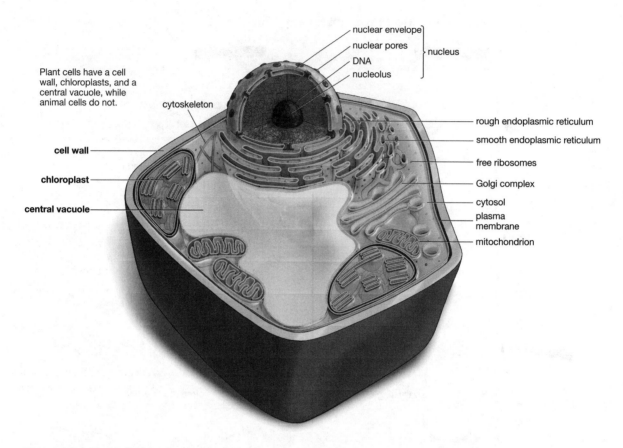

Plant cells have a cell wall, chloroplasts, and a central vacuole, while animal cells do not.

nuclear envelope
nuclear pores } nucleus
DNA
nucleolus

cytoskeleton

cell wall

chloroplast

central vacuole

rough endoplasmic reticulum
smooth endoplasmic reticulum
free ribosomes
Golgi complex
cytosol
plasma membrane
mitochondrion

FIGURE 4.5 b. Plant cell. The cell wall, central vacuole, and chloroplasts are not in animal cells, but the other components are common to both plant and animal cells.

high-power objectives. Sketch what you see in Question 4.

6. *Elodea* is a common aquatic plant that grows in ponds and lakes. Using forceps, remove a green leaf from the tip of a piece of *Elodea* and make a wet mount for microscopic examination. Iodine is not necessary.

Questions

1. **What is the advantage to staining specimens?**

2. **Can you see any organelles in the *Elodea* that you did not see in the onion epidermis?**

3. **Why was the absence of these organelles in the onion not surprising?**

a. Break a piece of onion scale.

b. Peel off a small piece of the inner epithelial lining.

c. Place the epithelial lining flat onto the drop of water, and lower the cover slip.

FIGURE 4.6   Temporary wet mount of onion epithelial tissue.

4. **Sketch your onion specimen at low- and high- power magnification. Label the nucleus and cell wall, and indicate the total magnification near each sketch.**

Exercise 4.6   ANIMAL CELLS

Your mouth, like all hollow organs, ducts, and exposed surfaces of your body, is lined by epithelial tissue. Among their many functions, epithelial cells serve as the body's first line of defense against invading pathogens. In this exercise, you will prepare a stained wet mount of your epithelial cells.

1. Gently scrape the inside of your cheek with a toothpick.

2. Rub the accumulated material onto a dry slide, and discard the toothpick.

3. After 1 minute, add a drop of methylene blue. Wait for 1 minute, and then rinse the methylene blue off the slide with a few drops of water. Add a cover slip and observe with a compound microscope immediately. Draw a typical cheek epithelial cell in the space below.

Questions

1. **What parts of the cheek epithelial cell does methylene blue adhere to? Sketch the cells and label the stained structures.**

2. **Compare the shape of the cheek cells to the shape of the plant cells. What causes the different appearance of these cells?**

3. **Do you see the small "black dots" near the cheek epithelial cells? What do you think they are?**

Algae

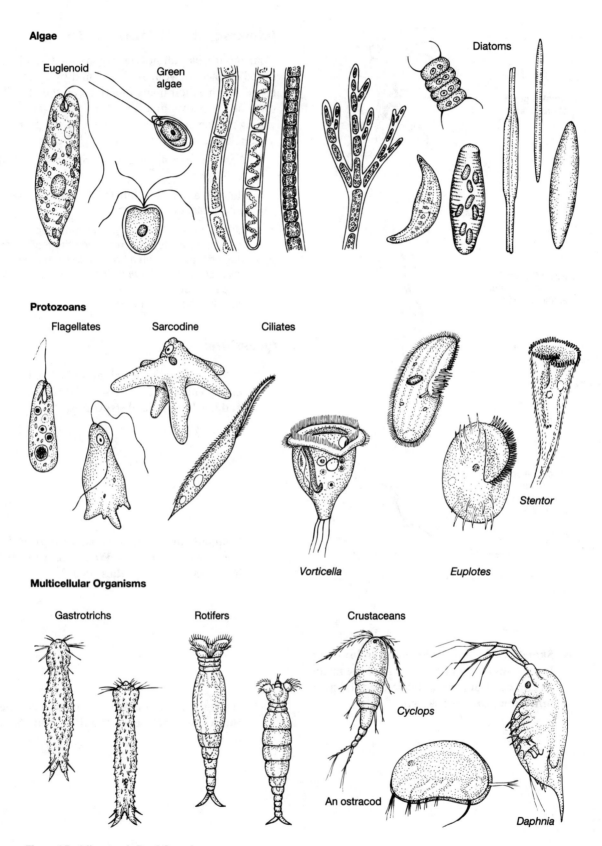

Figure 4.7 Microscopic Pond Organisms.

Exercise 4.7   POND WATER

1. Make a wet mount of pond water and examine the slide.

2. Most of the fascinating little critters that you see are single-celled organisms called protozoans. You may also see rotifers, gastrotrichs, and microcrustaceans, all of which are multicellular organisms with complex organs. Sketch what you find, and also try to identify some of the specimens by referring to Figure 4.7.

Sketches

1. **Sketch the pond organisms seen in the temporary wet mount.**

5

Diffusion and Osmosis

The maintenance of a stable internal environment in a cell or an organism is called **homeostasis**. This constant environment is necessary for cells to function efficiently. In order to maintain homeostasis, cells continually exchange materials with their environment. Ions and organic compounds such as sugars, amino acids, and nucleotides must enter the cell, while waste products and other substances must leave it. Regardless of the direction of movement (either into or out of the cell), the structure that regulates the movement of materials into and out of the cell is the cell membrane **[Krogh section 5.4]**.

Diffusion is the movement of molecules or ions from a region of their high concentration to a region of their low concentration (Figure 5.1). The exercises in this chapter will show you how diffusion enables a cell to maintain homeostasis. The rate of diffusion is affected by three factors:

1. **Temperature** At normal environmental temperatures, atoms, ions, and molecules are in constant motion, continuously colliding with one another. Molecules in a gaseous phase have greater kinetic energy (energy of movement) than molecules of the same substance in a liquid or a solid phase. As the temperature of any phase is raised, the speed of movement, frequency of collisions, and force of the collisions all increase.

2. **Molecule size** Molecules are always diffusing through a "medium" of other molecules, passing through the spaces in between. A medium can consist of a gas, a liquid, or even a solid. Larger molecules diffuse more slowly because of resistance from molecules of the medium. Smaller molecules can fit between the spaces in the medium more easily, and so can diffuse faster.

3. **Concentration gradient** If a high concentration of a substance is added to a solution, it will gradually diffuse until it is equally distributed throughout the solution. In any localized region of high concentration, the movement of molecules is, on the average, away from the region of highest concentration and toward the region of lowest concentration. The difference in concentration between high and low regions is called the **concentration gradient**. As diffusion proceeds, the diffusion gradient decreases, and diffusion slows **[Krogh section 5.2]**.

(a) Dye is dropped in   **(b)** Diffusion begins   **(c)** Dye is evenly distributed

water
molecules

dye
molecules

FIGURE 5.1   A few drops of red dye added to a beaker of water are at first heavily concentrated in one area. But then they begin to diffuse, eventually becoming evenly distributed throughout the solution.

Exercise 5.1   EFFECT OF MOLECULAR WEIGHT ON DIFFUSION

Wear gloves and goggles.

Molecular weight is an indication of the mass and the size of a molecule. Heavier molecules tend to be larger than lighter molecules. The purpose of this exercise is to determine the effect of molecular weight on the rate of diffusion through agar, which is a solid substance at room temperature. Agar is a gelatinous extract of red algae, used as a culture medium for small organisms and as a stabilizing agent in foods. The agar used in this experiment contains the pH indicator methyl red, which turns from yellow to pink in the presence of acids. By adding acids to the agar and measuring the portion of the agar that has turned pink, you can determine the rate of diffusion of the acids.

1. Obtain an agar-filled petri dish. Each petri dish contains agar with 0.02% methyl red (6 ml of the methyl red solution to 100 ml of agar). With a cork borer or a cut-off disposable pipette, slowly press two small holes in the agar. Make holes at equal distances from each other and from the edge of the petri dish. Label each hole on the bottom side of the petri dish.

2. Fill one hole with hydrochloric acid and the other with citric acid, one drop at a time, taking care not to overfill them. Record the start time.

3. At 15-minute intervals, measure the diameter of the acid that has diffused through the agar. Record your data in Table 5.1.

Questions

1. **Which acid exhibited the greatest rate of diffusion (mm/min)?**

smaller
pore size

TABLE 5.1 MEASUREMENT OF RATE OF DIFFUSION USING MOLECULAR WEIGHT		
Start time	**Distance moved (mm)**	
7:00 pm	**HCl**	**Citric acid**
1̶0̶ min	12 mm	10 mm
2̶0̶ 3̶0̶ m̶i̶n̶	14 mm	11 mm
3̶0̶ 4̶5̶ m̶i̶n̶	16 mm	13 mm
6̶0̶ m̶i̶n̶		

2. The molecular weight of citric acid is about five times that of hydrochloric acid. What is the relationship between molecular weight and rate of diffusion?

Exercise 5.2   OSMOSIS IN ANIMAL CELLS

Wear gloves and goggles for this experiment!
Do not inhale the acetic acid fumes. If
necessary, work under a fume hood.

If a solid substance, such as salt, is dissolved in water, the salt particles will diffuse through the water until they are evenly distributed. The salt is a **solute**, and the water is a **solvent [Krogh section 5.2]**. We are primarily interested in **osmosis**, the

diffusion of water molecules across a semiper-meable membrane (such as a cell membrane). Osmosis proceeds from an area of high water concentration (lower solute concentration) to an area of lower water concentration (higher solute concentration). See Figure 5.2. Water moves from a higher water concentration to an area of lower water concentration, following the rules of diffusion.

When comparing two solutions with different concentrations of solute, the solution with the higher solute concentration is termed **hypertonic**, and the solution with the lower solute concentration is **hypotonic**. Solutions with equal solute concentrations are termed **isotonic**. The terms _hypertonic_, _hypotonic_, and _isotonic_ are relative and are used only when comparing two or more solutions.

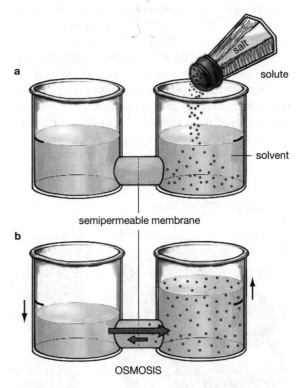

FIGURE 5.2   a. As salt is added to the solution on the right, the solution becomes hypertonic compared to the solution on the left of the selectively permeable membrane. b. Osmosis will proceed across the selectively permeable membrane from the hypotonic solution on the left to the hypertonic solution on the right.

solute - being dissolved
solvent -

Water will always diffuse from a hypotonic solution to a hypertonic solution. A hypotonic solution has a higher water concentration than a hypertonic solution.

A typical animal cell, for instance, has a solute concentration of 0.9% (Figure 5.3). If you put this cell in a hypotonic solution of distilled water (0% solute), then water will diffuse into the hypertonic cell. However, if you put this same cell into a solution of 5% solute, water will diffuse from the hypotonic 0.9% cell to the hypertonic 5% solution. *When two solutions are separated by a selectively permeable membrane, net movement of water will always be from the hypotonic solution to the hypertonic solution!* At **equilibrium**, note that water is still moving across the membrane; however, there is equal movement in both directions and thus no net movement of water. For example, a 0.9% cell placed in a 0.9% solution is at equilibrium with the solution.

A chicken egg is one of the largest living cells known. It is a single cell surrounded by a continuous cell membrane that lies directly underneath the eggshell. The shell of the egg is easily removed by soaking it in an acetic acid solution. Using chicken eggs, you are going to conduct an experiment to investigate the influence of one or more factors on the rate of osmosis. The eggs you will use have been soaked in a solution of 1 part acetic acid to 2 parts tap water (1:2).

osmotic pressure-pressure at equilibrium

1. **Wearing rubber or latex gloves and goggles,** obtain two eggs.

2. Dry the eggs with a paper towel.

3. Using a triple beam balance, weigh the eggs to the nearest 0.1 gram. Record these weights in the $Time_0$ row in Table 5.2.

4. Place one egg in a beaker of distilled water, and record the start time in the $Time_0$ cell in Table 5.2. Place the other egg in a beaker of 10% salt (NaCl) solution and record the start time.

5. While you wait, calculate the time intervals: if $Time_0$ is 8:35, then $Time_{15}$ is 8:50, and so on. Record these intervals in the "$Time_n$" column.

6. At 15-minute intervals, remove the egg from the beaker, dry off excess water, and weigh. After recording the weight, place the egg back into the beaker.

7. Record your results in Table 5.2 and complete the "Percent change" column. Percent change is calculated as follows:

$$\text{Percent change} = \frac{\text{weight}_{(n+15)} - \text{weight}_{(n)}}{\text{weight}_{(n)}} \times 100$$

where $time_{(n+15)}$ refers to the weight at the present time interval, and $weight_n$ refers to the weight at the previous time interval.

8. Graph your results in Figure 5.4, using different bars to represent the percent change in weight for the egg from each solution.

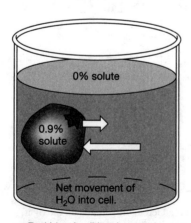

a. Red blood cell in a hypotonic solution.

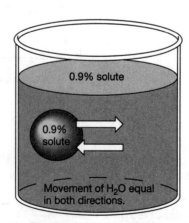

b. Red blood cell in a isotonic solution.

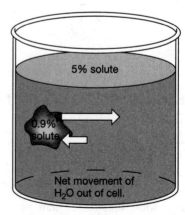

c. Red blood cell in a hypertonic solution.

FIGURE 5.3   Red Blood Cells in Hypotonic, Isotonic, and Hypertonic Solutions.

Hypotonic   Hypertonic

TABLE 5.2 PERCENT CHANGE IN WEIGHT OF EGGS					
Egg in distilled water			**Egg in 10% salt**		
Time$_n$	Weight (g)	Percent change	Time$_n$	Weight (g)	Percent change
0 7:10	~~82g~~	—	0	81.9g	—
15	82.8g	0.98 " or 97%	15	81.6g	~~0%~~
30	83.4g	.73	30	81.6g	—
45	83.6g	0.24%	45	81.9g	0.37%
~~60~~			~~60~~		

0.37 or 37%

Average = .65   Average = .25

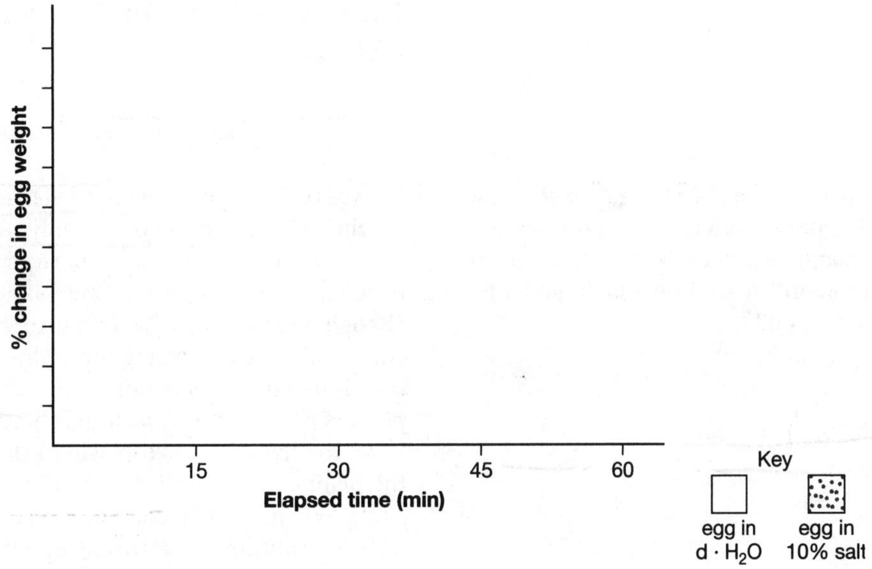

FIGURE 5.4   Bar Graph of Percent Change in Egg Weight. Use the symbols shown in the key.

Questions

1. **What happened to the weight of the eggs over time? Explain your results using the terms *hypotonic* and *hypertonic*.**

   water diffusing out of membrane because it (why it lost weight).

2. **Calculate the average percentage of change in weight for eggs in each solution.**

3. **Which egg exhibited the greatest weight gain? Explain your observations.**

   diffusing into the membrane (soaking it up)

4. **Was the percent weight change for the first 15-minute interval different from the percent weight change at intervals near the end of the experiment? If so, how much, and why does this happen?**

   trying to reach equilibrium concentration gradient

5. **Why did the rate of osmosis change over time?**

   as egg tries to reach equilibrium the water rate of osmosis decreases.

6. **Why do you think it is necessary to remove the shell to conduct this experiment?**

   shell blocks/prevents egg from absorbing water

7. **How could the rate of osmosis be increased in this experiment?**

   by heating up the water or getting rid of salt. (higher temp, higher rate diffusion)

Exercise 5.3   OSMOSIS IN PLANT CELLS

(Adapted from Mathis, 1996)

Leaves of the aquatic plant *Elodea* make excellent specimens for microscopic study. Aquatic plants tend to be at least slightly hypertonic to their environment, and this causes the plant to be turgid [**Krogh Figure 5.6**]. The pressure created by the continual influx of water through osmosis pushes the plasma membrane out against the cell wall. If a plant is placed in a hypertonic solution, water will leave the hypotonic cytoplasm of the cells, causing the plant to wilt.

When animal cells are placed in a dilute aqueous solution, they are extremely hypertonic to their environment. In such an instance, water will enter the cell by osmosis until the cells rupture or *lyse* (Figure 5.3a). Plant cells are encased by their cell walls, however; instead of rupturing the cell, the increase in water pressure would cause the water to flow back out of the cell. Eventually, equilibrium would be reached, in which case the flow of water into the cell due to the concentration differences would balance the flow out of the cell due to pressure differences. The pressure at equilibrium is called the **osmotic pressure** of the solution.

1. Tear three small leaves from the growing tip of an *Elodea* shoot with a pair of forceps.

2. Prepare a temporary wet mount of each leaf using pond water for one, distilled water for the second, and 20% NaCl for the third. Label each slide with a wax pencil.

3. Wait at least 10 minutes, then observe the cells in each leaf with a compound microscope. Compare the cells in distilled water and in 20% NaCl with the cells in pond water. The leaf in the pond water demonstrates the appearance of *Elodea* under "normal" conditions. Sketch the cells of each leaf in Question 1 below.

Questions

1. **Sketch the plant cells as seen on the temporary wet mounts:**

   Elodea **cell in pond water:**

   Elodea **cell in distilled water:**

   Elodea **cell in 20% NaCl:**

2. **What are the green organelles observed in the leaf, and what is their function?**

3. **What happened to the *Elodea* cells bathed in distilled water? Explain your answer using the terms *hypotonic* and *hypertonic*.**

4. **What happened to the cells when exposed to the salt solution? Explain your answer using the terms *hypotonic* and *hypertonic*.**

5. **Why is it useful to compare leaves in distilled water and salt solution to a leaf in pond water?**

6. **Of the three *Elodea* slides, which one exhibits the largest concentration gradient? Explain your answer.**

H

Exercise 5.4   THE EFFECTS OF OSMOSIS ON PLANT TISSUES

Design an experiment to demonstrate the effects of osmosis on potato pieces. Place equal-sized pieces of potato in hypotonic and hypertonic solutions that you can prepare yourself. After 45 minutes, compare the texture of the potato plugs in the two solutions.

Questions

1. Describe or sketch the setup you prepared, comparing or labeling the solutions inside and outside the potato tissue, using the terms *hypotonic* and *hypertonic*.

2. How do the two potato pieces differ in texture? Bend them to find out.

3. Imagine you are hosting your family's holiday dinner, and you notice that your celery and carrots are wilted. You don't want to exacerbate the holiday tensions, so to avoid a potential family feud, you need to find a quick way to restore the vegetables to their crisp texture. One relative suggests placing the vegetables in a bowl of salty water. She claims that the celery will not only be crisp again but it will also be presalted. Would you follow her advice, and why?

4. Another cousin suggests putting the celery in tap water with ice, then into the fridge for a quick chill. Would you follow his advice, and why?

5. While snooping under your sink, an in-law finds an old jug of distilled water that you use on rare occasions for your steam iron. He chuckles about using the distilled water, saying that you should just pour it on the celery and leave the celery on the table for everybody to eat anyway. Would you follow his advice, and why?

UNIT 2
Energy

E

Proteins and Enzymes

20 - 25 amino acids

Proteins are important organic molecules—they serve as nutrients, building materials, communicators, and catalysts.

A **protein** is a chain of amino acids linked together by peptide bonds (Figure 6.1). The sequence of amino acids comprising a protein determines its shape, and the shape determines its function **[Krogh section 3.7, Krogh Figure 6.9]**. Amino acids will interact in a specific way to cause the polypeptide chain of the protein to conform to a predictable three-dimensional shape. When the chemical bonds between amino acids of a polypeptide chain are disrupted, a change in shape results in a change in function.

Virtually all enzymes are proteins **[Krogh sections 6.5 & 6.6]**. As mentioned earlier, enzymes are organic **catalysts**, substances that increase the rate of a chemical reaction. Enzymes do not cause reactions to occur; they only speed up their rate and are not consumed in the reaction. For example, the enzyme carbonic anhydrase catalyzes the reaction of carbon dioxide with water to form carbonic acid:

$$CO_2 + H_2O \underset{\text{carbonic}}{\overset{\text{anhydrase}}{\rightleftharpoons}} H_2CO_3$$

The reaction will proceed without carbonic anhydrase, but at an almost imperceptible rate that would not be useful to an organism. The presence of carbonic anhydrase accelerates the reaction by 10 million times. At any given second, billions of complex biochemical reactions are occurring in your body and most are catalyzed by specific enzymes.

Each enzyme has an optimal pH and temperature. The optimal pH and temperature of an enzyme tends to be that of its natural environment. For instance, most human enzymes have an optimal temperature at or near body temperature (37° C). Salivary amylase has an optimum pH of 6.8, and it catalyzes the digestion of starch as food is chewed and swallowed. Pepsin is an enzyme that catalyzes the digestion of proteins in the food you eat. It works best at a pH of 2.0, which is the pH provided by hydrochloric acid in the stomach.

Because enzymes are proteins, their activity is determined directly by their shape. A change in the shape of an enzyme has a direct effect on its function, and it may no longer be able to catalyze a given reaction. An enzyme "fits" with its substrate

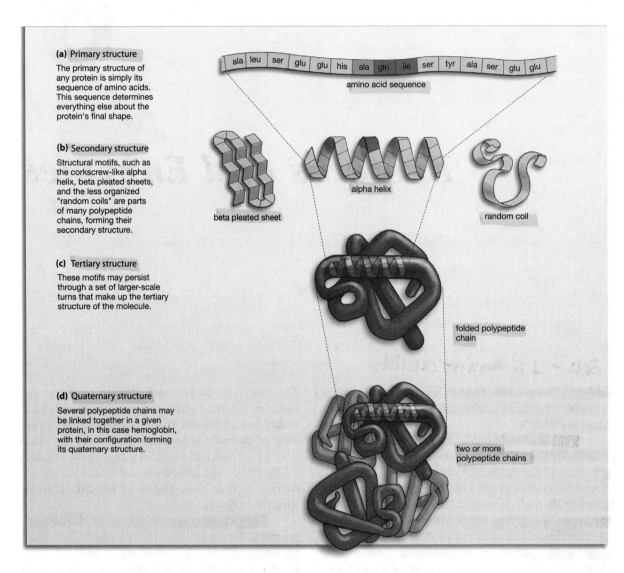

(a) Primary structure

The primary structure of any protein is simply its sequence of amino acids. This sequence determines everything else about the protein's final shape.

| ala | leu | ser | glu | glu | his | ala | gln | ile | ser | tyr | ala | ser | glu | glu |

amino acid sequence

(b) Secondary structure

Structural motifs, such as the corkscrew-like alpha helix, beta pleated sheets, and the less organized "random coils" are parts of many polypeptide chains, forming their secondary structure.

beta pleated sheet

alpha helix

random coil

(c) Tertiary structure

These motifs may persist through a set of larger-scale turns that make up the tertiary structure of the molecule.

folded polypeptide chain

(d) Quaternary structure

Several polypeptide chains may be linked together in a given protein, in this case hemoglobin, with their configuration forming its quaternary structure.

two or more polypeptide chains

FIGURE 6.1   Four Levels of Structure in Proteins.

much like a key fits into a lock. But when the temperature or pH in the environment of a protein is altered, chemical bonds may be broken and/or created, changing the shape of the protein molecule so that it no longer fits like a key into a lock. However, when the altering force is removed, the enzyme may regain its original shape and function. On the other hand, if the change in temperature or pH is too great, the enzyme completely and permanently loses it shape and function. In such a case, the protein or enzyme is **denatured** and usually "precipitates out" or solidifies. An example is the solidification of egg white when it is cooked or the curdling of milk when it is overheated.

Exercise 6.1a   CHEESE PRODUCTION USING ENZYMES

(© 1974, Carolina Biological Supply Company, used by permission.)

Wear gloves and goggles while transferring hydrochloric acid! Use caution near the boiling water bath.

Cheese and many other dairy products are produced by the breakdown of proteins in milk. Specifically, cheese is produced when the enzyme *rennin* catalyzes the conversion of the protein casein into a substance called **para-casein**. In this instance, peptide bonds are not being broken, but other bonds in the protein are being disrupted. This change in shape causes the proteins to coagulate and "fall out" of the solution. The para-casein is called the **curd** (the cheese) and the remaining liquid portion is called the **whey**. Remember the nursery rhyme:

> Little Miss Muffet,
> She sat on a tuffet,
> Eating her curds and whey;
> Down came a spider,
> Which sat down beside her,
> And frightened Miss Muffet away.

Now you know what little Miss Muffet was eating, but what is a tuffet?

1. Dissolve the powdered milk as directed on the package. Put 250 ml (1 cup) of milk in a 400-ml beaker.

2. Add 5 drops of 1 N HCl. Heat the milk to 32 °C (90 °F), stirring continuously.

FIGURE 6.2

3. Add 1 ml (10 drops) of Emporase™ (rennilase), stir, and remove from heat. Allow to stand until the milk coagulates (10–15 minutes).

4. Break up the curd with a glass rod, separating it from the whey.

5. Use another 400-ml beaker to catch the whey, and filter the mixture through cheesecloth. Save the curd and discard the whey. Keep the beaker.

Exercise 6.1b   CONVERTING CHEESE TO AMINO ACIDS

(© 1974, Carolina Biological Supply Company, used by permission.)

You are about to cut the cheese at a biochemical level. The enzyme *bacterial protease* catalyzes the breakdown of proteins in the cheese into individual amino acids. It catalyzes the breaking of peptide bonds holding the amino acids together. You will use the chemical ninhydrin, an indicator that turns purple in the presence of free amino acids, to determine whether you have successfully degraded the protein in your cheese. This digestion is very similar to digestion in your small intestine when you eat cheese. The proteins in cheese are broken down into amino acids, which are absorbed by the small intestine. These amino acids may be used in the synthesis of proteins in your body.

1. Label one 250-ml beaker "A" and one 250-ml beaker "B." To each beaker, add one-half of the curd and 100 ml of water. Stir well to break up the curd.

2. To beaker A, add 1 g (1/4 teaspoon) of the enzyme *bacterial protease*. Add no enzyme to beaker B. Stir both beakers and allow to stand for 5 minutes.

3. Cut two pieces of cheesecloth, place them over separate funnels, and filter 5 ml of each solution into separate test tubes marked A and B.

4. Add 1.0 ml (10 drops) of ninhydrin to each test tube and gently tap the tubes with your fingers to mix them.

5. Place test tubes A and B into a boiling water bath. After 10 minutes, remove the tubes and record any color change.

6. If time allows, take samples from beakers A and B at 10-minute intervals, filter, add ninhydrin reagent, and heat as before.

Questions

1. **What causes the proteins to coagulate (precipitate) in Exercise 6.1a?**

2. **Which test tube contains amino acids?**

3. **Which test tube, A or B, is the control tube? Explain.**

4. **If you completed Step 6, did the intensity of the color change increase as time elapsed? Why?**

Exercise 6.2 EFFECT OF pH ON ENZYME ACTIVITY

(Adapted from Glider, 1993)

Wear goggles and gloves.

Pineapple plants *(Ananas comosus)* belong to a group of tropical plants called **bromeliads** (Figure 6.3). Pineapples contain the proteolytic enzyme **bromelian**, which catalyzes the breakdown of proteins. You will allow bromelian to come in contact

with photographic film, which consists of a thin layer of gelatin mixed with silver particles that adhere to a strip of clear plastic. The silver particles give the film its dark color. Because gelatin is made of protein, it falls off the plastic strip when it comes into contact with bromelian. When this happens, the brown silver particles go with it and the film will look clear.

The pH of a solution refers to the concentration of H$^+$ ions in a solution **[Krogh section 3.2]**. These free ions are available to bond with atoms that comprise protein molecules and therefore have the ability to change the shape of the proteins enough to impair the protein's function, even though the peptide bonds between amino acids remain intact. Because bromelian is a protein, it has an optimal pH at which it works best. But if it comes into contact with solutions that are not within its optimal pH range, the enzyme will become impaired or deactivated. For this exercise, you will conduct an experiment to measure the activity of bromelian at various pH levels. You will need to measure the amount of time bromelian takes to catalyze the breakdown of gelatin on the film strips at various pH levels.

1. Obtain 10 clean test tubes, label them 1 through 10, and fill them according to Table 6.1.

2. Cover the mouth of each tube with parafilm and your thumb, mixing the contents of each tube by inverting it several times. Or, use a vortex mixer if one is available; be sure to direct the tubes away from your face.

3. Using pH paper and forceps, determine the pH of each tube and record this in Table 6.1. Be careful not to spill the solutions as you take the paper out of the tubes. You may need to use a broad-range

FIGURE 6.3 Harvesting pineapples at a plantation in Oahu, Hawaii.

 meat will denature because it is protein

TABLE 6.1	EFFECT OF pH ON BROMELIAN ACTIVITY						
Start time:							
	Experimental				**Control**		
Tube #	Filled with	pH	Clearing time	Tube #	Filled with	pH	Clearing time
1	2 ml 1 N HCl + 5 ml fresh juice			2	2 ml 1 N HCl + 5 ml boiled juice		
3	2 ml pH 3.0 buffer + 5 ml fresh juice			4	2 ml pH 3.0 buffer + 5 ml boiled juice		
5	2 ml pH 5.0 buffer + 5 ml fresh juice			6	2 ml pH 5.0 buffer + 5 ml boiled juice		
7	2 ml pH 7.0 buffer + 5 ml fresh juice			8	2 ml pH 7.0 buffer + 5 ml boiled juice		
9	2 ml 1 N NaOH + 5 ml fresh juice			10	2 ml 1 N NaOH + 5 ml boiled juice		

pH paper first, then a specific-range paper to pinpoint the pH.

4. Place the test tubes in a 40° C water bath and allow them to equilibrate for 5 minutes. While you wait, answer Questions 1 to 7.

5. Touching only the edges of the film, place a film strip in each test tube and record the start time.

6. Every 3 minutes, use forceps and vigorously shake the film in the test tubes, then check for clearing. A clear film will be completely transparent, with no brown residue. Record how long it takes for clearing to occur, in Table 6.1. If the film in the experimental or control tubes shows no signs of clearing after 60 minutes, record "no clearing" in the column labeled "Clearing time."

Questions

1. **Hypothesis:**

2. **Prediction (Which pH solution will allow bromelian to clear the film most quickly?):**

3. **Independent variable:**

4. **Dependent variable:**

O

5. **Controlled variables:**

6. **Experimental group:**

7. **Control group:**

8. **Was your prediction correct? (Do your data support your hypothesis?)**

9. **What is the optimal pH for bromelian activity? How do you know?**

10. **Why is boiled pineapple juice used for the control tubes?**

11. **Why doesn't bromelian catalyze the digestion of proteins in the lining of your stomach? (See Krogh section 26.7.)**

12. **Salivary amylase (Exercise 3.4) is an enzyme that catalyzes the digestion of starches in the mouth. What happens to the taste of a starchy food when a person chews it, and why?**

13. **What might happen to the effectiveness of salivary amylase if a person drinks a beverage while chewing food?**

14. **Do you think there might be an optimal pH and temperature for digestion of starch? How does this compare with the conditions present in our bodies?**

15. **Salivary amylase has to work quickly. Why might this be so?**

16. **How might certain substances such as the acetic acid (vinegar) in salad dressing affect salivary digestion?**

17. **What other foods may have a similar effect on enzymes?**

Exercise 6.3   EFFECT OF TEMPERATURE ON ENZYME ACTIVITY

Wear goggles and gloves.

Continuous with Exercise 6.2.

For this exercise, you will design and conduct an experiment to measure the effect of temperature on the activity of bromelian. Remember that the enzyme has an optimal temperature at which it works best, and your job is to figure out what temperature that is. Using methods described in Exercise 6.2, expose filmstrips to bromelian, and check for clearing. Be sure to use an experimental and a control group for each temperature tested; record these in Table 6.2. What solutions are you going to put in the experimental and control tubes? Hint: You need to use pH as a controlled variable.

These are the items you'll need:

At least 8 clean 15-ml test tubes

Wax pencil for numbering tubes

Fresh pineapple juice

Boiled pineapple juice

pH 3.0 buffer

Water baths of varying temperatures

Filmstrips

Disposable pipettes

Forceps

Recommendations:

1. Allow the temperature of the solutions in the test tubes to equilibrate in the various water baths for 5 minutes before exposing the filmstrips to bromelian. Answer Questions 1 to 7 while you wait.

2. Touch only the edges of the film.

3. Every 3 to 5 minutes, vigorously agitate the film, using forceps, and check for clearing. If the film in the experimental or the control tubes shows no signs of clearing after 60 minutes, record "no clearing" in the column labeled "Clearing time."

Questions

1. **Hypothesis:** The lower the temperature, the faster the bromelian will clear the film most effectively.

2. **Prediction (at which temperatures do you think bromelian will clear the film most effectively?):** 20°

3. **Independent variable:** Temperatures of bromelian (manipulated)

4. **Dependent variable:** Rate at which bromelian will clear the film.

TABLE 6.2   EFFECT OF TEMPERATURE ON BROMELIAN ACTIVITY

Start time: 7:09

Tube #	Filled with	Temp.	Clearing time	Tube #	Filled with	Temp.	Clearing time
1	fruit juice	20°	no clearing time	5	boiled f.j.	20°	7:41
2	fruit juice	30°	7:40	6	boiled f.j.	30°	—
3	fruit juice	60°	7:30	7	boiled f.j.	60°	—
4	fruit juice	70°	7:26	8	boiled f.j.	70°	—

5. Controlled variables:

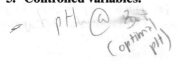

pH @ 3ost (optimal pH)

6. Experimental group:

The fesh juice juice

7. Control group:

Boiled fruit juice

8. Was your prediction correct? (Do your data support your hypothesis?)

No

9. What is the optimal temperature for the activity of bromelian?

70°

Exercise 6.4   EFFECTS OF ENZYMES ON YOUR DINNER

Use caution when handling boiling water.

Enzymes have desirable and undesirable effects on the food you eat. Next, you will prepare a steak and two varieties of gelatin dishes, one with fresh pineapple and one with canned pineapple, to illustrate this point.

1. Mix gelatin with 150 ml (1/2 cup) of boiling water per 3-ounce package. Stir to dissolve the gelatin, then add approximately 450 ml ($1\frac{1}{2}$ cups) of ice. Stir until all ice is melted.

2. In a bowl labeled "fresh pineapple," mix some fresh pineapple juice and fresh pineapple pieces with approximately 150 ml (1/2 cup) of the gelatin mixture.

3. Repeat Step 2, except use canned pineapple and a bowl labeled "canned pineapple."

4. Place both bowls into a refrigerator or freezer, and note the results just before you leave class.

5. Obtain two small pieces of fresh beefsteak. Apply meat tenderizer to one piece and pierce the meat with a scalpel or fork. Leave the other piece untreated.

6. After a few minutes, use a scalpel and cut through each steak. Note the texture of each.

Questions

1. **Explain the appearance of each food type. Why do they appear this way?**

   w/ tend = stringy / broken up / strained

   w/out tend = together, compact

2. **Look at the meat tenderizer label. What is the active ingredient?**

   Bromelain

3. **Explain how enzymes can be helpful in food preparation.**

   tenderizes / breakes up the meat tissue

4. **What foods might be a nuisance when preparing gelatin dishes?**

   fresh fruit pineapple

5. **What would happen to the bromelian if the steak were cooked?**

   — It would not tenderizer the meat
   — the heat denatures the bromelian

Exercise 6.5   ENZYME DEFICIENCIES

Web exercise

Investigate the effects of an enzyme deficiency in an animal (including humans). Try the Web sites listed below for ideas, but you should also conduct your own Internet search. Use a minimum of three reliable Web sites.

> http://www.medicinenet.com
> http://medlineplus.gov
> http://www.mic.ki.se/Diseases/index.html

Retype the questions below and provide your answers in double-spaced format.

Questions

1. **What search engine(s) did you use? What keywords?**

W

2. **What enzyme deficiency are you investigating? (Name the enzyme and the disease it causes.)**

3. **Normally, what biochemical reaction does the enzyme catalyze?**

4. **What are the effects of the enzyme deficiency on the individual's health?**

5. **Is there an effective treatment for the deficiency? If so, describe how it is applied, how it works, and its cost.**

6. **Are there any side effects of the treatment? Describe them. Might these side effects outweigh the benefits of treatment?**

7. **Cite your references alongside each answer above. See Appendix 1 for proper citation format.**

7

Deriving Energy from Food

Energy is required by living organisms for movement, transport, and growth. Nothing happens without energy! The Sun is the ultimate source of virtually all energy on the planet Earth. Solar energy is captured by plants through the process of photosynthesis. The glucose molecules holding this energy are then broken down by metabolic processes, creating usable energy for living systems. A food pyramid is provided in Figure 7.1 to help you understand the basic principles of the flow of energy through living systems.

➤ *Metabolism* refers to all chemical reactions that occur in the body of an organism. Aerobic cellular respiration is one very important metabolic pathway.

Cellular respiration is a series of reactions in which glucose molecules are broken down, releasing stored chemical bond energy (Figure 7.2). The released energy is used to make the energy-rich molecule ATP **[Krogh section 7.1]**. Carbon dioxide is released as a by-product of the breakdown of glucose. It is a crucial by-product from the perspective of plants, because they need CO_2 to perform photosynthesis. (Other organic compounds are also utilized in metabolic pathways. See **Krogh Figure 7.10**.)

Glycolysis is the first step in cellular respiration, and it results in the net production of two ATP

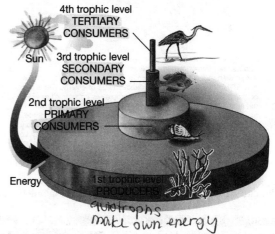

FIGURE 7.1 A Food Pyramid. The flow of energy through living systems starts with the sun. Sunlight energy is captured by autotrophic organisms and is stored in chemical bonds, in the process of photosynthesis. This energy is harvested by living organisms through cellular respiration. Then the energy is transferred up the pyramid, as organisms on upper levels consume those on lower levels. The number of organisms on each level decreases as energy flows upward, because energy transfer is not very efficient. Many organisms at the bottom of the pyramid support few organisms at the top.

molecules **[Krogh section 7.4]**. In glycolysis, the six-carbon glucose molecule is "split" into two, 3-carbon **pyruvic acid** (pyruvate) molecules. The pyruvic acid has three potential routes—aerobic

65

(a) In metaphorical terms

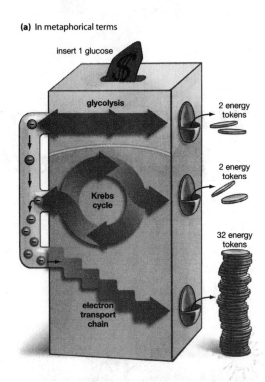

insert 1 glucose

glycolysis

2 energy tokens

Krebs cycle

2 energy tokens

electron transport chain

32 energy tokens

Just as the video games in some arcades can use only tokens (rather than money) to make them function, so our bodies can use only ATP (rather than food) as a direct source of energy. The energy contained in food—glucose in the example—is transferred to ATP in three major steps: glycolysis, the Krebs cycle, and the electron transport chain. Though glycolysis and the Krebs cycle contribute only small amounts of ATP directly, they also contribute electrons (on the left of the token machine) that help bring about the large yield of ATP in the electron transport chain. Our energy-transfer mechanisms are not quite as efficient as the arcade machine makes them appear. At each stage of the conversion process, some of the original energy contained in the glucose is lost to heat.

(b) In schematic terms

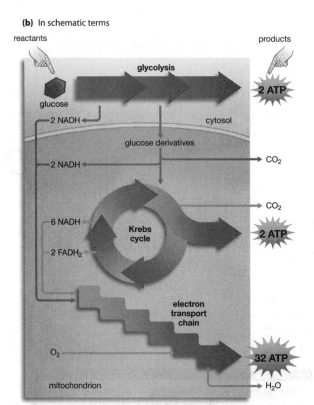

reactants

products

glycolysis

2 ATP

glucose

2 NADH

cytosol

glucose derivatives

2 NADH

CO_2

CO_2

6 NADH

Krebs cycle

2 ATP

2 $FADH_2$

electron transport chain

O_2

32 ATP

mitochondrion

H_2O

36 ATP maximum per glucose molecule

As with the arcade machine, the starting point in this example is a single molecule of glucose, which again yields ATP in three major sets of steps: glycolysis, the Krebs cycle, and the electron transport chain (ETC). These steps can yield a maximum of about 36 molecules of ATP: 2 in glycolysis, 2 in the Krebs cycle, and 32 in the ETC. As noted, however, glycolysis and the Krebs cycle also yield electrons that move to the ETC, aiding in its ATP production. These electrons get to the ETC via the electron carriers NADH and $FADH_2$, shown on the left. Oxygen is consumed in energy harvesting, while water and carbon dioxide are produced in it. Glycolysis takes place in the cytosol of the cell, but the Krebs cycle and the ETC take place in cellular organelles, called mitochondria, that lie within the cytosol.

FIGURE 7.2 Glycolysis and Aerobic Cellular Respiration.

cellular respiration, lactate fermentation, or alcoholic fermentation (Figure 7.3). In the course of aerobic cellular respiration, pyruvic acid is completely degraded and yields much ATP energy. In the course of both lactate and alcoholic fermentation, no additional ATP energy is derived, and when an organism is utilizing one of these pathways, the only source of ATP is glycolysis [**Krogh essay _Energy and Exercise_, page 144**]. The fermentative pathways get rid of the pyruvic acid and recycle the energy transfer molecules used in glycolysis.

Route 1 In **aerobic cellular respiration**, pyruvate enters the Krebs cycle and electron transport system.

$$C_6H_{12}O_6 + 6O_2 \rightarrow 6CO_2 + 6H_2O$$

+ energy (chemical + heat)

In this pathway, the pyruvic acid is broken down completely, and the high-energy electrons are stripped away and passed through a series of electron carriers. Energy is released at each transfer, and is used to make a net 34 ATP molecules. Oxygen is the final electron acceptor in the electron transport system, hence the name _aerobic_ cellular respiration. (If oxygen is absent, the whole pathway shuts down.) When oxygen atoms pick up these electrons and their associated hydrogen ions (protons), metabolic water is formed. Metabolic water is an important source of water to animals.

The kangaroo rat gains all of its water from its food. Ten percent of its water supply comes from moisture in the food, and 90 percent

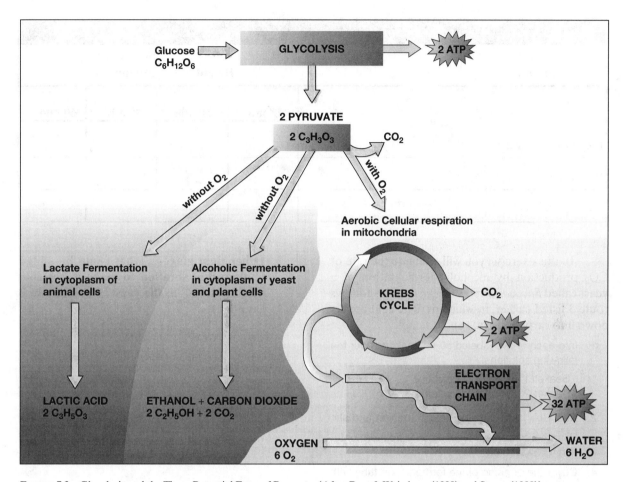

Figure 7.3   Glycolysis and the Three Potential Fates of Pyruvate. (After Bres & Weisshaar (1998) and Stryer (1988)).

comes from metabolic water *(Hickman et al., 1984)*.

Route 2 When oxygen is not available, respiration stops and aerobic organisms degrade pyruvic acid in lactate fermentation, with lactic acid being produced as a by-product. Buildup of lactic acid occurs as muscles become fatigued during exercise. In humans, this lactic acid is eventually transported to the liver, where it may be converted into glucose.

Route 3 Yeasts (single-celled fungi) are among the few organisms capable of converting pyruvic acid into ethanol through a process known as **alcoholic fermentation**. The yeast depends on the two ATP molecules generated from glycolysis as its primary energy source. The two pyruvic acid molecules are degraded into two molecules of carbon dioxide and two molecules of ethanol **[Krogh essay *When Energy Harvesting Ends at Glycolysis, Beer Can Be the Result*, page 140]**. No additional ATP

molecules are gained. When you consider that one of the unused "by-products" of this process (ethanol) can propel a race car around a track at over 200 miles per hour, the inefficiency of this route becomes apparent. Ethanol serves as an important source of clean-burning fuel for automobiles.

Exercise 7.1   ALCOHOLIC FERMENTATION: CARBON DIOXIDE PRODUCTION BY YEAST

(Adapted from Perry and Morton, 1995)

Human cells are not capable of carrying out alcoholic fermentation. Nevertheless, humans have utilized this process in some creative ways. Yeasts are used by vintners as a source of ethanol, by bakers as a source of carbon dioxide, and by brewers as a source of both to produce fine beers and lagers *(Raven et al., 1992)*.

TABLE 7.1 RATE OF CARBON DIOXIDE PRODUCTION BY YEAST				
Start time: 6:49	**Height of gas column (mm)**			
	~~15 min~~	30 min	~~45 min~~	~~60 min~~
1% glucose + yeast → left		83 mm		
1% starch + yeast → right		2 mm		

In this exercise, you will examine the rate of CO_2 production by alcoholic fermentation of a yeast called *Saccharomyces cerevisiae*. This follows route 3 listed earlier, in which pyruvic acid breaks down into carbon dioxide and ethanol.

1. Into each of two labeled 50-ml beakers, pour the following solutions:

   Beaker 1: 25 ml of 1% glucose
   Beaker 2: 25 ml of 1% starch

   20 mls of 10% glucose + starch

2. Add 1 g ($\frac{1}{4}$ tsp) of yeast to each beaker and stir.

3. Pour the contents of each beaker into fermentation tubes labeled "1" and "2."

4. Cover the opening of the fermentation tube with your thumb and invert the tube, so that the "tail" portion is filled. Place the tubes upright in the 37°C incubator, and record the start time in Table 7.1.

5. At 15, 30, 45, and 60 minutes, measure the height of the CO_2 column and record your results.

Questions

1. **Which solution had the greatest fermentation rate? Why didn't the other one work as well?**

   The glucose had greatest fermentation rate. The other didn't work as well because starch is more complex sugar?

2. **Other than glucose, what could be added to the other solution to make it a usable "food source" to the yeast? See Chapter 3 for a hint.**

   glucose (simple sugar) monosaccharide (3.4)
   glucose
   salivary amylase

Exercise 7.2   HEAT PRODUCTION BY LIVING ORGANISMS

In aerobic cellular respiration, only about 45% of the energy liberated is used in the synthesis of ATP. The remainder is lost in the form of heat. In endothermic (warm-blooded) organisms such as ourselves, this energy is not "wasted" because heat released as a result of cellular respiration is used for the maintenance of our body temperature.

On demonstration are three thermos bottles with thermometers, prepared about 24 hours ago. Thermos A contains germinating peas. The peas in thermos B were boiled and cooled to room temperature. Thermos C contains only water. Each thermos was filled at the same time with an approximately equal number of pea seeds and equivalent amounts of room-temperature tap water. Record the temperature of each thermos in Table 7.2.

TABLE 7.2 HEAT PRODUCTION BY PEAS	
Thermos	**Temperature**
A. Germinating peas	
B. Boiled peas	
C. Water	

FIGURE 7.4   A Volumeter.

Questions

1. In which bottle(s) is heat being produced?

2. What process is creating this heat?

3. In which bottle(s) is no heat produced? Why not?

Exercise 7.3   OXYGEN CONSUMPTION DURING CELLULAR RESPIRATION

During cellular respiration, organisms consume oxygen and release carbon dioxide. You will measure the rate of oxygen consumption of various organisms using a volumeter (Figure 7.4), which is a closed system. Carbon dioxide released by the animal will be absorbed by soda lime. As oxygen is consumed in the chamber, a bubble trapped in the attached pipette will move toward the chamber because the volume of gas in the chamber is decreasing.

The volume of gas in cubic centimeters (cc) is the same as measuring it in milliliters (ml).

1. Using the triple beam balance, determine the weight of each test organism to the nearest 0.1 gram. Record these amounts in Table 7.3. Answer Questions 1 to 3.

2. Place one organism into the inner cage. Do not allow the organism to touch the soda lime. Gently place the cage into the volumeter.

3. Moisten the inside of the calibrated tube with running water. Then push the tube into the rubber stopper. Put the rubber stopper in place on the volumeter.

4. Using the bubble wand, apply a drop of soap solution to the end of the tube until a bubble forms. Allow the bubble to stabilize.

5. Using a stopwatch, measure how much time it takes for the bubble to move a distance along the tube equal to 5 ml, or measure how many milliliters of oxygen the animal consumes in 10 minutes. Either way, you will record a rate of cc/min (ml/min).

    For example, an insect will consume oxygen more slowly than a mouse. In 10 minutes, a cockroach might consume 3 cc of oxygen. A mouse, on the other hand, may consume about 10 cc of oxygen in just 4 minutes.

    In this manner, practice measuring the amount of oxygen consumed, until your measurements are consistent. If several measurements from the same animal are not similar at all, ask your instructor for help.

TABLE 7.3 OXYGEN CONSUMPTION BY ANIMALS						
Organism: **Weight (g):**			**Organism:** **Weight (g):**		**Organism:** **Weight (g):**	
	O_2 consumed (cc)	Elapsed time (min)	O_2 consumed (cc)	Elapsed time (min)	O_2 consumed (cc)	Elapsed time (min)
Trial 1						
Trial 2						
Trial 3						
Total						
Average O_2 consumption rate (cc/min)						
Standardized O_2 consumption rate (cc/g/min)						

6. Record the results of three trials in Table 7.3. As soon as you are done with recording the trials for an organism, ask your instructor for help in removing it. Repeat Steps 2 to 6 for each organism.

7. Calculate the total cc of oxygen consumed and the total elapsed time for each organism. Record these in Table 7.3.

8. For each organism, calculate the average rate of oxygen consumption by dividing the total cc consumed during all three trials by the total time elapsed during all three trials.

9. For each organism, calculate the standardized consumption rate by dividing the average rate of oxygen consumption by the weight of the organism. This calculation will enable you to compare the rates between organisms, independent of their weights.

Questions

1. **Which of the organisms do you expect to consume the most oxygen?**

2. **Which organism do you expect to consume the least oxygen?**

3. **What factors might influence the rate of cellular respiration in organisms?**

4. **What is the advantage of reporting the metabolic rate of the various organisms in cc/gram/min rather than cc/min?**

5. **In order for this experiment to work, why must the carbon dioxide be removed by the soda lime?**

6. **Were your results consistent with your predictions? If not, explain any discrepancies.**

7. **How much oxygen is consumed by each organism in an hour? In a day? Give your answer in cubic centimeters (cc). Again, 1 cc = 1 ml.**

8

Photosynthesis

The human brain, so frail, so perishable, so full of inexhaustible dreams and hungers, burns by the power of the leaf.

Loren Eiseley, 1964

The flow of energy through living systems begins with plants, algae, and other chlorophyll-containing organisms that can make their own food. In other words, we depend on these organisms to live.

Autotrophic organisms make their own food. Heterotrophic organisms cannot make their own food and must get their energy from other sources.

In the process of **photosynthesis**, these autotrophic organisms capture solar energy and make glucose (Figure 8.1). These glucose molecules are then used both by autotrophic and heterotrophic organisms (including plants and animals) for cellular respiration.

In reality, glucose is only one product of photosynthesis. Most of the simple sugar resulting from photosynthesis is converted into starch (a glucose polymer) for storage, into sucrose for transport, or into structural components of plants [**Krogh section 8.5**].

The process of photosynthesis utilizes carbon dioxide, water, and light in the production of glucose. Oxygen is released as a by-product of this process. It is a very important by-product from the viewpoint of most organisms! You need to learn the balanced chemical equation for photosynthesis, given here. Does it look familiar? It is the opposite of aerobic cellular respiration.

$$6CO_2 + 6H_2O \xrightarrow{\text{light}} C_6H_{12}O_6 + 6O_2$$

The exercises in this chapter allow you to manipulate components of this equation. By manipulating

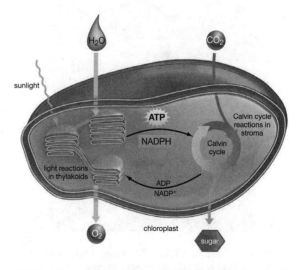

FIGURE 8.1   Summary of Photosynthesis in the Chloroplasts of Plant Cells.

73

the components, you can measure their effects on the rate of photosynthesis.

Exercise 8.1. EFFECTS OF CARBON DIOXIDE AND LIGHT ON PHOTOSYNTHESIS

On demonstration are four jars containing *Elodea* under different conditions, prepared in such a way that oxygen production by the plants can be measured. Two of the jars contain sodium bicarbonate (baking soda) as a source of carbon dioxide enrichment, and two of the jars contain tap water. One carbon dioxide–rich jar and one jar containing tap water were placed near a grow light. The two remaining jars were left in ambient light. Each jar is labeled with the start time of the experiment and the mass of the *Elodea* in that particular jar (Figure 8.2).

1. Answer Questions 1 through 5.

2. The size of the bubble is a direct measure in ml (cc) of oxygen production by the plants. When directed by your instructor, record the volume of oxygen trapped in each graduated cylinder. Record your results in Table 8.1.

FIGURE 8.2   Setup for Exercise 8.1. If the jar is tall enough, use a 50-ml beaker to hold the *Elodea* sprigs upright. The sprigs release oxygen, which is trapped by the funnel and graduated cylinder.

3. To standardize the oxygen consumption in each jar, divide the cc of oxygen consumed by the weight of the *Elodea*, and divide again by the number of hours elapsed during the exercise. Compare these numbers in all four jars.

Questions

1. **Construct a hypothesis regarding this experiment.**

   The elodea in sodium carbonate [bicarbonate] that's under the grow light will produce the greatest amount of oxygen

2. **State your prediction(s) regarding which *Elodea* sprigs will produce the greatest amount of oxygen.**

   The ones in the greatest light in sodium bicarbonate under the grow light.

3. **What are the independent variables?**

   The light (grow or ambient) and solutions in the jars (water or sodium bicarbonate).

4. **What is the dependent variable?**

   Amount of oxygen produced

$1 mL = 1 cc$

TABLE 8.1 OXYGEN PRODUCTION BY ELODEA				
Start time:				
Independent variables	**O₂ produced (cc)**	**Weight of Elodea (g)**	**Time elapsed (hour)**	**Standardized O₂ production (cc/g/hour)**
Water, grow light	1 / 0.05	2/2	12/45 min	
Water, ambient light	0.2 / 0.025	2/2	12/45 min	
Sodium bicarbonate, grow light	10 / 1	2/2	12/45 min	
Sodium bicarbonate, ambient light	1 / 0.05	2/2	12/45 min	

5. What is the source of the bubbles?

The oxygen comes from the elodea

7. According to your data, what has the greatest effect on the rate of photosynthesis—amount of carbon dioxide or amount of light? Explain.

The light
↳ produced significantly more oxygen

6. Which Elodea plants exhibited the greatest rate of oxygen production? Do the data support your hypothesis?

The sodium bicarbonate under the grow light. Yes this supports our hypothesis

8. Now consider what is lost when a tree is removed from an ecosystem (Figure 8.3). When plants are lost, a very important source of life-sustaining oxygen is lost. Deforestation (particularly of tropical forests) results not only in the loss of an oxygen source but also in the loss of one of the chief means by which CO_2 is removed from the atmosphere. Carbon dioxide is one of the most important of the greenhouse gases contributing to global warming [Krogh section 32.4]. To make matters worse, the burning of logging wastes releases excess CO_2 into the environment.

a   b

FIGURE 8.3   a. El Yunque Rainforest, Puerto Rico. It is the only tropical rainforest in the U.S. National Park System.
b. Destruction of Amazonian Rainforest.

Devise a way to estimate the amount of oxygen a tree can produce in a day.

10. How much *Elodea* (g) would be required to support each of the organisms in Exercise 7.3 for 24 hours?

9. On the basis of your calculation of oxygen consumption by the animals in Exercise 7.3, estimate how long the oxygen produced by an *Elodea* sprig would support each animal?

11. Assume that a resting adult human consumes about 1000 cc (ml) of oxygen per minute. At its current rate of photosynthesis, how long would it take the sprig of *Elodea* to produce enough oxygen to support a human for 1 minute? To support a human for 24 hours?

Exercise 8.2. CARBON DIOXIDE CONSUMPTION BY *ELODEA*

When in an aqueous solution, carbon dioxide reacts with water to form carbonic acid.

$$CO_2 + H_2O \rightleftharpoons H_2CO_3$$

This results in lowering the solution's pH. As CO_2 is consumed by aquatic plants through photosynthesis, the level of carbonic acid in a solution will decrease, leading to an increase in pH. Thus, monitoring pH provides an indirect measure of the amount of CO_2 consumed in the photosynthesis. This experiment uses bromothymol blue (BTB), a pH indicator that turns yellow at pH < 6.0, green at pH 6.0–7.6, and blue at pH > 7.6.

1. Place 75 ml of BTB solution into a 100-ml beaker. Blow exhaled air through a straw into the BTB solution just until it changes from blue to greenish-yellow, then stop.

2. Fill four test tubes three-fourths full with CO_2-rich BTB solution.

3. Wrap two of the test tubes with aluminum foil, leaving only the top open.

4. Obtain two 2-cm sprigs of *Elodea*, and place one into a wrapped tube and one into an unwrapped tube. The other two tubes will have no *Elodea* in them—use these for comparison. Immediately finish covering the two tubes wrapped in aluminum foil.

5. Place the unwrapped test tubes directly in front of the grow light, and the wrapped ones on your lab bench.

6. Allow the tubes to "incubate" for 1 hour. Answer Questions 1 to 5 while you wait. In Table 8.2 record any color changes that have occurred by marking an *X* in the appropriate space.

Questions

1. **State your hypothesis regarding this experiment.**

2. **State your prediction(s).**

3. **Dependent variable:**

4. **Independent variable:**

TABLE 8.2 CARBON DIOXIDE CONSUMPTION IN *ELODEA*			
	Yellow: pH < 6.0	**Green: pH 6.0–7.6**	**Blue: pH > 7.6**
Elodea, light			
Elodea, dark			
No *Elodea*, light			
No *Elodea*, dark			

5. **Why did the BTB turn yellow as you blew through the straw?**

6. **What is responsible for the color change of the solutions containing *Elodea* in front of the grow light?**

FIGURE 8.4   *Coleus* is a common houseplant. Use a *Coleus* plant with leaves that are red in the center, green on the margins, and white in between.

7. **Why might wrapped tubes with *Elodea* exhibit an increase in pH level?**

8. **Why might wrapped tubes with *Elodea* exhibit a decrease in pH level?**

Exercise 8.3.   PHOTOSYNTHETIC PIGMENTS

Wear goggles. Be very careful with the boiling water and ethanol!

A variety of pigments found in plants provide the brilliant colors observed in the botanical world. In this exercise you will examine the pigments of the multicolored leaves of a common houseplant called *Coleus*, whose leaves are green and red

(Figure 8.4). As photosynthesis proceeds, glucose is produced, converted to starch, and then stored in the leaves. By looking at the distribution of starch in *Coleus* leaves, you can determine which pigments facilitate photosynthesis and which do not. The green color is a result of chlorophylls *a* and *b*, and pigments called *anthocyanins* impart the red color.

Anthocyanins may be extracted in boiling water, and chlorophyll may be extracted in boiling 95% ethanol. The distribution of starch may be determined by conducting an iodine test. Recall that starch reacts with iodine to produce a very dark blue or black color.

1. Obtain a leaf from each of the *Coleus* plants. One of the plants has been maintained under a grow light and the other has been kept in the dark for several days.

2. Sketch the leaves (see Question 1), indicating the distribution of the different colors.

3. Using tongs, place leaves from each plant into boiling water. After 1 to 2 minutes, remove the leaves and note their appearance.

4. Using tongs, place the leaves into boiling 95% ethanol.

Exercise great caution at this step. Ethanol is VERY FLAMMABLE!

After 1 to 2 minutes, remove the leaves and note their appearance. Remove the ethanol from the hot plate, using an oven mitt.

Chapter 8 Photosynthesis 79

5. Place the leaves into petri dishes and cover them with iodine. After a reaction occurs, remove the leaves and note the distribution of starch.

Questions

1. **Sketch the color distribution in the leaves.**

2. **What happens to the water as the leaves are boiled in it?**

3. **Describe or sketch the appearance of the leaves after boiling them in water.**

4. **What happens to the ethanol as the leaves are boiled in it?**

5. **Describe or sketch the appearance of the leaves after boiling them in ethanol.**

6. **In what manner is starch distributed throughout the leaves?**

7. **Are anthocyanins photosynthetic pigments? How do you know?**

8. **How does the appearance of leaves from the plant kept in the dark differ from those kept in the light? What caused this difference?**

UNIT 3
Genetics

9

Cell Division

[handwritten notes:]
- know bold faced words
- know differences between Mitosis + Meiosis
- know steps of Mitosis

Cell division is the process cells use to reproduce themselves. **Mitosis** results in the formation of two daughter cells that are genetically identical to each other and the original parent cell. Mitosis occurs in somatic cells, which are body cells not involved in sexual reproduction. Binary fission of single-celled organisms is a form of mitosis. **Meiosis** occurs in germ or sex cells and produces haploid gametes (sperm by *spermatogenesis*, and eggs by *oogenesis*). Both mitosis and meiosis are only one phase of the cell cycle (Figure 9.1), and both

are followed by cytoplasm division, or **cytokinesis [Krogh Chapters 9 & 10]**.

During interphase, DNA molecules are long and threadlike, wrapped around proteins. Each DNA molecule is called a **chromosome**. Once a chromosome duplicates, it consists of two **chromatids** joined at the **centromere** (Figure 9.2). As long as the two chromatids remain together, they are referred to as **sister chromatids**. Each sister chromatid will become a chromosome in the new daughter cell. **Centrosomes** are organelles that

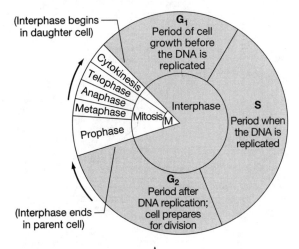

FIGURE 9.1   The Cell Cycle.

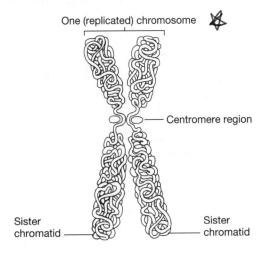

FIGURE 9.2   A Replicated Chromosome.

83

organize microtubule structures called **spindle fibers**, along which the chromosomes will migrate during mitosis. (Centrosomes are often called *centrioles*.) **[Krogh Figure 9.6]** A helpful link for studying cell division may be found at *http://www .biology.arizona.edu/cell_bio/tutorials/cell_cycle/main.html*.

Exercise 9.1   MITOSIS SIMULATION

Note that the process of mitosis is continuous, without rigid divisions between stages; however, it is most easily studied when divided into arbitrary stages. You can guess that the organism doesn't care whether its cells are in prophase or metaphase. In any event, **the end product of mitotic cell division is two genetically identical, diploid daughter cells**. To make sure you understand the processes of mitosis, you will use pop beads to construct model chromosomes involved in each phase of mitosis. Your hypothetical cells are from an organism that has a **diploid** (*2n*) chromosome number of 4 (with 2 pairs

FIGURE 9.3   Hypothetical Cell with Four Chromosomes.

of "matching" homologous chromosomes). Let the paternal chromosomes be blue and the maternal chromosomes be red. Use a different bead color for the centromeres and centrosomes.

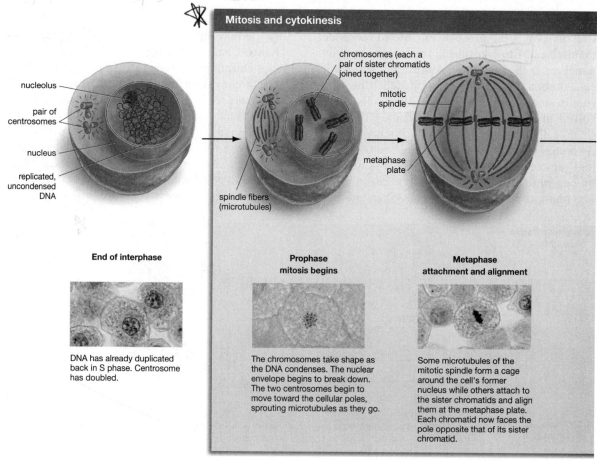

Mitosis and cytokinesis

nucleolus

pair of centrosomes

nucleus

replicated, uncondensed DNA

chromosomes (each a pair of sister chromatids joined together)

spindle fibers (microtubules)

mitotic spindle

metaphase plate

End of interphase

DNA has already duplicated back in S phase. Centrosome has doubled.

Prophase mitosis begins

The chromosomes take shape as the DNA condenses. The nuclear envelope begins to break down. The two centrosomes begin to move toward the cellular poles, sprouting microtubules as they go.

Metaphase attachment and alignment

Some microtubules of the mitotic spindle form a cage around the cell's former nucleus while others attach to the sister chromatids and align them at the metaphase plate. Each chromatid now faces the pole opposite that of its sister chromatid.

FIGURE 9.4   Mitosis and Cytokinesis.

1. With red pop beads, make maternal chromosomes. Make chromosome number 1 long with 12 pop beads (6 on each side of a centromere). Make chromosome number 2 shorter with 6 pop beads (3 on each side of a centromere). Refer to Figure 9.3 to get started.

2. Repeat Step 1 and make paternal chromosomes using blue pop beads instead of red. You should end up with four chromosomes.

3. Using four pop beads of a third color (yellow, perhaps), make a pair of centrioles, with two pop beads in each centriole.

4. With a piece of chalk (or a pencil and paper), draw a large "cell" on your lab bench and place the four chromosomes and centrosomes inside it. Assume that the nucleus is present, but don't draw it. Following the next steps, mimic what happens during the cell cycle. Use Figure 9.4 to guide you. Get a wet paper towel (or additional paper) to erase and redraw the cell as needed.

5. **S phase** The chromosomes replicate. Make four more chromosomes just like your originals.

6. **G$_2$ phase** The centrosomes replicate. *Centrosomes are lacking in plant cells*, although plant cells still have points of organization of the spindle fibers called *microtubule organizing centers* **[Krogh sections 4.6 & 9.4]**.

7. **Prophase** The centrosomes migrate to opposite ends of the cell, and spindle fibers appear between the two pairs of centrosomes (draw them). Just before metaphase, microtubules of the spindle apparatus become attached to the sister chromatids of each chromosome at the centromere. The chromosomes condense, becoming shorter and thicker.

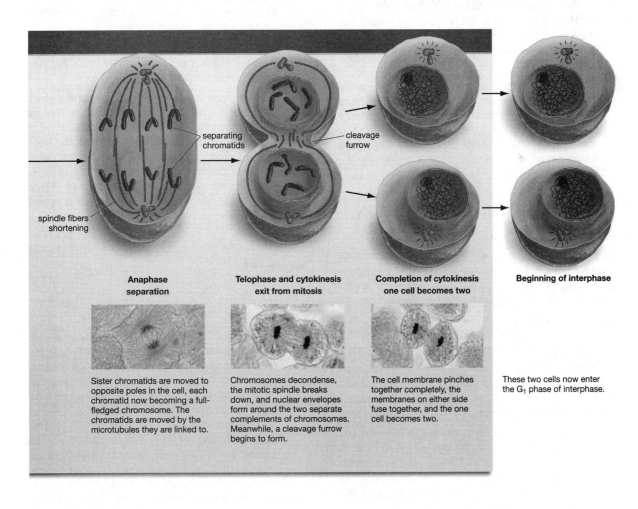

separating chromatids

cleavage furrow

spindle fibers shortening

Anaphase separation

Telophase and cytokinesis exit from mitosis

Completion of cytokinesis one cell becomes two

Beginning of interphase

Sister chromatids are moved to opposite poles in the cell, each chromatid now becoming a full-fledged chromosome. The chromatids are moved by the microtubules they are linked to.

Chromosomes decondense, the mitotic spindle breaks down, and nuclear envelopes form around the two separate complements of chromosomes. Meanwhile, a cleavage furrow begins to form.

The cell membrane pinches together completely, the membranes on either side fuse together, and the one cell becomes two.

These two cells now enter the G$_1$ phase of interphase.

FIGURE 9.4   *(Continued)*

8. **Metaphase** The centromeres line up precisely along the cell's central axis (equator), called the **metaphase plate**, with the arms of the chromatids trailing off randomly in various directions.

9. **Anaphase** During anaphase, the sister chromatids separate at the centromere. The two former sister chromatids are now considered daughter chromosomes. They move along the spindle fibers to opposite poles.

10. **Telophase** Telophase technically begins once the daughter chromosomes arrive at opposite spindle poles and stop moving. The spindle fibers disappear. Shortly following telophase, and to some extent overlapping it, cytokinesis occurs. **Cytokinesis** refers to division of the cytoplasm and is characterized by the appearance of a *contractile ring*, which forms a **cleavage furrow** in animal cells. The cleavage furrow deepens until division is complete. In plant cells, cytokinesis is characterized by the appearance of a **cell plate** instead of a cleavage furrow. Because of the rigid cell walls, it is impossible for a cleavage furrow to form in plant cells [**Krogh Figure 9.12**].

Because it is nearly impossible to determine the point at which the daughter chromosomes have stopped moving, you should consider telophase to begin with the appearance of the cleavage furrow or cell plate for animal cells and plant cells respectively.

Exercise 9.2 MEIOSIS SIMULATION

Meiosis produces haploid gametes that may take part in fertilization, which results in a new organism [**Krogh Chapter 10**]. An organism has a specific number of chromosomes—humans have 46, or 23 pairs. Twenty-three of the 46 chromosomes are paternal, contributed by the father's sperm, and the other 23 are maternal, contributed by the mother's egg. Obviously there must be a reduction in the chromosome number of gametes. If not, offspring resulting from sexual reproduction would each have 46 × 2, or 92 chromosomes! This reduction of chromosome number is from the **diploid** ($2n$) state of 46 chromosomes to the **haploid** (n) state of 23 chromosomes. It is important to bear in mind that each gamete will have all 23 chromosomes represented, but with only one copy of each, as opposed to two copies of each, as in the diploid ($2n$) condition.

The process of meiosis resembles mitosis in many ways, but the outcome is very different. In meiosis, **the cells proceed through two consecutive divisions (meiosis I and meiosis II)**, which end with the production of four haploid daughter cells. DNA replication does not occur between meiosis I and meiosis II. Using pop beads, you will construct "model chromosomes" illustrating each phase of meiosis I and meiosis II. Your model cells will represent the same hypothetical organism described in Exercise 9.1.

1. With red pop beads, make maternal chromosomes. Make chromosome number 1 long with 12 pop beads (6 on each side of a centromere). Make chromosome number 2 shorter with 6 pop beads (3 on each side of a centromere). Refer to Figure 9.3 to get started.

2. Repeat Step 1 and make paternal chromosomes using blue pop beads instead of red. You should end up with four chromosomes.

3. Using four pop beads of a third color (yellow, perhaps) make a pair of centrioles, with two pop beads in each centriole.

4. With a piece of chalk (or a pencil and paper), draw a large "cell" on your lab bench and place the four chromosomes and centrosomes inside it. Assume that the nucleus is present, but don't draw it. Following the next steps, mimic what happens during the cell cycle. Get a wet paper towel (or additional paper) to erase and redraw the cell as needed. Refer to Figure 9.5 as you proceed through the steps of meiosis.

Meiosis I

5. **S phase** The chromosomes replicate.

6. **G$_2$ phase** The centrosomes replicate.

7. **Prophase I** Each replicated chromosome lines up with its homologous partner. For instance, paternal chromosome 1 joins with maternal chromosome 1 to form a **tetrad**. (It's called a tetrad because the two homologous chromosomes have a total of four chromatids.) Homologous chromosomes exchange genetic material in a process called **crossing over**, a form of *recombination*. Crossing over may result in major shuffling of genetic material and is one way to provide genetic variation between siblings. Simulate this by exchanging two or three pop beads of a blue chromosome with two or three pop beads on the

adjacent red homologous chromosome. You'll end up with chromosomes of mixed colors.

8. **Metaphase I** Each tetrad lines up at the metaphase plate. Notice that the arrangement of the maternal and paternal chromosomes is random at the metaphase plate, meaning that they can line up with either the maternal or paternal chromosomes on either side. This also provides much of the variation between siblings. The probability that a gamete would wind up with all maternal or all paternal chromosomes is $(1/2)^{23}$ = 1 in 8,388,608. In other words, either chromosome of a pair has an equal chance at winding up at either pole. This is a principle called **independent assortment**, a concept Mendel figured out long before anyone knew about chromosomes! **[Krogh section 10.3]**.

9. **Anaphase I** The chromosomes comprising the homologous pairs separate from each other and migrate to opposite poles. It is important to note that the sister chromatids do not separate at the centromeres, as they do in mitosis.

10. **Telophase I** During telophase and subsequent cytokinesis, the two daughter cells are formed. Each resultant daughter cell of meiosis I has one-half the number of chromosomes as the parent cell. **Each new cell is haploid, but the chromosomes are still replicated!**

MEIOSIS II These stages proceed almost identically to mitosis, but no DNA replication occurs in the S-period of the cell cycle this time.

11. **Prophase II** Centrosomes migrate to opposite sides of each cell.

12. **Metaphase II** Replicated chromosomes line up at the metaphase plate.

13. **Anaphase II** Sister chromatids break apart and each becomes a separate, unreplicated chromosome.

14. **Telophase II** Four haploid daughter cells appear after cytokinesis.

15. Compare mitosis and meiosis side by side now. The fundamental differences between them are illustrated in Figure 9.6.

Questions

1. **Name two ways by which genetic diversity is accomplished during cell division. At what cell division stages do these events occur?**

2. **How do the chromosomes of an anaphase cell of mitosis differ from those of an anaphase cell of meiosis I?**

3. **What major event happens just before prophase I but does not occur before prophase II?**

4. **Which results in the change in chromosome number from diploid to haploid, meiosis I or meiosis II?**

5. **Do the two things mentioned in Question 1 happen in the stages of mitosis? Why or why not?**

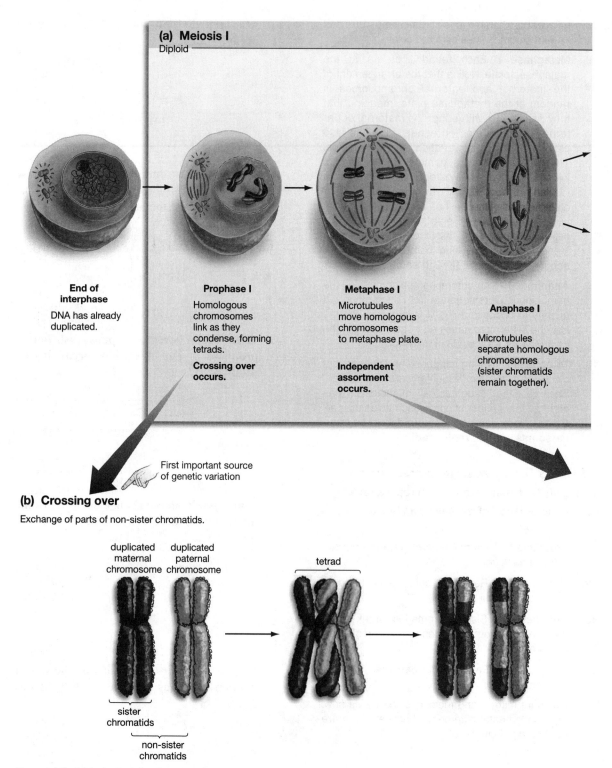

(a) Meiosis I
Diploid

End of interphase

DNA has already duplicated.

Prophase I

Homologous chromosomes link as they condense, forming tetrads.

Crossing over occurs.

Metaphase I

Microtubules move homologous chromosomes to metaphase plate.

Independent assortment occurs.

Anaphase I

Microtubules separate homologous chromosomes (sister chromatids remain together).

First important source of genetic variation

(b) Crossing over

Exchange of parts of non-sister chromatids.

duplicated maternal chromosome

duplicated paternal chromosome

tetrad

sister chromatids

non-sister chromatids

FIGURE 9.5   Meiosis, Crossing Over, and Independent Assortment.

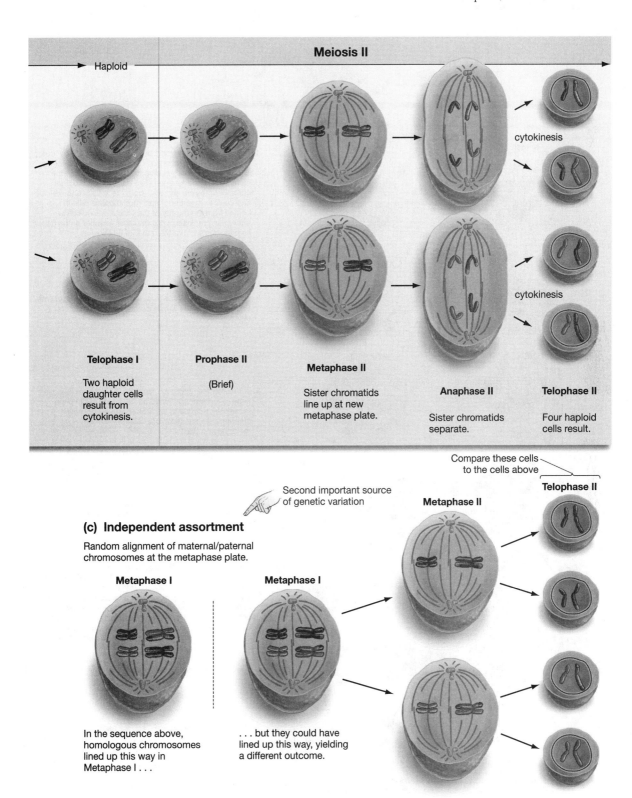

Meiosis II

Haploid

cytokinesis

cytokinesis

Telophase I

Two haploid daughter cells result from cytokinesis.

Prophase II

(Brief)

Metaphase II

Sister chromatids line up at new metaphase plate.

Anaphase II

Sister chromatids separate.

Telophase II

Four haploid cells result.

Compare these cells to the cells above

Telophase II

Second important source of genetic variation

Metaphase II

(c) Independent assortment

Random alignment of maternal/paternal chromosomes at the metaphase plate.

Metaphase I

Metaphase I

In the sequence above, homologous chromosomes lined up this way in Metaphase I . . .

. . . but they could have lined up this way, yielding a different outcome.

FIGURE 9.5   (*Continued*)

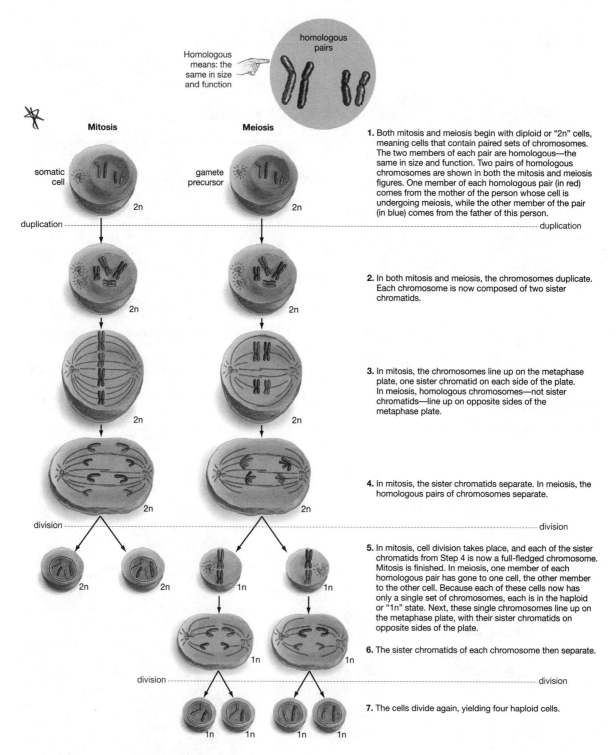

Homologous means: the same in size and function

homologous pairs

Mitosis

Meiosis

somatic cell

2n

gamete precursor

2n

duplication - duplication

2n

2n

2n

2n

2n

2n

division - division

2n

2n

1n

1n

1n

1n

division - division

1n

1n

1n

1n

1. Both mitosis and meiosis begin with diploid or "2n" cells, meaning cells that contain paired sets of chromosomes. The two members of each pair are homologous—the same in size and function. Two pairs of homologous chromosomes are shown in both the mitosis and meiosis figures. One member of each homologous pair (in red) comes from the mother of the person whose cell is undergoing meiosis, while the other member of the pair (in blue) comes from the father of this person.

2. In both mitosis and meiosis, the chromosomes duplicate. Each chromosome is now composed of two sister chromatids.

3. In mitosis, the chromosomes line up on the metaphase plate, one sister chromatid on each side of the plate. In meiosis, homologous chromosomes—not sister chromatids—line up on opposite sides of the metaphase plate.

4. In mitosis, the sister chromatids separate. In meiosis, the homologous pairs of chromosomes separate.

5. In mitosis, cell division takes place, and each of the sister chromatids from Step 4 is now a full-fledged chromosome. Mitosis is finished. In meiosis, one member of each homologous pair has gone to one cell, the other member to the other cell. Because each of these cells now has only a single set of chromosomes, each is in the haploid or "1n" state. Next, these single chromosomes line up on the metaphase plate, with their sister chromatids on opposite sides of the plate.

6. The sister chromatids of each chromosome then separate.

7. The cells divide again, yielding four haploid cells.

FIGURE 9.6   Comparison of Mitosis and Meiosis.

Exercise 9.3 PLANT MITOSIS

Now you will observe various stages of mitosis in cells located at the tip of an onion root, an area called the apical meristem. The root grows because of the production of new cells at the apical meristem. Because this is an area of pronounced growth, many cells in this region will be exhibiting mitosis (Figure 9.7).

1. Obtain a slide of onion *(Allium)* root tip. Using the scanning objective on a compound microscope, focus on the pink or purple-stained specimen.

2. On low power, focus on the cells that have dark spots in them. The spots are condensed chromosomes of dividing cells. Remember that cells in mitosis have no visible nucleus, but interphase cells will have a nucleus with very distinct nucleoli (darker structures within the nucleus).

3. With the high-power objective in place, scan the entire apical meristem, and locate cells in interphase and in various stages of mitosis.

Questions

1. **Sketch cells you find on the slide that are undergoing each stage of mitosis.**

   Interphase:

   Prophase:

   Metaphase:

Anaphase:

Telophase:

2. **What are the characteristics of each cell that allow you to identify its cell division stage? Label them in your sketches.**

Exercise 9.4 ANIMAL MITOSIS

During this exercise, you will observe various stages of mitosis in animal cells. One of the best places to find cells actively dividing by mitosis is the early embryo, where cells are large and divide rapidly with a short interphase. You will observe animal cell mitosis from animal embryos that have been preserved on a slide (Figure 9.8).

1. Obtain a slide of animal cells undergoing mitosis. Find an embryo section with the scanning objective, and then use the low-power objective. You should now be able to see individual cells in the embryo, although it may be difficult to see many details.

2. Click the high-power objective into place. The dark-stained materials are the chromosomes, and their positions should allow you to identify the stage of mitosis the cell is undergoing.

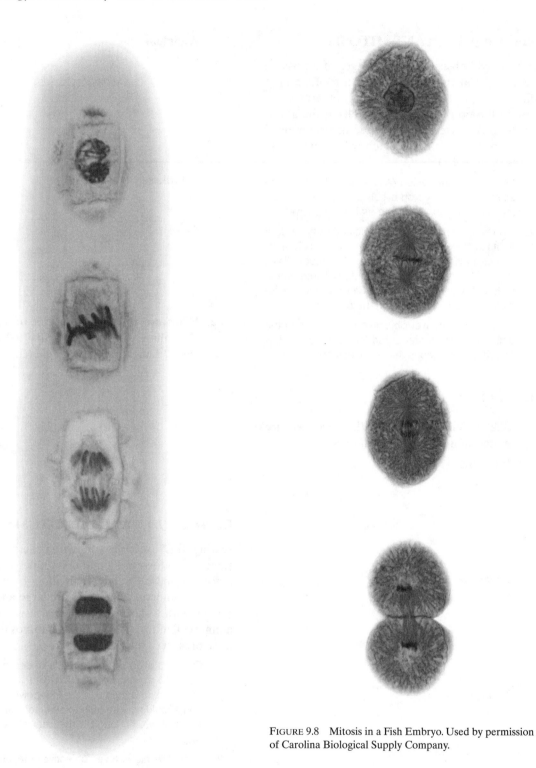

FIGURE 9.8   Mitosis in a Fish Embryo. Used by permission of Carolina Biological Supply Company.

FIGURE 9.7   Mitosis in an Onion Root Tip. Used by permission of Carolina Biological Supply Company.

Questions

1. **Sketch cells that are undergoing each stage of mitosis.**

   Prophase:

   Metaphase:

   Anaphase:

   Telophase:

2. **What are the characteristics of each cell that allow you to identify its stage of mitosis? Label them in your sketches.**

10

Mendelian Genetics

The deep language of the genes is an Esperanto of biological design which can be understood by a Babel of organisms.
R. Fortey, 2000

Gregor Johann Mendel was a monk who elucidated the basic principles of genetics, the laws governing the transmission of characteristics from parents to offspring. He conducted breeding experiments with pea plants, and his findings were published in 1866 **[Krogh Chapter 11]**. Mendel meticulously observed the offspring of controlled matings, counting the numbers of each type of offspring. Although Mendel could be considered the "Father of Genetics," his work was largely ignored until three botanists independently confirmed his findings at the beginning of the twentieth century.

Mendel proposed that discrete particles "carried" hereditary information from one generation to the next. This was in direct opposition to the view of his day, which stated that parental traits somehow "blended" together to form an offspring that was intermediate between the parents. Mendel determined that each individual has two of these particles for each trait (that we now know as genes). He knew that only one factor is contributed during gamete formation, and that the offspring received only one factor from each parent. This concept became known as Mendel's **law of segregation**. Additionally, Mendel determined that the segregation of particles for one trait (such as flower color) was independent of that of other traits (such as height). This concept became known as Mendel's **law of independent assortment**.

Likewise, many twentieth-century scientists are famous for their work with genetics. Thomas Hunt Morgan is recognized for his work with fruit flies *(Drosophila melanogaster)*. One thing Morgan proposed was gene linkage, which occurs when genes appear not to separate independently, as Mendel's law states! Morgan's work at first appeared to be a monkey wrench in genetics knowledge. But a few years later, Barbara McClintock used corn *(Zea mays)* to provide strong evidence about how crossing over occurs. McClintock's work explained how some traits appeared to be linked to others, thus solving that mystery and allowing Mendel's principles to stand. Refer to Figure 9.5 for a review of crossing over and independent assortment.

Crossing over refers to the exchange of parts of chromosomes during prophase I of meiosis.

The exercises in this chapter will give you firsthand experience with the same kinds of organisms these pioneering geneticists used.

95

Exercise 10.1   THE MONOHYBRID CROSS

(Adapted from Carolina Drosophila Manual, @1988, Carolina Biological Supply Company, used by permission.)

Extended five-week exercise

In this exercise, you will conduct a monohybrid cross, a cross in which only one trait is studied. To study patterns of inheritance, the trait that you will observe is the eye color of fruit flies (Figure 10.1), which may be red (wild type) or brown (sepia) in our experiment. You will cross fruit flies from two pure-breeding lines. One line is true-breeding red-eyed and the other is true-breeding brown-eyed.

> *Pure breeding* or *true breeding* means that when individuals having a specific trait are crossed, they always produce progeny like themselves, generation after generation. For example, when red-eyed flies from a pure-breeding line are crossed, their offspring all have red eyes.

Week 1

1. Prepare a fresh vial for your own flies. Using a small plastic cup, pour equal parts *Drosophila* medium and tap water into a clean vial. Drop 5 to 10 grains of yeast onto the medium. On a new label, write your initials and the date, and place the label on the vial. Get a clean foam stopper, and set the new vial and stopper aside. Look at the culture vials with the adult flies in them. Answer Question 1.

2. You will begin by placing three male and three female flies together in fresh vials and allowing them to mate. Study Figure 10.1b to differentiate male and female flies. Half of the groups will use brown-eyed males and red-eyed females. The other groups will use red-eyed males and brown-eyed females. Adult females are ready to mate within 12 hours of emerging from the pupa stage, but you must be sure to use virgin females! Males and females must be separated by the time they are 12 hours old. Ask your instructor for virgin female flies or for flies that are less than 12 hours old. (Promptly thank the person who separated them for you.)

3. You must anesthetize the flies before separating males and females. You may do this in their culture vial or in a new vial that you have transferred the flies to; ask your instructor which method to use. To transfer the flies into a new vial, you must tap the culture vial on your lab bench to move the

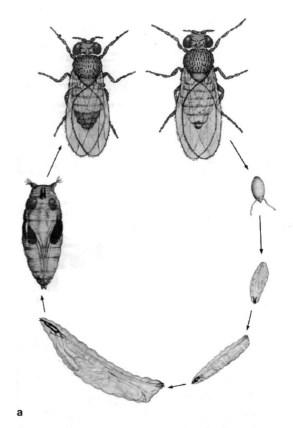

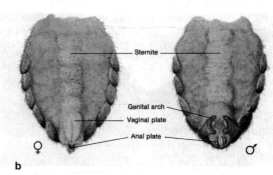

FIGURE 10.1   a. Life Cycle of *Drosophila melanogaster*. b. Female (left) and male (right) flies. © 1988, Carolina Biological Supply Company, used by permission.

flies downward, remove the stopper, and quickly place an empty vial over the culture vial. Then invert the vials and tap the flies into the empty one.

> *Don't tap the culture vial too hard—you don't want to transfer the medium, too.*

Place stoppers into both vials, and don't let any flies get away! Dip the absorbent end of a wand into the Flynap™ bottle. Gently tap the vial containing the flies on your lab bench, push the stopper aside with your index finger, and quickly place the absorbent end of the wand below the stopper. Try to avoid getting Flynap™ on the foam stopper.

4. Watch the flies closely, and remove the stopper immediately after all the flies have stopped moving. If you are anesthetizing the flies in the original culture vial, do not leave the wand in it longer than four minutes. If you are using an empty vial (without culture medium), do not leave the wand in it longer than two minutes.

DO NOT OVERANESTHETIZE THE FLIES!

---

5. Pour the sleeping flies onto an index card. Use a dissecting microscope and soft-bristle paintbrush to separate males from females of each eye color. Place three male flies of one eye color and three female flies of the other eye color in the new culture vial you prepared in Step 1. Note that it doesn't matter which sex has which eye color, as long as all the females have one color, and all the males have the other color. Put the stopper in place, and store the vial in a safe place until next week. Answer Questions 2 and 3.

Week 2

6. Retrieve the fly vials that you set up last week, and anesthetize your flies. Once the flies have stopped moving, place them into the morgue. (They have finished their work.) Put the vial (which now contains eggs and larvae of the F_1 generation) back in a safe place, and you will have new flies by next week.

Week 3

7. Compare your F_1 flies with the parental types. Answer Questions 4 to 6. Following the protocol from Week 1, anesthetize the flies and place three male and three female flies together in a fresh vial. Since all the flies have the same eye color this time, the sorting process will be simpler. You no longer need the vial that the F_1 flies came from. Ask your instructor what to do with the old vial.

Week 4

8. Retrieve your fly vials that you set up last week and anesthetize your flies. Once the flies have stopped moving, place them into the morgue. They have also finished their work. Put the vial (which now contains the eggs and larvae of the

F_2 generation) back into the incubator until next week. Answer Question 7.

Week 5

9. Retrieve your vial containing the F_2 generation of flies. Anesthetize the flies as before; but this time, allow sufficient time for the anesthesia to kill the flies. This will make it easier to count them without having to worry about having the animals fly away in the process. Be sure to retain your vial containing the larvae, in case your instructor asks you to perform an additional cross. Observe and count the number of each phenotype that you have among your F_2 generation and record these numbers in the "Observed" column in Table 10.1.

10. Determine whether the ratio that you observed is consistent with the ratio you expected for the two possible phenotypes, following the procedure described in Exercise 10.2. Answer Question 8.

Questions

1. **What are the phenotypes of the parental flies?**

2. **Why is it important to have virgin female flies?**

3. **What are the genotypes of the parental flies? (Assign letters for their traits.)**

TABLE 10.1	CHI-SQUARE ANALYSIS OF *DROSOPHILA* RATIOS				
Phenotype	**Observed (o)**	**Expected (e)**	**o − e**	**$(o − e)^2$**	**$(o − e)^2 \div e$**
Red eyes					
Brown eyes					
$\Sigma(o) =$				$\Sigma X^2 =$	

Level of significance: 0.05

Degrees of freedom: 1

4. **Which type of parental fly do the F_1 offspring look like?**

5. **Which eye color is the dominant trait?**

6. **On the basis of what you have learned in lecture and the observations that you have made thus far in lab, write a prediction concerning** ratios of phenotypes among your F_2 generation. The "expected" values in Table 10.1 will be generated from these ratios.

7. **Why is it not important to have virgin female flies for the F_2?**

8. Is the ratio of phenotypes observed what you predicted from Question 6? Explain.

Exercise 10.2 CHI-SQUARE ANALYSIS

Continuous with Exercise 10.1

As you may guess, it is very unlikely that the numbers that you observed in the F_2 generation show the exact ratio that you expected. When you compare these numbers, you need to know how much difference is acceptable. Is the difference in the numbers that you observe and the numbers that you expect "real" or a result of random deviation away from the expected values (due to pure chance)? In other words, is the difference great enough to show that our prediction is wrong?

It is useful to have a statistical test to measure the "goodness of fit" of your data. A commonly used test to measure the goodness of fit is the **chi-square test**, in which the observed numbers in each category are compared to the expected numbers in

accordance with a given prediction. Here is the formula for this test:

$$X^2 = (observed - expected)^2 \div (expected)$$

1. Perform a chi-square test on your F_2 data: The observed numbers are the actual flies that you have counted for each phenotype. Add up the "observed" numbers to get the total number of flies observed [$\Sigma(o)$]. Write this number in Table 10.1.

2. To calculate the expected number of flies for each phenotype, multiply the total number of flies observed by the ratios predicted by a Mendelian monohybrid cross (3:1). For example, if you counted 400 total flies, the expected number of red-eyed flies would be $400 \times 3/4 = 300$. Fill in the corresponding column in Table 10.1.

3. Calculate the individual chi-square values for each phenotype, using Table 10.1 as a guide.

4. Calculate the overall chi-square using the formula

$$\Sigma X^2 = X^2_1 + X^2_2$$

where you add up the individual chi-square values of each phenotype. Record your calculations in Table 10.1.

5. The ΣX^2 is then interpreted using Table 10.2. *Degrees of freedom* refers to the number of categories minus 1. In this case, you have two phenotypes, hence two categories. Thus, you have

TABLE 10.2 CHI-SQUARE VALUES AND PROBABILITIES *(Bliss, 1967)*					
Confidence interval	**70%**	**80%**	**90%**	**95%**	**99%**
Chi-Square **Degrees of Freedom**	**30%**	**20%**	**10%**	**5% (sig.)**	**1%**
1	1.074	1.642	2.706	3.841	6.635
2	2.408	3.219	4.605	5.991	9.210
3	3.665	4.642	6.251	7.815	11.345
4	4.878	5.989	7.779	9.488	13.277
5	6.064	7.289	9.236	11.070	15.086

one degree of freedom for this experiment. Using the 5% column for one degree of freedom, compare your ΣX^2 to the value in the table. Should ΣX^2 be less than 3.841 (for one degree of freedom), this means that your results do not differ significantly from the predicted 3:1 ratio. If the number you obtain is greater than 3.841, then your outcome does differ significantly from the prediction, and the "hypothesis" is rejected.

By choosing the 5% column, we are utilizing a 95% confidence interval for this test. When using a 95% confidence interval, only 5 times in 100 would a difference giving a X^2 value greater than 3.841 be the result of random deviation. In instances where there is a "strong" penalty attached to "being wrong," a much higher confidence interval might be applied. For instance, if we were conducting a study in which we were testing the effectiveness of a drug with potentially lethal side effects, we would want to know with a great degree of certainty that the drug works. In other words, we would apply a very high burden of statistical confidence, maybe at the 99% interval. In these instances, in order for a difference to be accepted as being "real," not just chance deviation from the expected value, it would have to have a X^2 value of 6.635 or greater to be considered significantly different. When using the 99% confidence interval, only one time in 100 would a difference giving a chi-square value greater than 6.635 be the result of random deviation. See Table 10.2.

6. If the difference between your outcome and what you predicted is too great to be explained away by "random chance," then you must reject your hypothesis. What this means from our standpoint is probably not going to rock Mendel's boat of laws. Rather, we must consider whether our crosses were performed correctly. Did you start with virgin females? Did you count all the adults in your vial? Do red- or brown-eyed flies die more easily than the other, hence preventing us from counting them?

Questions

1. **Are your F_2 data consistent with your prediction in Exercise 10.1?**

2. **Does statistical analysis support your prediction? Explain how your chi-square values help determine this.**

Exercise 10.3   THE DIHYBRID CROSS AND CHI-SQUARE ANALYSIS

In this exercise you will utilize corn to study the results of a dihybrid cross, a cross that involves two traits. Each kernel (seed) on a ear of corn represents a distinct individual. Thus, corn makes an excellent classroom model for genetic studies because each ear of corn may represent a large sample of offspring from a specific cross. The traits you will be examining are color and texture of the kernel. The kernels will be either purple or yellow, and either smooth (starchy) or wrinkled (sweet). See Figure 10.2. Corn with purple, smooth kernels was crossed with corn having yellow, wrinkled kernels. The resultant F_1 individuals were all purple and smooth. Plants from the F_1 generation were crossed, and the corn grown from them makes up the F_2 generation.

1. Answer Questions 1 through 4.

2. Obtain an ear of F_2 corn and mark your starting point with a pin. Count and record the phenotypes for kernels on your ear of corn using a laboratory counter (or good old-fashioned hash marks on a piece of paper.) Record your data in Table 10.3.

3. The next step will be to determine whether the ratio that you observed is close to the 9:3:3:1 ratio that you expected for the four possible phenotypes. To determine the expected number of purple, smooth kernels, multiply the total number of kernels by 9/16 (0.5625). To calculate the expected number of purple, wrinkled kernels for each phenotype, multiply the total number of kernels counted by 3/16 (0.1875). Continue this for the other two phenotypes, using predicted Mendelian ratios. Record these in the "Expected" column in Table 10.3.

FIGURE 10.2 Genetic Corn. Purple, starchy corn is on the left, and yellow, sweet corn is on the right.

TABLE 10.3 CHI-SQUARE ANALYSIS OF *ZEA MAYS* RATIOS					
Phenotype	**Observed (o)**	**Expected (e)**	**o − e**	**(o − e)²**	**(o − e)² ÷ e**
Purple, smooth					
Purple, wrinkled					
Yellow, smooth					
Yellow, wrinkled					
$\Sigma(o) =$				$\Sigma X^2 =$	
Level of significance: 0.05					
Degrees of freedom: 3					

4. Calculate the individual chi-square values for each phenotype, using Table 10.3 as a guide.

5. Next you will calculate the overall chi-square using the formula

$$\Sigma X^2 = X^2_1 + X^2_2 + X^2_3 + X^2_4$$

where you add up the individual chi-square values of each phenotype. Record your calculations in Table 10.3. The ΣX^2 is then interpreted using Table 10.2. Recall that *degrees of freedom* refers to the number of categories minus 1. In this case, you have four phenotypes, hence four categories. Thus, you have three degrees of freedom for this experiment. Compare your chi-square value to the value in the 5% column for three degrees of freedom, which you will see is 7.815. If the number you obtain is greater than 7.815, then your outcome does differ significantly from the prediction and you must reject your hypothesis. If your number is less than 7.815, then your findings are consistent with your hypothesis.

Questions

1. **What are the phenotype(s) of the F$_1$ individuals?**

2. **Assign genotypes to each parent and to the F$_1$ individuals.**

3. **Identify the dominant traits.**

4. **List the phenotypes that you will expect to see in the F$_2$ generation, and give the ratio at which you will expect the phenotypes to occur. (The "expected" values in Table 10.3 were generated from these ratios.)**

5. **Are your data consistent with your prediction?**

6. **Does statistical analysis support your prediction? Explain how your chi-square values help determine this.**

11

Biotechnology

*The stakes are high in this biological game
of unparalleled consequences.*
M.A. Palladino, 2002

Biotechnology is the use of living organisms to create useful products or to facilitate processes—mostly used to increase human health and comfort. In the following exercises, you will gain hands-on experience with methods used in modern applications of biotechnology. Carefully study **Chapter 15** in your Krogh textbook before you begin.

The myriad processes of biotechnology allow us to manipulate organic molecules such as DNA and proteins. You will compare DNA fragments from different sources, using **restriction enzymes** to cut DNA at particular base sequences. Then you will examine how these sequences separate by size during **gel electrophoresis**.

Phoresis means "to be carried."

This will enable you to identify a disease-causing virus by its *DNA fingerprint*, in a process called **DNA typing**. You will also learn how to make **recombinant DNA** by splicing together fragments of DNA from different organisms. You will then insert this recombined DNA into another organism that expresses the foreign genes, a process called **transformation** (or **transgenic biotechnology**, as your textbook calls it). Then you will conduct cloning, allowing the organism to multiply asexually, producing exact genetic copies

of the desired DNA. These clones express the desired trait.

As you cut up, splice together, move, and separate DNA, keep the applications of this technology in mind. Why are we interested in certain fragments of DNA? Why do we insert them into other organisms? Why do we need many clones of these organisms?

Exercise 11.1   RESTRICTION ENZYMES AND ELECTROPHORESIS OF DNA

(© 1997, University of Reading. Used by permission of Carolina Biological Supply Company and the University of Reading.)

Bacteriophages ("bacteria-eaters") are viruses that invade bacteria. To reproduce, bacteriophages (or phages) must take over the molecular machinery of their bacterial hosts. The phage lambda (λ) preys upon *Escherichia coli*. Lambda can insert its DNA into the bacterial chromosome, which replicates the foreign DNA. Then the host cell manufactures more λ phages, which lyse the cell as the phages are released (Figure 11.1). However, bacteria have a defensive mechanism to prevent bacteriophages from doing their harm: **restriction enzymes**. They are made by bacteria to *restrict* the

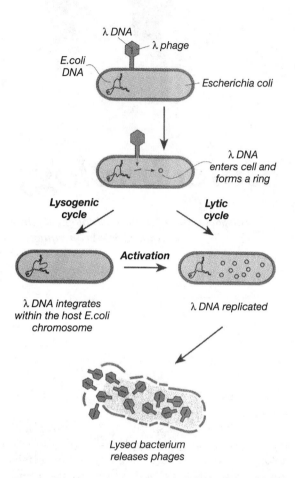

FIGURE 11.1   Bacteriophage Lambda Infecting Bacterial Cell. © 1997, University of Reading. Used by permission of Carolina Biological Supply Company and the University of Reading.

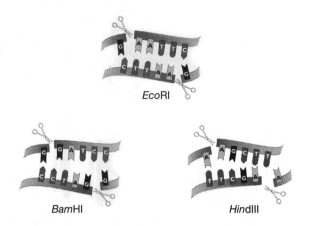

FIGURE 11.2   Recognition Sites of Three Restriction Enzymes. © 1997, University of Reading. Used by permission of Carolina Biological Supply Company and the University of Reading.

proliferation of invading viruses, cutting foreign DNA at specific base sequences within the DNA molecule **[Krogh section 15.2]**.

In this exercise, you will use three restriction enzymes to cut up DNA: *Eco*RI, *Bam*HI, and *Hind*III. Each restriction enzyme recognizes a specific sequence of nitrogenous bases, and cuts the DNA molecule at that **recognition site** (Figure 11.2). The DNA fragments that result will be of different sizes. You will then separate the DNA fragments by **gel electrophoresis** (Figure 11.3) **[Krogh section 15.5]**.

First, a gel is cast of agarose, a very pure form of agar, which is obtained from seaweed. At one end of the slab of gel are several small wells, made by the teeth of a comb that is placed in the gel before it sets. A buffer solution is poured over the gel, so that it fills the wells and makes contact with the electrodes at each end of the gel. Ions in

the buffer solution conduct electricity. The buffer also stops the gel from drying out.

The DNA fragments are mixed with a small volume of leading dye. The dye is dissolved in a dense sugar solution, so that when it is added to the wells, it sinks to the bottom, taking the DNA with it. An electrical current is applied to the electrodes, setting up an electrical field across the gel. Phosphate groups give the DNA fragments a negative electrical charge, so that the DNA migrates through the gel toward the positive electrode. Small fragments move quickly through the porous gel—larger fragments travel more slowly. In this way, the pieces of DNA are separated by size. The loading dye also moves through the gel, so that the progress of the electrophoresis can be seen. After electrophoresis, the gel is stained to reveal the DNA bands more clearly.

Here are some helpful hints:

Hint 1

Do not touch the pipette tips with your bare skin, or you may contaminate the samples.

Hint 2

Practice using the pipette before doing the "real thing" with DNA. Ask your instructor for dye, an extra precast gel, water or buffer solution, a micropipette, and pipette tips. Pour the buffer solution over the gel to a depth of 2 mm. Place a pipette tip onto the pipette. After drawing 20 μl (microliters) of

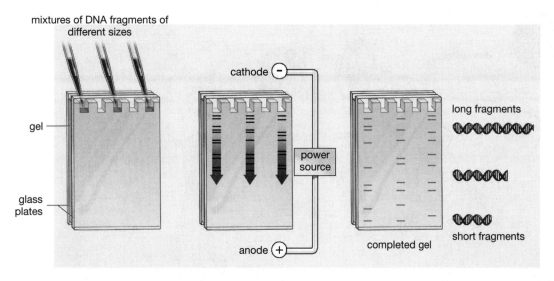

FIGURE 11.3   Banding Patterns of DNA.

dye into the pipette tip, hold the pipette at eye level, slowly insert the tip into a well in the gel, and very gently fill the well with the dye. Repeat this for more practice.

Hint 3

When transferring solutions with a micropipette, the tips may have tiny droplets of solution left in them. It is important for you to transfer as much solution as you can, so remove the droplets by touching the pipette tip to the inner wall of the microtube.

Rehydration of the Lambda DNA

1. Refer to Figure 11.4 as you proceed. Add 100 μl of distilled water to the λ DNA tube. Cap the tube lightly and allow it to stand for five minutes.

2. Hold the closed tube firmly at the top, then flick the side repeatedly with your finger to mix the contents. Do this for one full minute. If you find that drops of liquid are scattered inside the tube, tap it firmly on the bench several times to return the liquid to the bottom.

3. Allow the tube to stand for another five minutes. To be certain that the DNA is thoroughly mixed in the solution, pull the solution up and eject it from the pipette several times.

Cutting the DNA With Restriction Enzymes

4. Add a fresh tip to the micropipette. Put 20 μl of λ DNA solution into an enzyme tube of your choice.

Mix the liquid and the dried enzyme by carefully drawing the liquid up and down into the tip a few times. The liquid in the enzyme tubes should have a distinct blue hue, but there should be no concentration of dye at the bottom of the tube.

5. Repeat this for each enzyme tube and the yellow "control" tube, using a fresh pipette tip each time to prevent cross-contamination between the tubes.

6. Cap each tube with an appropriately colored lid.

7. Put the tubes into the foam rack provided. Incubate the tubes at 37°C, in a water bath or incubator, for 30 to 45 minutes. While you wait, complete Steps 9 to 13.

8. Place the tubes in a hot-water bath at 65°C for 10 minutes to denature the enzymes.

Casting the Agarose Gel

9. Melt the agarose gel (0.8% made up in Tris-Borate-EDTA (TBE) buffer). The molten agar may be stored in a hot-water bath at 55–60°C until needed.

10. Place the gel box on a level surface, where you can leave it undisturbed for the next 15 minutes. This is necessary because if the gel sets at an angle, the DNA fragments will not run evenly through the gel. Slot the comb in place at one end of the gel box.

11. Pour about 10 ml of molten agarose into the gel box so that it fills the central cavity and flows under and between the teeth of the comb. Try not to spill liquid into the areas at either end (if some gel does flow over, just leave it to set—you can scoop it out later).

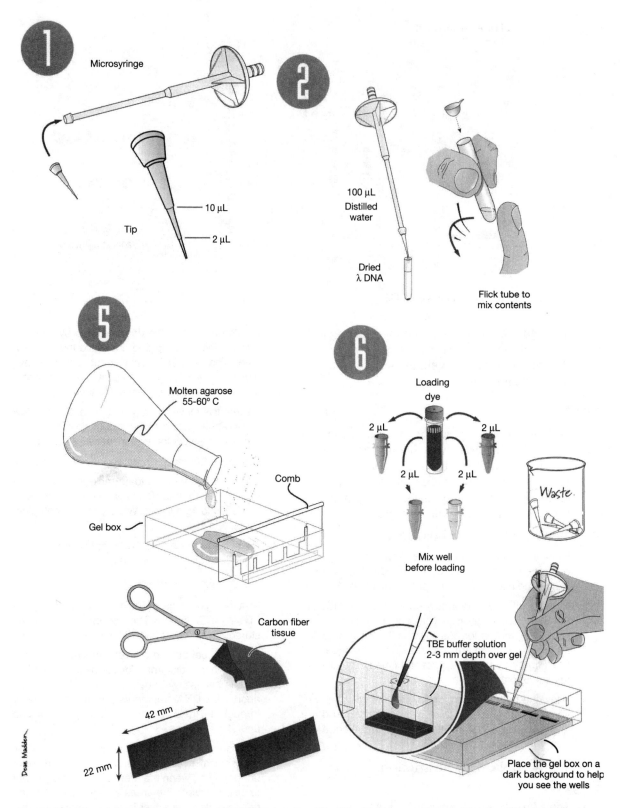

FIGURE 11.4   Restriction Analysis Procedure. © 1997, University of Reading. Used by permission of Carolina Biological Supply Company and the University of Reading.

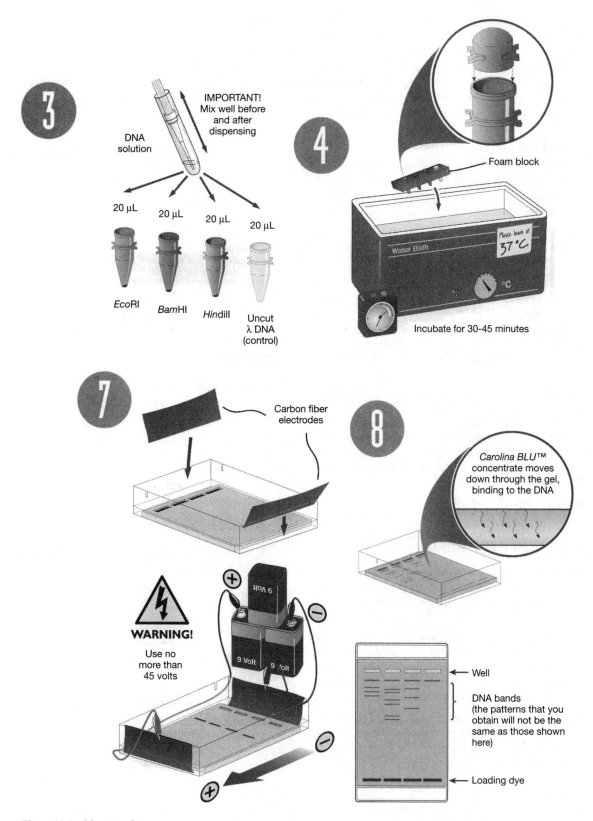

3

DNA solution

IMPORTANT! Mix well before and after dispensing

20 μL
*Eco*RI

20 μL
*Bam*HI

20 μL
*Hin*dIII

20 μL
Uncut λ DNA (control)

4

Foam block

Water Bath

Please leave at 37°C

°C

Incubate for 30-45 minutes

7

Carbon fiber electrodes

WARNING!
Use no more than 45 volts

9 Volt
9 Volt 9 Volt

8

Carolina BLU™ concentrate moves down through the gel, binding to the DNA

Well

DNA bands (the patterns that you obtain will not be the same as those shown here)

Loading dye

Figure 11.4   (*Continued*)

12. Do not disturb the box until the gel has set hard. The gel will exhibit an opaque appearance when set.

13. Cut two pieces of the carbon fiber tissue, each about 42 mm × 22 mm. These will be the electrodes at either end of the gel box. Put the electrodes aside until you have loaded the gel. (Don't forget to complete Step 8 above.)

Loading the Gel

14. Pour just over 10 ml of TBE buffer solution into the gel box. The liquid should cover the surface of the gel and flood into the reservoirs that will hold the electrodes.

15. Very gently ease the comb from the gel, allowing the buffer solution to fill the wells left behind. Take care not to tear the wells.

16. Put the gel box on a dark surface, where it will remain undisturbed while the gel is running.

17. Put a clean tip on the micropipette. Add 2 μl of loading dye to the DNA you wish to load. Mix the dye and the DNA sample very thoroughly by drawing the mixture up and down in the microsyringe tip.

18. Expel any air from the pipette tip, without expelling the loading dye/DNA mixture! Pipette the mixture into one of the wells, holding the tip above the well but under the buffer solution. Take great care not to puncture the bottom of the well with the micropipette tip. (Remember, you should practice this with dye and an extra gel first!)

19. Make a note of which DNA you have put into the well. Repeat Steps 17 to 19 for the control and remaining tubes, using a new tip each time.

Running the Gel

20. Fit one electrode at each end of the box. Your instructor will assist you in attaching the power supply (alligator chips) to the electrodes. Turn the switch on.

21. Check that contact is made between the buffer solution and the electrodes. (If after a few minutes, the DNA doesn't appear to be moving, then the current isn't flowing through the buffer solution.) Add a little more buffer if necessary.

22. Disconnect the batteries or power supply once the blue loading dye has reached the end of the gel. If you leave it connected, the DNA will run off the end of the gel.

23. Rinse the alligator clips in tap water and dry them thoroughly to prevent corrosion.

Staining the DNA

24. Remove and discard the electrodes and pour off the buffer solution. The buffer solution may be reused.

25. With a gel spatula or flexible plastic ruler, **slowly** remove the gel from the gel box, separating it from the edges of the gel box first. Place the gel (top side up) in a staining tray.

26. While wearing gloves, pour 10 ml of DNA-staining solution onto the gel.

27. After exactly four minutes, pour the stain back into its bottle.

28. Carefully rinse the gel with distilled water three times to remove excess stain. During each rinse, "soak" the gel for a few minutes.

29. Observe the bands on your gel using the light board provided. If you cannot see clear DNA bands, put the soaking gel and staining tray into a zippered plastic bag to prevent desiccation, and allow the gel to destain until next week.

Questions

1. **How many bands do you see? Sketch the bands of the gel, and indicate the DNA source for each well.**

2. **Which enzyme produces the smallest fragment? The largest?**

3. **What does the undigested DNA control show us?**

4. **Did the enzymes digest all the DNA they were incubated with? How do you know?**

5. **Which enzyme has the most restriction sites on lambda DNA? How do you know?**

6. **How difficult was it to manipulate the DNA? Are you surprised by this?**

7. **How do you know the DNA was cut if you can't even see the DNA fragments? How do biotechnicians (and chemists, physicists, and astronomers, for that matter) know how to make conclusions if they can't even "see" what they are working with?**

Exercise 11.2   OUTBREAK! FINGERPRINTING VIRUS DNA

(© 1998, Carolina Biological Supply Company. Used by permission.)

This exercise also uses electrophoresis, but this time the banding patterns will help you identify whether or not a fictitious virus is pathogenic. Review the "helpful hints" given in Exercise 11.1. Follow the story and instructions given below. Note that the DNA has already been prepared for you and is ready to electrophorese—a restriction enzyme has already been used to cut the DNA samples.

You are a molecular biologist working for the Centers for Disease Control and Prevention (CDC). Your job is to help track epidemics and to monitor emerging diseases.

Emerging diseases are those that are increasing in frequency.

Five years ago, a cluster of cases of hemorrhagic fever occurred in an isolated town in northeastern Alabama. The disease killed approximately 30% of those who caught it. The most alarming aspect of the new disease was that it was highly contagious from human to human, thus posing the threat of an epidemic. Medical authorities, including your office, believe the only reason the outbreak did not erupt into a major epidemic was that the Alabama town was so small and isolated. A CDC team traced the disease to a virus carried by the numerous local squirrels. An extensive trapping campaign was carried out in an attempt to eliminate the virus by eliminating the squirrels infected with it. No further cases have been reported in Alabama.

Three years ago a suspicious outbreak occurred in Pennsylvania. Several people fell ill with a hemorrhagic fever. The symptoms of the disease were the same as those of the Alabama fever, but no one died. The Pennsylvania fever was apparently less contagious than the Alabama fever since the Pennsylvania patients were exposed to many people who did not come down with the disease. However, the Pennsylvania virus was also traced to the local squirrel population.

You were asked to compare the viruses that caused the two outbreaks. You found that the virus particles looked similar. Both viruses had a DNA genome, but the base sequences of the genomes were different in many places.

Now, three people in Missouri have fallen ill with a hemorrhagic fever. Their symptoms are similar to the symptoms of the Alabama and the Pennsylvania fevers. Local medical personnel were alarmed and called the CDC to determine if the dangerous Alabama virus had reappeared. The patients have been placed in quarantine. You are flown to the scene.

You must immediately determine whether the Missouri patients are infected with the highly contagious and deadly Alabama virus, the Pennsylvania virus, or some other agent.

One of the techniques you decide to use is examination of viral DNA by restriction analysis. Wearing protective clothing, you enter the patient's isolation rooms in the Missouri hospital, carefully draw samples, place them on ice, and rush back to the biological containment laboratory at the CDC for a variety of tests. For the restriction analysis, you isolate virus particles, extract DNA, and subject the samples to restriction digestion and agarose gel electrophoresis. In your gel, you include samples of DNA from the Pennsylvania and the Alabama viruses for comparison.

Cast Agarose Gel

1. Place a gel-casting tray on a level surface, where you can leave it undisturbed for the next 15 minutes. This is necessary because if the gel sets at an angle, the DNA fragments will not run evenly through the gel. Slot the comb in place at one end of the gel box.

2. Pour about 10 ml of molten agarose into the gel box so that it fills the central cavity and flows under and between the teeth of the comb. Try not to spill liquid into the areas at either end (if some gel does flow over, just leave it to set—you can scoop it out later).

3. Do not disturb the box until the gel has set hard. The gel will exhibit an opaque appearance when set.

4. Pour just over 10 ml of TBE buffer solution into the gel box. The liquid should cover the surface of the gel and flood into the reservoirs that will hold the electrodes.

5. Very gently, ease the comb from the gel, allowing the buffer solution to fill the wells left behind. Take care not to tear the wells.

6. Put the gel box on a dark surface, where it will remain undisturbed while the gel is running.

Loading the Gel

7. Using a fresh pipette tip, draw one of the three DNA samples into your pipette. Expel any air from the pipette tip, without expelling the DNA! Pipette the DNA mixture into one of the wells, holding the tip above the well but under the buffer solution. Take great care not to puncture the bottom of the well with the micropipette tip.

8. Make a note of which DNA you have put into the well. Repeat Step 7 for the remaining DNA samples, using a new tip each time.

Running the Gel

9. Cut two pieces of the carbon fiber tissue, each about 42 mm × 22 mm. These will be the electrodes at either end of the gel box. Fit one electrode at each end of the box. Your instructor will assist you in attaching the power supply to the electrodes. Turn the switch on.

10. Check that contact is made between the buffer solution and the electrodes. (If after a few minutes, the DNA doesn't appear to be moving, then the current isn't flowing through the buffer solution.) Add a little more buffer if necessary.

11. The loading dye will separate into two bands of color. The faster-moving, purple band is the dye bromophenol blue; the slower-moving, aqua band is xylene cyanol. Allow the DNA to electrophorese until the bromophenol blue band is about 2 cm from the end of the gel.

12. Disconnect the batteries or power supply. If you leave it connected, the DNA will run off the end of the gel!

13. Rinse the alligator clips in tap water and dry them thoroughly to prevent corrosion.

Staining the DNA

14. Remove and discard the electrodes and pour off the buffer solution. The buffer solution may be reused.

15. With a gel spatula or flexible plastic ruler, **slowly** remove the gel from the gel box, separating it from the edges of the gel box first. Place the gel in a staining tray.

16. While wearing gloves, pour 10 ml of DNA-staining solution onto the gel.

17. After exactly four minutes, pour the stain back into its bottle. (If you are using "Final Stain Solution," the gel will need to be stained for 15 minutes.)

18. Carefully rinse the gel with distilled water three times to remove excess stain. During each rinse, "soak" the gel for a few minutes to remove the stain.

19. Observe the bands on your gel using the light board provided. If you cannot see clear DNA

bands, put the soaking gel and staining tray into a zippered plastic bag to prevent desiccation, and allow the gel to destain until next week.

Questions

1. **Compare the DNA fingerprints of the Pennsylvania, Alabama, and Missouri virus isolates. What can you conclude about the virus infecting the Missouri patients?**

2. **Why would the restriction fragment patterns from the two viruses be different?**

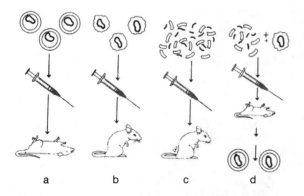

FIGURE 11.5 Griffith's Transformation Experiment. A. When the virulent S-strain was injected into a mouse, the mouse died. B. When the non-virulent R-strain was injected into a mouse, the mouse lived. C. When the heat-killed S-strain was injected into the mouse, the mouse lived. D. When the heat-killed S-strain was mixed with living R-strain bacteria, the mouse died, and living S-strain bacteria were cultured from the mouse. This indicates that the living R-strain bacteria were transformed into a virulent type, like the S-strain. © 1999, Maria Rapoza and Helen Kreuzer, Carolina Biological Supply Company. Used by permission.

Exercise 11.3 TRANSFORMATIONS

Extended two-week exercise

(© 1999, Maria Rapoza and Helen Kreuzer, Carolina Biological Supply Company. Used by permission.)

Transformation is a process in which foreign DNA is taken up by an organism that expresses the foreign genes. In a classic 1928 experiment, a physician researcher named Frederick Griffith was the first to demonstrate transformation (Figure 11.5). Griffith worked with two strains of the *Pneumococcus* bacterium: an encapsulated S-strain that is virulent (meaning that is pathogenic or disease-causing) and a nonencapsulated R-strain that is nonvirulent. When dead bacterial cells of the S-strain were exposed to living R-strain cells, Griffith found that something from the dead S-strain cells *transformed* the R-strain cells into virulent agents.

> Bacteria often have an outer protein covering (capsule) that helps them to adhere to substrate and that also provides some defense against the immune system of the host it is infecting. This accounts for the virulence of the S-strain and the nonvirulence of the R-strain.

It wasn't until 1944 that DNA was implicated as the transforming agent. Unfortunately, Griffith died in a World War II bomb blast while working in his laboratory in London, so he never lived to recognize the significance of his work. In retrospect, this simple experiment turns out to be one of the more important in the history of biology because it led to the implication of DNA as genetic material. This provided the conceptual basis for the important work by others many years later, including James Watson and Francis Crick.

In this exercise, you are going to create "glow-in-the-dark" *Escherichia coli* bacteria by using recombinant DNA to transform the *E. coli* cells **[Krogh section 15.2]**. You are going to use **plasmids**, extra pieces of DNA found in some types of bacteria. The plasmids contain DNA that encode for **antibiotic resistance**, which allows bacteria to survive when exposed to antibiotics that normally kill bacterial cells (cells without the plasmids that carry the gene for resistance). See Chapter 14 for information about antibiotic resistance.

For this transformation experiment, plasmids containing an antibiotic resistance gene were recombined with a section of DNA that encodes for bioluminescence. **Bioluminescence** is a fascinating feature of some organisms that emit light, such as fireflies, algae, and jellyfish.

If you ever go boating in the sea at night, look in the wake of the boat. You may notice a glow as the water is disturbed. Or, if you walk on a beach at night, kick the

wrack line (the high tide line of washed-up sea "junk") and you may see a momentary glow.

To create the recombined plasmid, DNA was taken from a jellyfish, *Aequorea victoria*, which has the GFP (Green Fluorescent Protein)

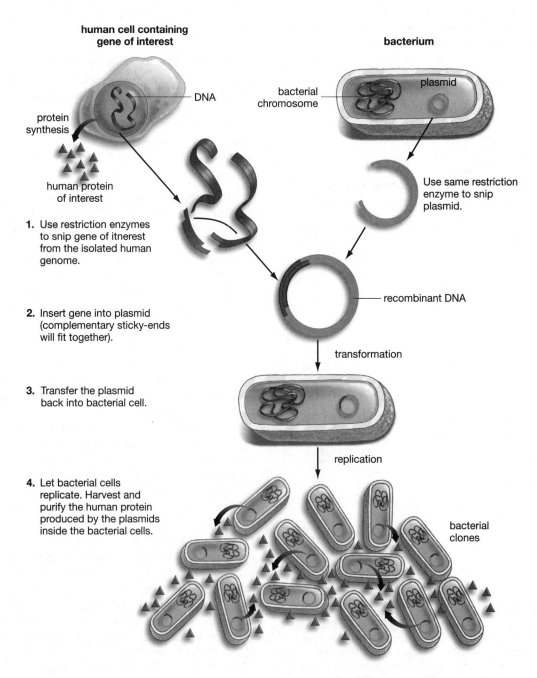

human cell containing gene of interest

DNA

protein synthesis

human protein of interest

bacterium

bacterial chromosome

plasmid

Use same restriction enzyme to snip plasmid.

1. Use restriction enzymes to snip gene of itnerest from the isolated human genome.

2. Insert gene into plasmid (complementary sticky-ends will fit together).

recombinant DNA

transformation

3. Transfer the plasmid back into bacterial cell.

replication

4. Let bacterial cells replicate. Harvest and purify the human protein produced by the plasmids inside the bacterial cells.

bacterial clones

FIGURE 11.6   Recombinant DNA, Transformation, and Cloning: Using Bacteria to Produce a Needed Human Protein.

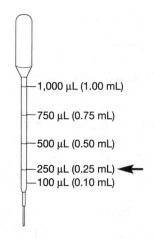

FIGURE 11.7   Pipette Graduations.
© 1999, Maria Rapoza and Helen
Kreuzer, Carolina Biological Supply Company. Used by permission.

gene, encoding for a protein that makes the jellyfish bioluminescent. The same restriction enzyme is used to create "sticky ends" on the jellyfish DNA and on plasmid DNA. The sticky ends of the two DNA sources were then spliced together, creating a new, recombined plasmid. The same technology can be used to produce plasmids containing human DNA, or DNA of other animals (Figure 11.6).

The first genetically engineered pets arrived on the U.S. market in January 2004—zebra fish that glow in the dark, thanks to the jellyfish GFP gene.

Your job is to apply the plasmids to a solution of *E. coli* bacteria, and if the bacteria successfully "take up" the recombined plasmid, they too will become bioluminescent. You will also see that they are able to grow on media (nutrients in a petri dish) containing the antibiotic ampicillin (Amp). The recombined plasmid will enable the bacteria to survive, even in the presence of Amp. Refer to Figures 11.7 and 11.8 as you complete these steps.

Week 1

1. Mark one sterile 15-ml tube "+". Mark another "−". (Plasmid DNA will be added to the + tube; none will be added to the − tube.)

2. Use a sterile transfer pipette to add 250 μl of ice-cold calcium chloride to each tube. Calcium chloride makes the bacterial cell membranes more permeable, so that plasmids can enter the cells.

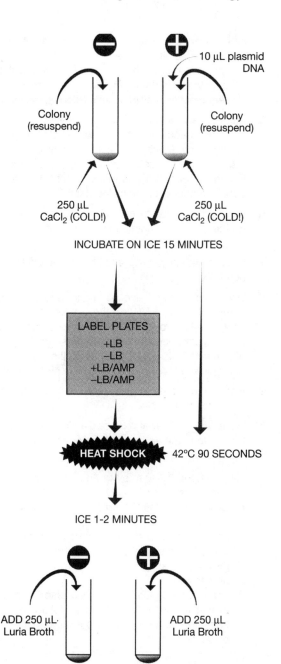

FIGURE 11.8   Transformations Procedure. © 1999, Maria Rapoza and Helen Kreuzer, Carolina Biological Supply Company. Used by permission.

3. Place both tubes on ice.

4. Use a sterile plastic inoculating loop to transfer a cell mass about the diameter of a pencil eraser from isolated colonies of *E. coli* cells from the starter plate into the + tube.

   a. Be careful not to transfer any agar from the plate along with the cell mass.

b. Immerse the cells on the loop in the calcium chloride solution in the + tube and vigorously spin the loop in the solution to dislodge the cell mass. Hold the tube up to the light to observe that the cell mass has fallen off the loop.

5. Immediately suspend the cells by repeatedly pipetting in and out with a sterile transfer pipette. Examine the tube against light to confirm that no visible clumps of cells remain in the tube or are lost in the bulb of the transfer pipette. The suspension should appear milky white.

6. Return the + tube to ice. Transfer a mass of cells to the − tube and suspend as described in Steps 4 and 5 above.

7. Return the − tubes to ice. Both tubes should now be on ice.

8. Use a sterile plastic inoculating loop to add one loopful of plasmid DNA to the + tube. When the DNA solution forms a bubble across the loop opening, its volume is 10 μl. Immerse the loopful of plasmid DNA directly into the cell suspension and spin the loop to mix the DNA with the cells.

9. Return the + tube to ice and incubate both tubes on ice for 15 minutes.

10. While you wait, label your media plates with your lab group name and date. Write on the perimeter of the petri dish with small print.

a. Label one LB/Amp plate "+." This is an experimental plate.
b. Label the other LB/Amp plate "−." This is a control.
c. Label one LB plate "+." This LB plate is a control to test the viability of the cells after they have gone through the transformation procedure.
d. Label one LB plate "−." This is also a control to test the viability of the cells following the transformation procedure, but these have no plasmids.

11. Following the 15-minute incubation on ice, "heat-shock" the cells. Remove both tubes directly from ice and immediately immerse them in the 42°C water bath for 90 seconds. Gently agitate the tubes while they are in the water bath. Return both tubes directly to ice for one or more minutes.

12. Use a sterile transfer pipette to add 250 μl Luria broth (LB) to each tube. Gently tap the tubes with your finger to mix the LB with the cell suspension. Place the tubes in a test tube rack at room temperature for a 10-minute recovery.

13. Now you will remove some cells from each transformation tube and spread them on the plates. Do one plate at a time, from start to finish. Cells from the − tube should be spread on the − plates, and cells from the + tube should be spread on the + plates.

a. "Clamshell" (slightly open) the lids and carefully pour 4 to 6 glass beads onto each plate. Your instructor may have done this for you already.
b. Use a sterile transfer pipette to add 100 μl of cells from the - transformation tube to each of the - plates.
c. Use a back-and-forth and up-and-down shaking motion, not swirling round and round, to move the glass beads across the entire surface of the plate. This should evenly spread the cell suspension all over the agar surface.
d. When finished spreading, let the plates rest for several minutes to allow the cell suspensions to become absorbed into the agar.
e. To remove the glass beads, hold each plate vertically over a container, clamshell the lower part of the plate, and tap out the glass beads into the container. Do not touch or disturb the beads.

14. Use another sterile transfer pipette to add 100 μl of cell suspension from the + DNA tube onto the appropriate plate.

15. Immediately spread the cell suspensions as described in Step 13.

16. Wrap the plates together with tape and place the plates upside down either in the incubator or at room temperature. Incubate them for 24 to 36 hours in a 37°C incubator, or 48 to 72 hours at room temperature. After the incubation period, the plates will be moved to the refrigerator to stop bacterial growth. Answer Question 1.

17. Predict the results. Write "growth" or "no growth" in Table 11.1, depending on whether you think the plate will show growth. Give the reasons for your predictions.

Week 2

18. Examine the plates for bacterial growth and bioluminescence. Your instructor will assist you in using a UV light, if available.

19. Observe the colonies through the bottom of the petri plate. Do not open the plates. Record your observed results in Table 11.1. If your observed results differed from your predictions, explain why you think that may have occurred.

TABLE 11.1 GROWTH ON TRANSFORMATION PLATES	
A. Experimental	**B. Control**
LB/Amp + Prediction: Reason: ( ) Observed result: # of colonies:	LB/Amp − Prediction: Reason: ( ) Observed result: # of colonies:
C. Control	**D. Control**
LB + Prediction: Reason: ( ) Observed result: # of colonies:	LB − Prediction: Reason: ( ) Observed result: # of colonies:

20. Count the number of individual colonies and, using a permanent marker, mark each colony as it is counted. If the cell growth is too dense to count individual colonies, record "lawn."

Questions

1. **What does the phenotype of the colonies tell you? Explain.**

2. **Compare and contrast the number of colonies on each of the following pairs of** plates. **What does each comparison tell you about the experiment?**

   a. **LB + and LB −**

   b. **LB/Amp − and LB −**

c. **LB/Amp + and LB/Amp −**

6. **Why is there no growth on the LB/Amp − plates?**

d. **LB/Amp + and LB +**

7. **Why does growth occur in only certain spots on the LB/Amp + plate?**

3. **What are you selecting in this experiment? (That is, what tells you that the transformation has been successful and the plasmids are in the recipient cells?)**

8. **Which plate(s) glows most brightly? Why?**

9. **Which plates glow only a little? Why?**

4. **In biotechnology research, what is the value of placing a gene for antibiotic resistance into recombinant plasmids?**

10. **What one plate would you first inspect to conclude that the transformation occurred successfully? Why?**

5. **Why is there a "lawn" of growth on the LB + and LB − plates?**

UNIT 4
Evolution and Classification

12

Evolution

This statement by Yale University's famous pale-ontologist was a bold one, but it emphasized the excitement over the principles of evolution in the late 1800s. **Evolution**, simply put, refers to change over time. *Organic evolution* is the process by which living organisms change over time, and *natural selection* is its main driving force. **Natural selection** occurs when some individuals are more successful than others in surviving and hence re-producing, because of traits that give them a better "fit" with their environment.

The theory of organic evolution through nat-ural selection was formulated by Charles Robert Darwin and Alfred Russel Wallace (Figure 12.1).

(a)

(b)

FIGURE 12.1   a. Charles Darwin in 1881. b. Alfred Russel Wallace.

Darwin's 1859 treatise, *The Origin of Species*, provided the foundation of modern biology **[Krogh Chapter 16]**. Others had previously considered the notion that an evolutionary process is at the heart of the diversity of life, but there were two features that set *The Origin of Species* above all preceding works. First, it provided overwhelming and well-articulated evidence that evolution occurs; and most importantly, with natural selection it provided a mechanism by which evolution may proceed.

The essence of the theory of natural selection is as follows:

1. **Variation exists among individuals of a population in nature.** This inherent variation provides the material for the process of natural selection.
2. **Populations exhibit phenomenal reproductive potential.** Darwin got this idea from *An Essay on the Principle of Population* by T. R. Malthus, which stated that animal and plant populations tend to increase beyond the environment's capacity to support them **[Krogh section 16.4]**.
3. **Environments have limited resources, which leads to a struggle for survival.** Individuals having variations that are not beneficial are more likely to be eliminated. Likewise, those with advantageous characteristics survive and reproduce. Thus there is a "struggle for survival," and a process of natural selection operates. This results in what is sometimes called "survival of the fittest," or as Darwin called it, *the preservation of favored races*.
4. **Populations in nature are continually evolving.** Eventually, enough change occurs that a new species will be the result. This process is called **anagenesis** (Figure 12.2). Anagenesis can account for the change exhibited by a group of organisms over time; however, it cannot account for the tremendous diversity of life that exists on Earth. When a population is subdivided such that two or more components of the population are reproductively isolated, accrued changes may lead to **cladogenesis**.

Darwin formulated a strong theory explaining the process by which evolution occurs, but he lacked the knowledge of how characteristics were passed from one generation to the next. As you know,

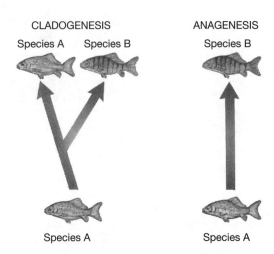

FIGURE 12.2 Modes of Speciation. Speciation can occur either by the branching of one lineage into others (cladogenesis) or by gradual change along a single lineage (anagenesis).

DNA is the material that provides the phenotypes, or characteristics, upon which natural selection may act. In other words, genetics lies at the heart of evolution. Darwin had no knowledge of genes and chromosomes; genetic knowledge was the one element needed to make the theory complete. Ironically, after Darwin's death, a copy of Mendel's 1866 paper on his pea plant experiments was found among Darwin's effects. The reprint was unread.

Exercise 12.1 NATURAL SELECTION

The peppered moth, *Biston betularia*, nicely demonstrates the phenomenon of natural selection. This moth exists in two phases—light and dark. Under natural circumstances the light phase is the most common because moths of the light phase are able to blend in with their background, especially if the background is covered with lichens (Figure 12.3a).

See a discussion about lichens in Exercise 16.4.

The birds that eat peppered moths are less likely to see the light moths, and are less likely to eat them. However, when the black moths rest on the light-colored background, the contrast makes a highly advertised meal for a bird. It is as if the black moths are emitting a flashing neon sign that says

a

b

FIGURE 12.3   Evolution in Action. a. Light and dark moths on a tree covered with light-colored lichens. b. Light and dark moths on a soot-covered tree.

EAT ME! And the birds do just that. However, if the environment is affected by industrial soot as in Figure 12.3b, the opposite happens, with light moths being detected and eaten more often.

In this exercise, you will examine the role that environment can play in determining how a specific phenotype can provide a selective advantage to an organism. Color patterns of peppered moths involve complete dominance of a single gene with two alleles. Pigment production leads to the dark phase and is a dominant trait. Possession of at least one copy of the dominant allele (such as in the *AA* or *Aa* genotypes) results in a dominant phenotype (black moth). Homozygous recessive individuals (*aa*) have the light color.

Traits are encoded on genes. An **allele** is an alternative form of a gene; each individual carries two alleles for most traits.

You are going to play the role of a predatory bird. Model peppered moths, represented by beads, will be placed on a soot-covered resting spot, represented by sand. The experiment will begin with 90% light moths (white beads) and 10% dark moths (red beads). The following proportions of genotypes will be used for the parental generation (P):

1 homozygous dominant (*AA*) dark moths
39 heterozygous (*Aa*) dark moths
360 homozygous recessive (*aa*) light moths

1. Predict what will happen as predation occurs. Answer Questions 1–3.

2. Separate into groups of 4 students. Two "setup students" from each group will be responsible for setting up the exercise and conducting the mating of the moths. Two students will play the role of "birds," and should leave the classroom until it is time for them to eat "moths."

3. Instructions for setup students: Your instructor will give you a cup containing 100 moths that have randomly been drawn from the 400 moths comprising the parental generation. The 100 moths should be randomly distributed into a container of red sand (this is the soot-covered tree). Gently shake the container until the moths are partially or fully submerged in the sand.

4. Instructions for birds: Enter the room, and when prompted, use forceps to remove 50 moths from the "tree" (the cup) one at a time. Don't worry about what color to pick up—just eat. You may shake the cup if you like, to expose more moths (beads). Remember you are a bird and you are hungry. After you have finished eating, leave the room to nap while your meal is digested.

5. Setup students: After the 50 "moths" have been eaten, retrieve the surviving moths by pouring the sand through a strainer. The 50 moths will be combined with surviving "moths" from all other student groups and placed into a container. The container will be passed around and each student group will randomly pull 25 pair of "moths" from the container.

6. Mate each of the 25 pair you selected. Assume that each breeding pair produces 4 offspring. As a shortcut, the possible offspring have been listed below. Record the total number of each genotype and phenotype represented in your 100 offspring in Table 12.1. These represent the F_1 generation.

    $AA \times AA = 4\ AA$ (All Red)

    $AA \times Aa = 2\ AA$ (Red); 2 Aa (Red)

    $AA \times aa = 4\ Aa$ (Red)

    $Aa \times Aa = 1\ AA$ (Red); 2 Aa (Red); 1 aa (White)

    $Aa \times aa = 2\ Aa$ (Red); 2 aa (White)

    $aa \times aa = 4\ aa$ (White)

7. In Table 12.1, record the class data of the 400 offspring generated by moths from all student groups.

8. Acquire 100 moths from the stock cups to represent **your** 100 F_1 offspring. You will have to choose the exact number that you listed for each genotype in Table 12.1.

9. Your instructor will collect your F_1 moths in order to pool all 400 offspring from the 4 class groups.

10. Repeat Steps 2–7 to acquire data for the F_2 generation, but this time record your data in Table 12.2. After the birds have finished eating the second time, they will remain to assist the others process the F_2 data.

11. All students: Now you must run a control for this experiment, so that you can compare what happens in the absence of selective pressure. Your instructor will give you a cup of 100 moths from the original parental generation. Without peeking, randomly eat (remove) 50 moths from the cup and set them aside.

12. Birds: Mate the remaining moths in the cup as described in step 6 above. Record your data for the control F_1 generation in Table 12.3.

13. Prepare the control F_1 generation, as described in Step 8 above, except use data from Table 12.3. Your instructor will pool all control F_1 individuals.

14. Repeat Steps 11 and 12 to acquire data for the control F_2 generation. Record your data for the control F_2 generation in Table 12.4.

Questions

1. **If the "resting spot" is heavily affected by industrial soot, what color moths will be eaten most readily?**

2. **How do you predict this will affect the genotypic ratios and the phenotypic ratios?**

TABLE 12.1   F₁ GENERATION WITH SELECTIVE PRESSURE

Your mating pair(s)	F₁ from your mating pair(s)		Total F₁ from all mating pairs in class		Total F₁ from all mating pairs in class	
	Genotype	Number	Genotype	Number	Phenotype	Number
	AA		AA		Dark	
	Aa		Aa			
	aa		aa		Light	

TABLE 12.2   F₂ GENERATION WITH SELECTIVE PRESSURE

Your mating pair(s)	F₂ from your mating pair(s)		Total F₂ from all mating pairs in class		Total F₂ from all mating pairs in class	
	Genotype	Number	Genotype	Number	Phenotype	Number
	AA		AA		Dark	
	Aa		Aa			
	aa		aa		Light	

TABLE 12.3   F₁ GENERATION WITHOUT SELECTIVE PRESSURE (CONTROL)

Your mating pair(s)	F₁ from your mating pair(s)		Total F₁ from all mating pairs in class		Total F₁ from all mating pairs in class	
	Genotype	Number	Genotype	Number	Phenotype	Number
	AA		AA		Dark	
	Aa		Aa			
	aa		aa		Light	

TABLE 12.4   F₂ GENERATION WITHOUT SELECTIVE PRESSURE (CONTROL)

Your mating pair(s)	F₂ from your mating pair(s)		Total F₂ from all mating pairs in class		Total F₂ from all mating pairs in class	
	Genotype	Number	Genotype	Number	Phenotype	Number
	AA		AA		Dark	
	Aa		Aa			
	aa		aa		Light	

3. Is predation by the birds random or nonrandom? Explain.

4. How does the background affect prey item selection? (Is your prediction correct?) Explain.

5. Is the selection of moths in the "Control" (Steps 9–11) random or nonrandom?

6. What happens to the proportions of the genotypes after each predation? Compare the experimental to the control.

7. What happens to the proportions of the phenotypes after each predation? Compare the experimental to the control.

8. Is natural selection occurring? Explain your answer.

9. Is evolution occurring? Explain your answer.

10. Explain how the processes described in this exercise could lead to the appearance of a new species.

Exercise 12.2   DETECTING EVOLUTION: THE HARDY-WEINBERG PRINCIPLE

Continuous with Exercise 12.1

Exercise 12.1 illustrates how gene frequencies can fluctuate when a selective pressure (predation of moths with a particular phenotype) is applied, using Punnett squares and allele combinations. But how can we measure how much these allele combinations have changed? This shift in allele frequencies is called **microevolution**, and it can be measured with the Hardy–Weinberg equation. Before you proceed with this exercise, carefully read the essay Detecting Evolution: The Hardy–Weinberg Principle on page XX of your Krogh textbook.

Remember the two main reasons why the Hardy–Weinberg equation is important. First, it can approximate the frequencies of alleles in a population. We can't determine this simply by looking at the phenotype. Why not? Because we don't know whether an organism with a dominant phenotype (such as a dark moth) has a homozygous dominant (AA) or heterozygous (Aa) genotype—the dominant phenotypes will be expressed, and recessive alleles will not. Using the Hardy–Weinberg equation, we can calculate the ratios of dominant *and* recessive alleles in a gene pool. Second, it helps us measure just how much a population is evolving, by comparing the frequencies of alleles in an evolving population (an experimental group) to a population that is not evolving at all (a control group).

In order for the Hardy–Weinberg equation to be applicable to a population, the following conditions must be met:

1. No genes are undergoing mutation
2. The population is very large
3. The population is isolated from other populations of the same species
4. All members of the population survive and reproduce
5. Mating is random.

Given this ideal set of circumstances, frequencies of alleles in the population remain relatively constant over time, and are not affected by crossing over and independent assortment. In other words, sexual reproduction by itself does not

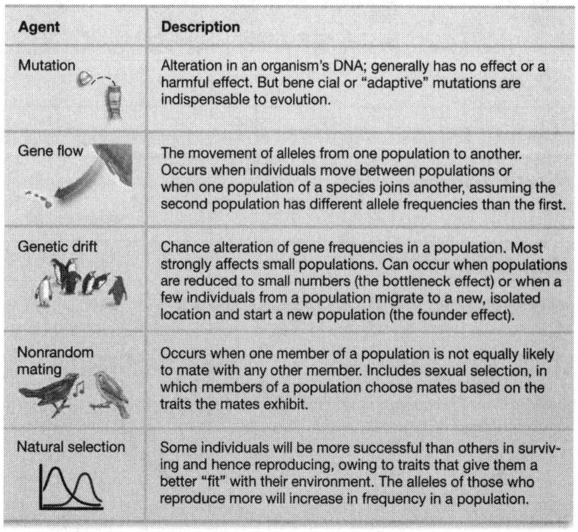

Agent	Description
Mutation	Alteration in an organism's DNA; generally has no effect or a harmful effect. But bene cial or "adaptive" mutations are indispensable to evolution.
Gene flow	The movement of alleles from one population to another. Occurs when individuals move between populations or when one population of a species joins another, assuming the second population has different allele frequencies than the first.
Genetic drift	Chance alteration of gene frequencies in a population. Most strongly affects small populations. Can occur when populations are reduced to small numbers (the bottleneck effect) or when a few individuals from a population migrate to a new, isolated location and start a new population (the founder effect).
Nonrandom mating	Occurs when one member of a population is not equally likely to mate with any other member. Includes sexual selection, in which members of a population choose mates based on the traits the mates exhibit.
Natural selection	Some individuals will be more successful than others in surviving and hence reproducing, owing to traits that give them a better "fit" with their environment. The alleles of those who reproduce more will increase in frequency in a population.

FIGURE 12.4   Agents of change: five forces that can bring about change in Allele frequencies in a population.

change the frequency of alleles within the population. However, these ideal circumstances are not realistic, and you know that other forces have a profound impact on the gene pool (Figure 12.4).

The Hardy–Weinberg Equilibrium gives you a control group—what the population would look like if it weren't evolving. Under the ideal circumstances of the Hardy–Weinberg Principle, allele frequencies may be described by the following equations:

$$p^2 + 2pq + q^2 = 1$$

where a gene exists as two alleles, so that

$$p + q = 1$$

You may recognize this as a standard quadratic equation that you learned to use in algebra class! For example, if the frequency of A is 70% (0.70), then the frequency of allele a must be 30% (0.30). If the Hardy–Weinberg Equation is applied to a gene that exists in alleles designated A and a, then

$$p^2 = \text{the genotypic frequency of } AA$$
$$2pq = \text{the genotypic frequency of } Aa$$
$$q^2 = \text{the genotypic frequency of } aa$$

Table 12.5 shows you how to calculate the Hardy–Weinberg Equilibrium, using the parental generation from Exercise 12.1 as an example.

TABLE 12.5 HARDY–WEINBERG CALCULATIONS FOR THE PARENTAL MOTH POPULATION			
Equations	$p^2 + 2pq + q^2 = 1$ and $p + q = 1$		
Genotype	p^2	$2pq$	q^2
Solve for q	-----	-----	$q^2 = 0.9000$ $q = 0.9487$
Solve for p	If $p + q = 1$ then $p + 0.9487 = 1$ $p = 0.0513$ $p^2 = (0.0513)^2$ $p^2 = 0.0026$	-----	-----
Solve for $2pq$	-----	$2pq = 2\,(0.0513)$ (0.9487) $2pq = 0.0973$	-----
Check your work: $p^2 + 2pq + q^2 = 1$	0.0026+	0.0973+	0.90 = 1
Determine the genotype distribution for 400 moths	$AA = 400\,(.0026)$ $AA = 1$	$Aa = 400\,(0.0973)$ $Aa = 39$	$aa = 400\,(0.90)$ $aa = 360$
Are these the genotypic frequencies we **started** with?	YES	YES	YES
Percent change between these frequencies and the parental genotypic frequencies	—	—	—
Is microevolution occurring?	NO	NO	NO

Study it carefully, noting how to solve for each variable (each allele). Since the Hardy–Weinberg principle states that allele frequencies stay constant generation after generation, the data in Table 12.5 can be used as a control.

In the following exercise, you will demonstrate how the Hardy–Weinberg Equilibrium can estimate the frequency of alleles, using only the recessive phenotype given in the F_2 moth population you sampled in Exercise 12.1. You will test the Hardy–Weinberg principle by comparing how much your calculations differ from those you actually obtained for the F_2 generation.

1. Calculate the Hardy–Weinberg equilibrium for the experimental F_2 peppered moth population on the basis of the **phenotype** for light colored moths you listed in Exercise 12.1, Table 12.2. (Use the total phenotype from all mating pairs in the class.) Show your work and record your calculations in Table 12.6. You must start with the recessive phenotype. You already know that the light phase is a recessive phenotype, so you can deduce that the frequency of light moths from the F_2 represents the homozygous recessive (*aa*) individuals. Remember that q^2 = the genotypic frequency of *aa*. Plug this frequency into the equation in Table 12.6. Answer question 1.

TABLE 12.6 HARDY–WEINBERG CALCULATIONS FOR THE EXPERIMENTAL F_2 MOTH POPULATION			
Equations	$p^2 + 2pq + q^2 = 1$ and $p + q = 1$		
Genotype	p^2	$2pq$	q^2
Solve for q	-----	-----	$q^2 =$ $q =$
Solve for p	If $p + q = 1$ then	-----	-----
Solve for $2pq$	-----	$2pq =$	-----
Check your work: $p^2 + 2pq + q^2 = 1$			
Determine the genotype distribution for 400 moths	$AA =$	$Aa =$	$aa =$
Are these the genotypic frequencies we **ended** with?			
Percent change between these frequencies and the parental genotypic frequencies			
Is microevolution occurring?			

2. Determine the value for q.

3. Your next step is to plug q into the equation $p + q = 1$, and solve for p. Remember that $p^2 =$ the genotypic frequency of *AA*.

4. Now solve for $2pq$. Remember that $2pq =$ the genotypic frequency of *Aa*.

5. Check your work by adding up the values for $p^2 + 2pq + q^2 = 1$. If the components don't add up to 1, find where you have miscalculated.

6. Determine the genotype frequencies for a population of 400 moths (multiply each component in the equation by 400).

7. Compare the frequencies you just generated with the Hardy–Weinberg equilibrium to the **genotypes actually obtained** in Table 12.2. The frequencies should match closely. This shows you that the Hardy–Weinberg really works—and that it can be used to calculate genotype ratios, starting only with the knowledge of a recessive phenotype frequency. Answer Question 2.

8. How much have the genotypic frequencies changed between the parental generation and the F_2 generation in Exercise 12.1? Report the differences using the percent change formula (See Exercise 5.2) for each genotype, and also record these in Table 12.6.

Questions

1. Why can't all three genotypic ratios be estimated solely by examining the phenotypic ratios?

2. How do these estimated ratios of *AA* and *Aa* genotypes (according to the Hardy-Weinberg method) compare to those that you actually obtained in the F_2 generation in Exercise 12.1?

3. Compare the percent change between the parental generation and the F_2 generation in Exercise 12.1. Is this difference in allele frequencies consistent with your expectations, based on your knowledge of the evolutionary process? Explain.

4. What statistical test might you use to determine whether the genotypic ratios predicted by Hardy–Weinberg ("expected" values) are consistent with the genotypic ratios generated in the experiment ("observed" values)?

5. Recall that five conditions must be met in order for the Hardy–Weinberg equilibrium to apply to a natural population. Which (if any) of the five conditions are violated in our model? How might this/these violation(s) cause our results to deviate from our expected values for the control group?

6. How could you determine the frequencies of the individual *A* and *a* alleles (not just the genotypes) for the moth populations?

Exercise 12.3   GENETIC DRIFT: THE FOUNDER EFFECT

Natural selection is one of many ways that allele frequencies can change in a population. See Figure 12.4. **Genetic drift** is another way, and it occurs when populations are reduced to small numbers (the bottleneck effect) or when a few members of a population migrate to a new location, forming a separate population (the **founder effect**). The founder effect can, in part, explain regional variations among members of a species **[Krogh Figure 17.6 and Figure 18.4.]**.

1. Again, using the original genotypic ratios, place *1 AA*, *39 Aa*, and *360 aa* moths into a container. (This is the P generation.) Each group of students should randomly select 4 moths (2 mating pairs) from the container.

TABLE 12.7 OFFSPRING OF FOUNDER EFFECT MOTHS				
First mating pair	**F₁ from first mating pair**			
	Genotype	**Number**		
	AA			
	Aa			
	aa			
Second mating pair	**F₁ from second mating pair**		**Total F₁ genotypes from your mating pairs**	
	Genotype	**Number**	**Genotype**	**Number**
	AA		*AA*	
	Aa		Aa	
	aa		*aa*	

2. Mate your selected moths, using the Punnett squares in Table 12.7. Calculate the genotypic ratios for your new "founder" population by totaling the number of each phenotype found in all 8 F₁ offspring.

3. Compare these genotypic ratios to the ratios of the P generation.

Questions

1. **Compare the total genotypic ratios of the F₁ generation in Table 12.1 to the total genotypic ratios of the F₁ in Table 12.7. What has happened to the frequencies of genotypes in the founder population?**

2. **Is genetic drift occurring in this situation? Explain your answer.**

3. **Is evolution occurring in this situation? Explain your answer.**

Exercise 12.4   HUMAN EVOLUTION: QUANTITATIVE SKULL ASSESSMENT

(Adapted from Gunstream, 2001)

The hypothesized evolutionary relationships of primates are given in Figure 12.5. Many evolutionary changes have occurred in the lineage connecting contemporary humans to our nonhuman ancestors. The purpose of this exercise is to measure these changes and critically examine model skulls of primates, including humans **[Krogh section 19.8]**.

The analysis of skull morphology is an exceedingly complex topic, and much debate surrounds the interpretation of some of the evidence. However, it has been firmly established that our closest extant (living) primate relatives are the chimpanzees, followed by the gorillas.

For similarities between primate DNA, see
http://news.bbc.co.uk/hi/english/sci/tech/newsid_1333000/1333730.stm

If we share a common ancestor with these primates, we should share skull features, and intermediate forms should be observable in our common ancestors (Figure 12.6). Here are some of these features (Figure 12.7).

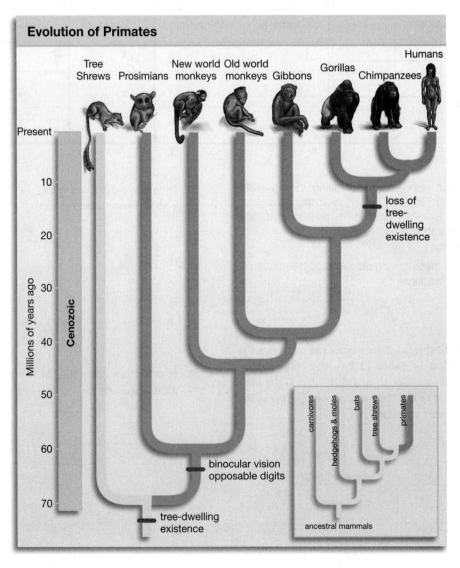

FIGURE 12.5   Evolution of Primates.

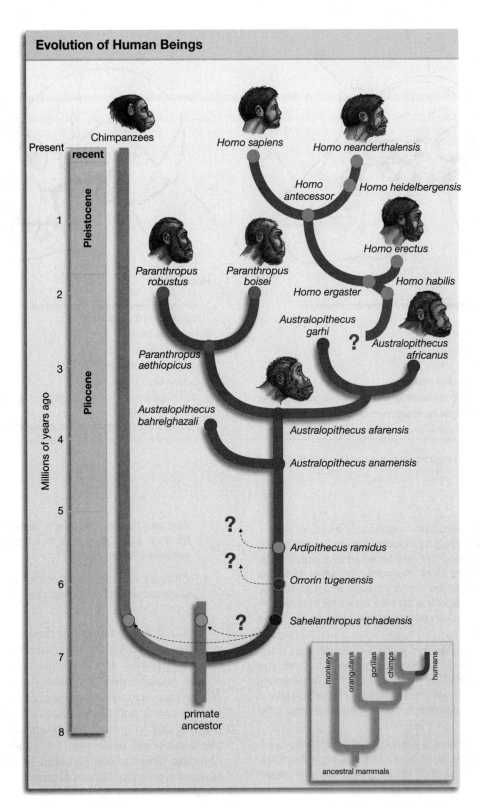

FIGURE 12.6 Evolution of Humans.

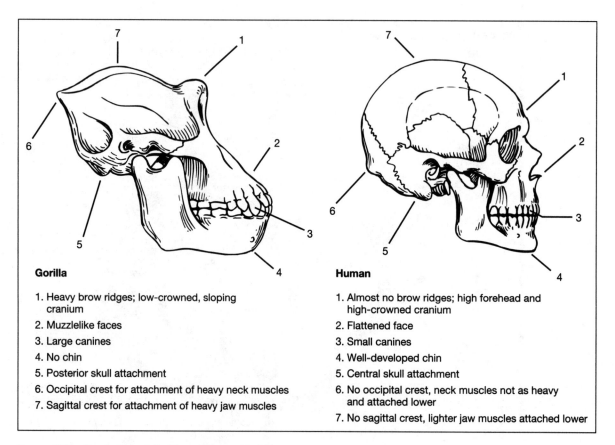

FIGURE 12.7   Comparison of Gorilla and Human Skulls.

Gorilla

1. Heavy brow ridges; low-crowned, sloping cranium
2. Muzzlelike faces
3. Large canines
4. No chin
5. Posterior skull attachment
6. Occipital crest for attachment of heavy neck muscles
7. Sagittal crest for attachment of heavy jaw muscles

Human

1. Almost no brow ridges; high forehead and high-crowned cranium
2. Flattened face
3. Small canines
4. Well-developed chin
5. Central skull attachment
6. No occipital crest, neck muscles not as heavy and attached lower
7. No sagittal crest, lighter jaw muscles attached lower

1. One measurable evolutionary trend in the lineage leading to modern humans is a flattening of the face. This may be linked to a switch in dependence from the sense of smell to visual senses as the primary means of detecting food and predators. A flat face allows for binocular vision, in which both eyes can focus on the same image.

2. Also, very evident is the increase in cranial volume, which facilitates an increase in brain size and complexity. Enhanced cerebral development correlates with the appearance of a vertical face in modern humans as the forehead and brain grew larger.

3. Another trend is a reduction of the chewing apparatus. This led to drastic restructuring of many features of the skull, including the chewing muscles.

With less muscle mass there would be less need for thick bone structures to withstand chewing stresses *(Poirier, 1990)*.

4. Changes in morphology of the teeth would be expected as dietary habits changed. See Exercise 21.3.

On demonstration are model skulls of the primates *Australopithecus afarensis, Australopithecus africanus, Homo habilis, Homo erectus, Homo heidelbergensis, Homo neanderthalensis, Homo sapiens* (modern human), *Pan troglodytes* (chimpanzee), and *Gorilla gorilla* (lowland gorilla). Working in groups, you will gather quantitative anatomical data from the skulls; these are objective observations gathered by precise measurement, given in numerical terms.

The cranial volume for each species is given in Table 12.8.

1. Using Figures 12.8 and 12.9 as your guide, use calipers to make the following measurements to the nearest millimeter for each skull. Record your data in Table 12.8.

2. **Cranial breadth:** maximum width of the cranium.

3. **Cranial length:** maximum distance between the posterior surface and the small prominence, which is called the **glabella**, between the brow ridges.

4. **Facial breadth:** maximum distance between the lateral surfaces of the cheekbones, which are called **zygomatic arches**.

5. **Facial projection length:** distance between the anterior margins of the auditory canal and the upper jaw, which is called the maxilla.

6. **Skull length:** maximum distance between the posterior surface of the cranium and the anterior margin of the maxilla.

7. Calculating indices: Indices are important comparative tools because they eliminate differences resulting from the size of the specimens being analyzed. An index is a ratio obtained by dividing one measurement by another, then multiplying by 100. Calculate values for the following indices and record them in Table 12.8.

8. **Cranial index** = cranial breadth ÷ cranial length × 100. The cranial index gives information concerning the shape of the head. A cranial index of less than 75 is considered a long-headed condition, and an index of 80 or greater is considered a round-headed condition.

9. **Skull proportion index** = cranial breadth ÷ facial breadth × 100. The skull proportion index is a measurement of the size of the cranium relative to the face. The greater the index, the larger the cranium.

10. **Facial projection index** = facial projection length ÷ total skull length × 100. The facial projection index is a measure of the protrusion of the lower face. The greater the index, the more muzzle-like the face is.

TABLE 12.8 QUANTITATIVE PRIMATE SKULL MEASUREMENTS									
	Cranial volume (cc)	Cranial breadth (mm)	Cranial length (mm)	Facial breadth (mm)	Facial projection length (mm)	Skull length (mm)	Cranial index	Skull proportion index	Facial projection index
A. afarensis	400–500								
A. africanus	450–550								
H. habilis	500–750								
H. erectus	750–1250								
H. heidelbergensis	875–1450								
H. neanderthalensis	1200–1750								
H. sapiens	1300–1700								
P. troglodytes	300–500								
G. gorilla	350–750								

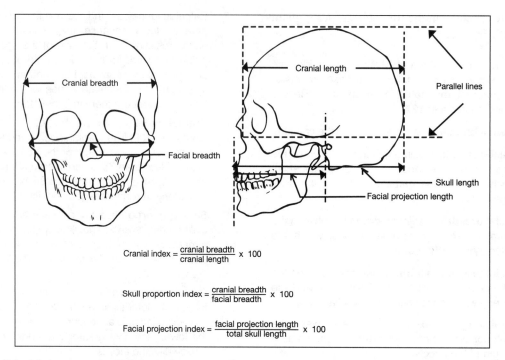

FIGURE 12.8   Measurements for Skull Analysis.

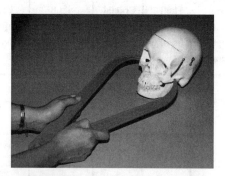

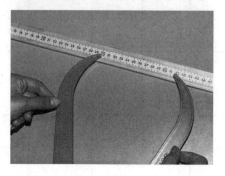

FIGURE 12.9   Measuring with Calipers.

Questions

1. **According to the cranial indices, which species is most round-headed? Which is most long-headed? List them in order, starting with the most long-headed species.**

2. According to the skull proportion indices, which species has the largest cranium? Which has the smallest cranium? List them in order.

6. What are the implications of cranial volume in primates?

3. According to the facial projection indices, which species is most muzzle-like? Which is the least? List them in order.

Exercise 12.5   HUMAN EVOLUTION: QUALITATIVE SKULL ASSESSMENT

(Adapted from Gunstream, 2001)

Continuous with Exercise 12.4

In this exercise, you will gather qualitative observations that are more subjective because they require some judgment on your part. For example, instead of making a measurement, you will be asked to compare various skull features, and then to rate your responses on a numerical scale [**Krogh section 18.5**].

1. Use Figure 12.7 to guide you while conducting measurements for the qualitative features listed in Table 12.9. Rate each feature 1 through 5.

2. After you have gathered all data, discuss your data with your classmates, and answer the following questions. Make inferences about the evolutionary relationships of the species represented. Attempt to interpret each skull morphology feature observed in terms of its potential evolutionary and behavioral significance.

4. Which traits appeared earlier in primates? Which appeared later?

Questions

1. What major trends in skull morphology are represented as evolution proceeds from *Australopithecus afarensis* to modern humans?

5. Is your assessment consistent with the lineage found in Figure 12.7? List any differences you find.

2. Based on the differences in dentition (teeth) among all the skulls represented, what

TABLE 12.9 QUALITATIVE SKULL ASSESSMENT					
	Skull and vertebra attachment	**Brow ridges**	**Length of canines**	**Forehead**	**Chin**
	1 = most posterior 5 = most central	1 = most pronounced 5 = least pronounced	1 = longest 5 = same as incisors	1 = most sloping 5 = best developed	1 = no chin 5 = most developed
A. afarensis					
A. africanus					
H. habilis					
H. erectus					
H. heidelbergensis					
H. neanderthalensis					
H. sapiens					
P. troglodytes					
G. gorilla					

might you infer about the dietary habits of each species?

3. What might you infer from the position of the skull and vertebra attachment? (What effect might this have on the animal's posture?)

4. How are the skulls of chimpanzees and gorillas similar and how are they different?

5. Is it the gorilla's or the chimpanzee's skull that is the most similar one to that of modern humans (*H. sapiens*)?

6. **Why do you think the chimpanzee and gorilla skulls are included for these analyses?**

7. **Which type of data was most valuable—quantitative or qualitative? Why?**

8. **When you view your quantitative and qualitative data together, what conclusions do you make regarding human evolution?**

9. **One area of contention in the evolutionary relationships of humans is that of the relationship of *Homo sapiens*, *Homo neanderthalensis*, and *Homo heidelbergensis*. Some contend that modern humans arose directly from *H. neanderthalensis*. Another more contemporary view suggests that *H. sapiens* did not arise directly from *H. neanderthalensis*, but shares a common ancestor with *H. neanderthalensis* in *H. heidelbergensis*. New data contradict both of these views (Figure 12.6), suggesting that both *H. sapiens* and *H. heidelbergensis* share a common ancestor in *Homo antecessor*. Compare your data for *H. sapiens*, *H. neanderthalensis*, and *H. heidelbergensis* (unfortunately model skulls are not yet available for *H. antecessor*). Do you observe**

anything that would support one of the above views over the others? Explain your answer.

Exercise 12.6   SHAKING OUR FAMILY TREE

Figure 12.6 gives one of several contemporary hypotheses explaining recent events in human evolution; however, it is important to realize that hominid paleontology (the study of human fossils) is one of the most contentious and dynamic subdisciplines in natural history. Recent discoveries have dramatically altered our view of human evolution (Gibbons, 2002; Gore, 2002; Lemonick and Dorfman, 2002; Pickrell, 2002).

For instance, the fossil remains of a very chimp-like hominid dating to nearly 7 million years ago (mya) found in Chad, *Sahelanthropus tchadensis*, is considered by some anthropologists to have been more similar to modern humans than Lucy (*A. afarensis*) who dates back to just over 3 mya. Others contend that instead of being a hominid, *S. tchadensis* was an early ape. Its discoverer, Michel Brunet, contends that *S. tchadensis* is **the point** of divergence of our lineage with that of chimps (Wong, 2003). Have we uncovered the quintessential "missing link"? Hopefully, future fossil finds will resolve this debate. Molecular data suggests that the lineages leading to humans and chimps diverged some 5 to 7 mya. Could *S. tchadensis* be a true "missing link"? Another interesting note about *S. tchadensis* is that it seriously questions the title long-held by the Great Rift Valley of Eastern Africa as the cradle of human evolution. Chad is located in West Africa, some 1,500 miles west of the Great Rift Valley. One of the most interesting questions regarding human evolution is whether humans arose in a small confined area or whether the origin of modern humans was a more multiregional event (Stringer, 2003).

One prevailing idea is that early hominids (*H. heidelbergensis*) arose exclusively in Africa. Then, one group of *H. heidelbergensis* moved north out of Africa into Europe and Asia some 200,000 years ago and gave rise to the Neanderthals. Modern humans then came *north* out of Africa into Europe and Asia a mere 100,000 years ago. The population of *H. heidelbergensis* remaining in Africa gave rise to us (*H. sapiens*).

An alternative view based on the finding of fossil remains in northern Spain in 1994 suggests another possible scenario. These remains, named *Homo antecessor,* date to about 1 mya. It has been proposed that *H. antecessor* gave rise to *H. sapiens* and *H. heidelbergensis* in Africa, and that *H. heidelbergensis* migrated to Europe where they gave rise to *H. neanderthalensis*. If this is the case, *H. heidelbergensis* is not in the direct lineage leading to *H. sapiens* at all. The recent finding of human (*H. sapiens*) remains in Ethiopia dating to over 150,000 years ago (White et al., 2003) lends credence to the "out of Africa" hypothesis (Stringer, 2003).

An alternative view holds that *H. heidelbergensis*, *H. neanderthalensis*, and *H. erectus* were really all different varieties of an early human species (which would be referred to as *H. erectus* if you accept this viewpoint). Other researchers consider *H. heidelbergensis* to be a subspecies of *H. sapiens*, distinct from *H. erectus*, and have proposed that the ancestral species is *H. rhodesiensis* (White et al., 2003). In either event, the arrival of hominids in Europe and Asia was probably a recent event (within the past 250,000 years) involving relatively modern hominids.

A recent fossil find in the Republic of Georgia now shakes even this "modern" branch of our family tree. Paleontologists uncovered the remains of a chimp-like *H. habilis* dating to 1.7–1.8 mya, placing hominids in Eurasia some 1.5–1.6 million years earlier than previously believed. Furthermore, tremendous variation among fossils from the same "bed" have led some researchers to suggest that all hominids arising after *H. habilis* should be considered *H. erectus*, making *H. erectus* our immediate ancestor. A growing body of evidence suggests that Neanderthals were in Europe a very long time ago (Stringer, 2003). The early inhabitation of Eurasia by hominids is consistent with the observation that European fossils exhibiting distinctive anatomical features of the Neanderthals have been found to be over 350,000 years old (Klein, 2003).

As you can see, we are far from reaching a consensus concerning our shared genealogy. Fortunately, new fossil finds create ripples of excitement in the media, and so there are plenty of materials available that can help us understand their implications. Popular scientific journals such as *National Geographic*, *Natural History*, and *Discover* contain articles detailing recent finds in hominid paleontology. Your job is to summarize an article's main points about a recent fossil find and its implications to the current paradigms of human evolution.

Here are the requirements for writing your manuscript:

> The article upon which the paper is based must have been published within the last two years.

> Your paper must be 2 to 5 pages, typed and double-spaced.

> The paper should contain a *minimum* of three references other than your source article (these references are not subject to publication date restrictions).

> You should provide competing interpretations of the findings.

In the course of your paper, answer the following questions:

1. Are the data consistent with the tree given in Figure 12.6?

2. What is the proposed hypothesis?

3. What predictions may be generated from the proposed hypothesis?

4. Outline a series of observations that might be conducted to test the predictions.

Exercise 12.7   EVOLUTION AND CREATIONISM

Web exercise

[Krogh sections 16.8–16.9]

For an interesting study in religion and science, investigate the evolution versus creationism controversy. It deals mainly with the teaching of evolution in public schools, which some religious groups have objected to because of their belief in a literal biblical account of creation as given in the book of Genesis. Investigate the body of literature available on the subject (see the

Chapter 12 references), and start by visiting a few Web sites. You will quickly learn that there are not just two sides to this issue; a variety of opinions exist.

Retype the questions listed below, and provide your answers in double-spaced format. Be sure to cite your references; see Appendix 1 for proper citation format. For additional literature and more information about this topic, refer to the Chapter 12 literature cited section at the end of this book.

Here are some Web sites to get you started:

http://catalaw.com/dov/docs/dw-scicr.htm
http://www.ncseweb.org
http://www.ncseweb.org/resources/articles/1593_the
_creationevolution_continu_12_7_2000.asp
http://www.nap.edu/html/creationism
http://www.icr.org
http://www.christiananswers.net/q-eden
/genetic-mutations.html
http://www.talkorigins.org
http://www.emagazine.com/may-june_2003
/0503feat2.html
http://www.sciencemag.org/cgi/content/full
/299/5612/1523
http://www.answersingenesis.org

Questions

1. **What type of experiences have you had regarding the topic of evolution versus creationism? For example, how was the topic of evolution treated in your high school?**

2. **How have those experiences influenced your own opinion?**

3. **How has the topic been treated in your college experience so far? How has it been treated in your church?**

4. **Do you think this is an important issue? Explain.**

5. **Study the *www.ncseweb.org* homepage. How many states have considered passing antievolution legislation during the past 12 months? Using examples from two different states, determine what actions would be made illegal by such legislation.**

6. **Now, do you think this is an important issue? Explain.**

7. **List the arguments presented by the major groups involved in this controversy. (Study the** *www.ncseweb.org/resources/articles /1593_the_creationevolution _continu_12_7_2000.asp* **site.)**

8. **Defend the position of one group from each end of the continuum mentioned in Question 7. Give at least one reason why you might agree and why you might disagree with each group.**

9. **Use an Internet search engine to find one additional Web site that would be helpful for the general public to better understand creationism, evolution, and the controversy between them. Provide the following information in a separate, one-page, typed, double-spaced format:**

   Page title:

URL:

Author/organization:

Summary of content:

Educational value:

13

Classification and Taxonomic Keys

*The first step to wisdom, as the Chinese say,
is getting things by their right names.*

E.O. Wilson, 1994

Nature has enormous diversity—more than 1.5 million species of animals and 250,000 species of plants have been named *(Hickman et al., 1993)*. Yet, the species thus far described represent only a fraction of those that exist. This chapter examines the system of classification that biologists use to catalog and retrieve information about this vast array of organisms. These exercises provide an outline for you to understand how organisms are separated into groups, so that you will not have to swim in a sea of seemingly unrelated facts as you learn about the Archaea, Bacteria, and Eukarya. Rather, you will be able to make connections between traits of organisms, classification of organisms, and their evolution **[Krogh sections 18.4 & 18.5]**.

Our classification system was established by the great Swedish botanist of the eighteenth century, Carolus Linnaeus (Figure 13.1). Linnaeus appears to have had a natural talent for naming, describing, and classifying organisms, especially flowers. Such collection and classification was Linnaeus's passion, and it was his life's ambition to name and describe all of the known kinds of plants, animals, and minerals. Linnaeus named and described thousands of species of plants and animals in his lifetime.

Linnaeus made two great conceptual contributions to the sciences of **taxonomy** (the naming of organisms) and **systematics** (the classification of organisms and groups of organisms):

1. He gave us a system of **binomial nomenclature.** (*Binomial* means "two names.")

2. He arranged groups of organisms in a hierarchical fashion, ranking organisms on different levels.

FIGURE 13.1   Carolus Linnaeus.

141

FIGURE 13.2   Virginia Opossum, *Didelphis Virginiana.*

Each species is given a latinized name composed of two words. For instance, the scientific name of the only marsupial that occurs in North America, the Virginia opossum, is *Didelphis virginiana* (Figure 13.2). The first word is the **genus**, and it always begins with a capital letter and is a noun. The second term always begins with a lowercase letter, and it is usually an adjective. The first and second terms together refer to the **species**. Scientific names, or species names, are always italicized because they are written in a different language. When writing in "longhand," however, we cannot italicize words, and so genus and species names should be underlined, which denotes italics. Thus, if you were writing in longhand, the scientific name of the Virginia opossum would be written as Didelphis virginiana.

What is a species? "Species" is another one of those elusive concepts in biology, but it always refers to a group of organisms, and a single species includes *all* individuals making up that group. Therefore, you do not catch a species and you cannot eat a species, unless you intend to catch or eat every representative of the group. For instance, the scientific name of the domestic cow is *Bos taurus.* If you say, "I went to McDonald's for lunch and ate *Bos taurus,*" you are saying that you ate every single cow on the face of this planet for lunch.

Ernst Mayr elaborated the most common concept of species: **Species** are groups of actually or potentially interbreeding natural populations, which are reproductively isolated from other such groups.

Occasionally, biologists will abbreviate a species name after the name has already been mentioned. For example, since the Virginia opossum has already been discussed here, we can now refer to it as *D. virginiana.* Likewise, when several species of the same genus are being discussed, it is appropriate for a biologist to refer to the group as *Didelphis* spp. instead of listing all the species in that group.

Exercise 13.1   CLASSIFICATION

(Adapted from Burns, 1968)

Linnaeus's scheme arranges organisms into groups called **taxa** (singular, *taxon*) with an ascending series of increasing inclusiveness. This allows us to classify all living organisms into a meaningful system, so that information concerning any given group of organisms can be easily stored and retrieved.

The most commonly recognized taxa in contemporary systematics are listed, in order, following this paragraph. Beside each, the complete classification of the Virginia opossum is given to illustrate the hierarchical nature of the system. The species *Didelphis virginiana* contains all Virginia opossums. The genus *Didelphis* encompasses not only the Virginia opossum but also other species of opossum in the genus *Didelphis.* Family Didelphidae contains several genera, of which *Didelphis* is one. Order Marsupialia, in addition to encompassing the opossum, contains kangaroos and all other marsupials. Class Mammalia, in addition to opossums and kangaroos, includes a vast assortment of other animals including dogs, cats, mice, whales, and humans. Phylum Chordata, in addition to the mammals, includes birds, amphibians, reptiles, and fish. Kingdom Animalia, in addition to those listed already, includes sponges, jellyfish, worms, insects, and all other animals. Finally, Domain Eukarya includes all eukaryotic organisms, whose cells have a nucleus and membrane-bound organelles. Thus, you can see how the classification system is arranged in a series of categories of **ascending levels of inclusiveness**. Taxonomic names above the genus level (family, order, class, phylum, kingdom, and domain) begin with a capital letter but are not italicized.

The taxa that zoologists refer to as phyla (singular, *phylum*) are called **divisions** by botanists when referring to plants.

DOMAIN Eukarya

KINGDOM Animalia

PHYLUM Chordata

CLASS Mammalia

ORDER Marsupialia

FAMILY Didelphidae

GENUS *Didelphis*

SPECIES *Didelphis virginiana*

A strong argument supporting the theory of organic evolution is that *a unified classification system can be produced for living organisms.* As one moves from bacteria to animals, the number of properties in common changes, and a hierarchical framework is naturally produced. If those concepts, or groups, that biologists place organisms into were not somehow related in very real ways, it would be impossible to construct an orderly classification scheme.

In this exercise, you will use the Linnaean classification system to classify "organisms," and you will discover some of the basic methods (and problems) involved with classification.

1. Obtain a "classification" envelope that has 20 different objects in it. Each object represents a different species. Study these organisms carefully.

2. Are there similar characteristics among some of them? Use all available taxonomic characters (traits) to determine the relationships between the organisms and to arrange them in an orderly scheme that reflects these relationships. Assume that all of these organisms belong to Domain Eukarya and Kingdom Animalia. You must classify the organisms in the remaining taxonomic categories: phylum, class, order, family, genus, and species.

3. Make a classification system that includes all of the species. Be sure to indicate the characteristics of each taxonomic group. For instance, what characteristics must an organism have to fit into an *order* or *family* taxon?

4. Name your organisms. Be sure to write the scientific names properly.

5. Construct a dichotomous key for identification of all species (see Exercise 13.2 for an example).

6. Exchange your key and specimens with that of another laboratory group. Attempt to identify all of the "species" using their dichotomous key.

Questions

1. **Compare your key with the one composed by another group. Are you able to identify their organisms? Decide what characteristics are most useful in their classification.**

2. **Different characteristics have probably been utilized by student groups in your class. There may be some disagreement about what characteristics are most important. List the traits that different groups have used as phylum-level characteristics (i.e., what characteristics have been used to separate phyla?)**

3. **What problems might biologists encounter when classifying new organisms?**

4. **What solutions or problems might DNA technology provide in classifying organisms?**

Exercise 13.2   A TAXONOMIC KEY

For this lab, you will be identifying trees near your campus using a taxonomic key, comparing characteristics listed in the key to the morphological (anatomical) characteristics seen in the trees. At each step in the key, you will choose between two characteristics listed (hence the term **dichotomous key**), and then continue to the next couplet as directed, until the name for the tree is finally given.

Start with the first couplet (1a and 1b), read both choices, and make a decision about which choice to follow, based on your observations of the tree's anatomy. When measurements are given in the key, measure the characteristic you are observing with a ruler. For instance, the diameter of a leaf is sometimes given in a couplet. In this case, simply place your ruler across the widest part of the leaf and compare your measurement with that given in the key. Because there are natural variations in any population of living organisms, base your decisions on numerous specimens (leaves) whenever possible. Most of the characteristics listed in the key describe leaf morphology, so you must first know some basic leaf characteristics.

LEAF MORPHOLOGY

(Adapted from Mathis, 1996)

The technical term for a leaf is a **blade**, and the **petiole** connects a leaf to its branch at the **node** (Figure 13.3). The space between nodes is called an **internode**. There may or may not be a **terminal bud** at the end of the branch, but you will almost always see an **axillary bud** at a node. Pay careful attention to the position of the axillary bud; a **simple leaf** has only one blade (or leaflet) per axillary bud, and a **compound leaf** has several blades (leaflets) per axillary bud. See Figure 13.4.

> The axillary bud is an important feature! Find it before looking for other features on a tree.

Once you determine whether the leaf is simple or compound, you can determine its pattern of venation. Figure 13.5 shows the venation of simple leaves. A **pinnate** leaf has a feather-like pattern; a **palmate** leaf has a hand-like pattern, with the veins branching out from a common point near the base of the leaf. Compound leaves also can have pinnate or palmate venation, so take care to identify these features carefully.

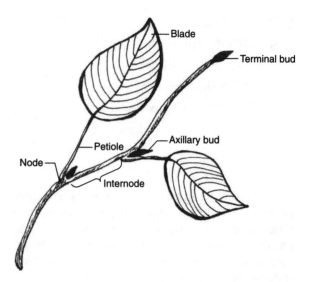

FIGURE 13.3   Leaf Morphology.

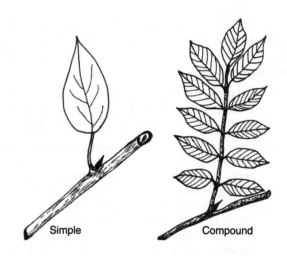

FIGURE 13.4   Simple versus Compound Leaves.

The leaf attachment on the branch is another important characteristic. See Figure 13.6. Two leaves are **opposite** if their nodes are directly across from one another, but are **alternate** if their nodes are offset. **Whorled** leaves have at least three nodes that are attached at the same position on a branch. Again, first look for the position of the axillary buds to identify these features so that you will not mistake leaflets on a compound leaf with a simple leaf.

FIGURE 13.5   Pinnate versus Palmate Venation.

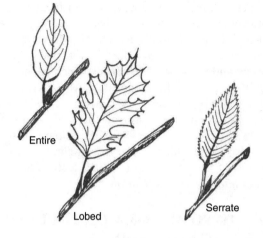

FIGURE 13.7   Leaf Margins.

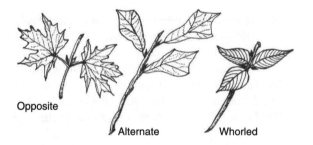

FIGURE 13.6   Leaf Arrangement.

The leaf edges, or margins, can be **entire** (smooth), **serrate** (toothed), or **lobed**, as in Figure 13.7. Some leaves have deep lobes that extend nearly to the main vein of the leaf. Some are "doubly serrate," meaning they have double teeth on their margins. Some leaves are serrate *and* lobed, and some are so deeply serrate that the teeth may be mistaken for lobes. Most evergreens have needle- or scale-like leaves with a waxy surface, but a few broadleaf trees such as mountain laurel and rhododendrons are evergreen, usually with thick leaves. (It usually helps to have a field guide with you to get you through the more difficult spots in the key.)

As you go out tromping through the woods, you should be wary of poison ivy (Figure 13.8). This vine produces a substance that may elicit an allergic reaction and a severe skin rash (dermatitis). Most such cases of dermatitis are acquired by direct

FIGURE 13.8   A Plant to Avoid. Poison ivy, with its serrate "leaves of three," can be seen with white or cream-colored berries in the fall.

contact with the plant; however, the rash can result from contact with shoes or clothes, which may have touched the plant months earlier. The vine is easily recognized by its compound leaves with three leaflets, and ivory or white berries, which appear in the fall and persist into the winter. Also be careful not to touch poison oak or poison sumac, which can elicit similar skin reactions (see couplets 57 and 58 in the key). Poison oak, which is common in the southern United States, looks very similar to poison ivy, except that it has up to seven leaflets and is hairy under the leaves *(Harlow, 1957).*

KEY TO SOME COMMON TREES OF THE UNITED STATES

Adapted from Trees of the Eastern and Central United States by W. M. Harlow (1957)

1a. Tree evergreen 2
1b. Tree not evergreen 7
2a. Leaves 1/2 in. or more in width, broad.......
............Mountain Laurel (*Kalmia latifolia*)
2b. Leaves less than 1/2 in. wide, needle-like or narrow, or small and scale-like 3
3a. Some leaves scale-like; others short needles. Eastern Red Cedar (*Juniperus virginiana*)
3b. Leaves needle-like 4
4a. Leaves in bundles 5
4b. Leaves not in bundles, but occur singly 1/3–1/2 in., arranged in flat spray, rounded on tip, 2 white bands on underside ....
........Eastern Hemlock (*Tsuga canadensis*)
5a. Needles in bundles of....................................5
.........................White Pine (*Pinus strobus*)
5b. Needles in bundles of 2 or 3 6
6a. Needles in 2s, about 5 in. long, snapping cleanly when bent double ....
............................Red Pine (*Pinus resinosa*)
6b. Needles stiff and twisted, often in tufts along trunk ....
................................Pitch Pine (*Pinus rigida*)
7a. Leaves opposite 8
7b. Leaves alternate.......................................17
8a. Leaves simple...9
8b. Leaves compound...................................13
9a. Leaves lobed 10
9b. Leaves not lobed; side veins parallel margin; leaf margin smooth, bark plate-like or blocky...
........Flowering Dogwood (*Cornus florida*)

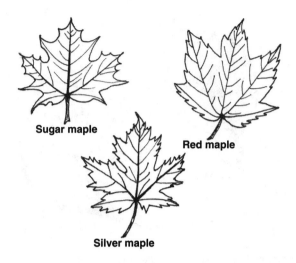

FIGURE 13.9   Common Maple Species.

10a. Margins of lobes wavy, not repeatedly toothed, green underneath (5 lobes) in Figure 13.9) ....
...................Sugar Maple (*Acer saccharum*)
10b. Margins definitely toothed 11
11a. Underside of mature leaf silvery 12
11b. Underside of mature leaf green; teeth fine and double; bark green with white stripes..
.......Striped Maple (*Acer pennsylvanicum*)
12a. Leaves 5-lobed; sides of terminal lobe diverge (V-shaped); fruit with widely spreading wings; bruised twigs have rank odor .....
...............Silver Maple (*Acer saccharinum*)
12b. Leaves variable, 3 or 5-lobed; terminal lobe usually pointed (Figure 13.9); bruised twigs without offensive odor; often with distinct red tint on petiole ....
...........................Red Maple (*Acer rubrum*)
13a. Leaves palmately compound; 5 leaflets, broadest in the middle and tapering both ways......................Buckeye (*Aesculus* spp.)
13b. Leaves pinnately compound....................14
14a. Maple with 5 or 7 (sometimes 3 or 9) leaflets; leaflets extremely variable, some deeply lobed or sometimes again compounded ....Boxelder (*Acer negundo*)
14b. Leaflets not deeply lobed 15
15a. Leaflets lacking a stem, joined directly to the midrib (main vein); 7 to 13 leaflets, usually 9 Black Ash (*Fraxinus nigra*)...........................
15b. Leaflets have a distinct stem (usually 7 leaflets, rarely 3–11) 16

16a. Leaflets sparingly toothed, elliptical to almost oval..
................White Ash (*Fraxinus americana*)
16b. Leaflets conspicuously toothed, narrow, elliptical to lance-shaped...............................
.........Green Ash (*Fraxinus pennsylvanica*)
17a. Leaves simple...18
17b. Leaves compound....................................56
18a. Leaves lobed 19
18b. Leaves unlobed 32
19a. Underside of leaf covered with silvery wool
.... White Poplar (*Populus alba*)
19b. Underside of leaf not silvery or woolly...20
20a. Fruit an acorn..21
20b. Fruit not an acorn 28
21a. Rounded lobes, lacking hairs or bristles....22
21b. Bristle-like or hair-tipped lobes 27
22a. Leaves serrate, or with rounded or scalloped teeth...23
22b. Leaves cut into lobes by distinct gulfs or sinuses 24
23a. Margin irregular, some sinuses much deeper than others, often almost lobed; brownish-velvety below ....
.........Swamp White Oak (*Quercus bicolor*)
23b. Margin regularly wavy or with rounded coarse teeth; smooth or inconspicuously hairy below ....
..................Chestnut Oak (*Quercus prinus*)
24a. Leaves basically shaped like a cross (may be highly variable in shape).............................
...........................Post Oak (*Quercus stellata*)
24b. Leaves not cross-shaped 25
25a. Sinuses reaching nearly to the midrib.........
...........................White Oak (*Quercus alba*)
25b. Sinuses reaching not more than halfway to the midrib (main vein) 26
26a. Leaves regularly lobed, smooth below........
...........................White Oak (*Quercus alba*)
26b. Leaves irregularly lobed, velvety below......
.........Swamp White Oak (*Quercus bicolor*)
27a. Lobes 7 to 11; sinuses tend to be about the same depth ....
...........Northern Red Oak (*Quercus rubra*)
27b. Lobes 5 to 7; more or less scruffy (like dandruff) below; sinuses irregular....................
....................Black Oak (*Quercus velutina*)
28a. Leaves and twigs have a spicy odor and flavor; leaves unlobed, mitten-shaped, and/or 3-lobed ....Sassafras (*Sassafras albidum*)

28b. Leaves and twigs not spicy.......................29
29a. Base of leaf stem hollow, enclosing the next year's bud; 4–7 in. in diameter; 3–5-lobed with margins coarsely toothed; bark mottled, creamy white and brown, the former where the old bark peels off........................
................Sycamore (*Platanus occidentalis*)
29b. Base of leaf stem solid..............................30
30a. Sap of broken leaves and twigs milky (cloudy); leaves rough above (like sand paper), hairy below....................................
..................................Mulberry (*Morus* spp.)
30b. Sap clear, not cloudy................................31
31a. Leaves mostly 4-lobed, the apex "chopped off" or indented with a wide notch.............
.............Tuliptree (*Liriodendron tulipifera*)
31b. Leaves 5- or 7-lobed, star-shaped ....
.........Sweetgum (*Liquidambar styraciflua*)
32a. Sap of broken leaves and twigs milky (cloudy); leaves rough above (like sandpaper) ....Mulberry (*Morus* spp.)
32b. Sap clear (not cloudy) 33
33a. Leaf stem about as long as the diameter of the blade, conspicuously flattened so that leaf trembles in the slightest breeze 34
33b. Leaf stem circular or grooved in cross section (not as in 33a)................................36
34a. Leaves circular in outline 35
34b. Leaves triangular; base of leaf, flat or slightly heart-shaped; twigs, stout; buds, large and resinous ....
..Eastern Cottonwood (*Populus deltoides*)
35a. Teeth fine; buds brown and shiny ....
...... Quaking Aspen (*Populus tremuloides*)
35b. Teeth coarse; buds grayish, powdery, and dull .....Bigtooth Aspen (*Populus grandidentata*)
36a. Leaf margin entire, not toothed in any way; leaves ovate to elliptical, long pointed, slightly thickened; shiny dark green, yellow in autumn ....
.............Persimmon (*Diospyros virginiana*)
36b. Leaf margin serrate or with rounded teeth 37
37a. Leaves average 4 in. or more in diameter, nearly circular, somewhat heart-shaped at base ....Basswood (*Tilia americana*)
37b. Leaves not circular or, if so, less than 4 in. in diameter...38
38a. Leaves with medium- to large-sized conspicuously single teeth on margin...........39

38b. Leaves with double teeth or such small ones that it is difficult to see whether they are single or double..42

39a. Leaves lopsided and more or less heart-shaped at the base (2 1/2 in. to 4 in. long; sharply serrate with a narrow, curved tip) .. .. Hackberry (*Celtis occidentalis*)

39b. Leaves more or less equal at the base, not heart-shaped...40

40a. Teeth ending in a hair or bristle (6–8 in. long); terminal bud lacking on twig............. ....Chestnut (*Castanea dentata*)

40b. Teeth without bristles 41

41a. Teeth sharp with a papery rattle (21/2–5 in. long); buds long and lance-shaped; terminal bud present ....Beech (*Fagus grandifolia*)

41b. Teeth slightly rounded; buds short and egg-shaped; terminal bud present ....Chinkapin Oak (*Quercus muehlenbergii*)

42a. Twigs armed with long thorns .... Thornapple or Hawthorn (*Crataegus* spp.)

42b. Twigs without thorns or spines.................43

43a. Leaves with conspicuous medium- to large-sized double teeth; very lopsided at base..44

43b. Leaves with very small single or double teeth...45

44a. Leaves oval, tend to look trough-like because of a fold at the midrib (main vein); doubly serrate, rough above ....Slippery Elm (*Ulmus fulva*)

44b. Leaves usually elliptical or broadest near apex; flat.. ....American Elm (*Ulmus americana*)

45a. Pith conspicuously triangular when sliced crosswise; shrub or small tree....................... ....Alder (*Alnus rugosa*)

45b. Pith circular or so small as not to be seen... 46

46a. Twigs with an intensely bitter quinine taste or bitter almond flavor............................47

46b. Twigs not as above 53

47a. Twigs bitter; leaf narrow; bud, covered by a single scale (a willow) 48

47b. Twigs with faint or strong bitter almond flavor; leaves 2–6 in. long, 1–11/2 in. wide, shiny and dark green above, paler below............. ....Black Cherry (*Prunus serotina*)

48a. Upper and lower leaf surfaces both shiny .. .. Shining Willow (*Salix lucida*)

48b. Lower leaf surface not shiny 49

49a. Both leaf surfaces green and smooth; apex, often curved ....Black Willow (*Salix nigra*)

49b. Upper leaf surface green; the lower with grayish or silvery bloom, hairy, or both....... 50

50a. Leaves lance-shaped 51

50b. Leaves narrowly elliptical........................52

51a. Tree with a drooping appearance ....Weeping Willow (*Salix babylonica*)

51b. Tree with an upright form ....White Willow (*Salix alba*)

52a. Veins sunken into the upper surface ....Bebb Willow (*Salix bebbiana*)

52b. Veins not sunken.. ....Pussy Willow (*Salix discolor*)

53a. Veins parallel to the margin ....Buckthorn (*Rhamnus cathartica*)

53b. Veins are not parallel to the margin 54

54a. Bark papery; spurshoots (dwarfed twigs bearing tufts of leaves) common................. ....Birch (*Betula lutea*)

54b. Bark smooth and blue-gray or finely shreddy; spurshoots lacking 55

55a. Bark smooth and blue-gray; trunk "muscular" appearing... ....American Hornbeam (*Carpinus caroliniana*)

55b. Bark shreddy (shredding off in narrow, scaly plates that curl at free ends)...................... ....Hophornbeam or Ironwood (*Ostrya-virginiana*)

56a. Margins of leaflets smooth 57

56b. Margins of leaflets serrate 58

57a. Five to 11 paired leaflets, somewhat broad at base, 1–2 in. long; small spines on leaf stem ....Pricklyash (*Zanthoxylum americanum*)

57b. Five to thirteen leaflets, elliptical, 2 1/2–3 1/2 in. long; restricted to swampy areas, or areas of standing water.. ....Poison Sumac (*Toxicodendron vernix*)

58a. Leaves or twigs, when broken, exude milky sap; leaflets 7–31 (1–4 in. long)..................... ....Staghorn sumac (*Rhus typhina*)

58b. Sap clear..59

59a. Leaflets 9–21, 2–5 in. long; second year's pith (center of twig cross section) shows chambers when sliced lengthwise ....Walnut (*Juglans* spp.)

59b. Leaflets 9 or fewer; pith solid 60
60a. Leaflets 5 (rarely 7), 3–7 in. long..............61
60b. Leaflets 7–9, 2–8 in. long 62
61a. Bark smooth or becoming furrowed with forking ridges ....
....................Pignut Hickory (*Carya glabra*)
61b. Bark separating into long, narrow, curved strips, loosely attached at middle ....
................Shagbark Hickory (*Carya ovata*)
62a. Leaflets slightly hairy beneath; twigs slender, ending in bright yellow, slightly flattened bud ....
......Bitternut Hickory (*Carya cordiformis*)
62b. Leaflets densely hairy beneath; twigs brown, stout, and hairy, ending in large hairy bud....
.....Mockernut Hickory (*Carya tomentosa*)

Questions

1. **Why do you think that keys use characteristics such as leaf shape rather than tree size?**

2. **Did you find more evergreens or deciduous trees?**

3. **Which tree species were most common?**

4. **Were some trees easier to identify than others? Why?**

5. **Were there any trees that you couldn't identify with this key? Sketch their leaf characteristics and describe or sketch the bark.**

6. **Where might you add changes to the key in order to find a place for these specimens?**

7. **Using a field guide for your area, identify the forest type near your campus.**

8. **Take two unrelated leaves, and two related leaves; for each pair, discuss the similarities and differences in morphology. Sketch their characteristics here:**

9. **Why might it be important to know what species an organism represents?**

UNIT 5
The Diversity of Life

14

Bacteria

... in every little particle of (the Earth's) matter, we now behold almost as great a variety of Creatures, as we were able before to reckon up in the whole Universe it self.

Robert Hooke, 1665

This chapter begins a survey of life on our planet, and the smallest of these organisms are bacteria. All bacteria are **prokaryotic**, which means that they possess no nucleus or membrane-bound organelles. Bacteria are very successful organisms. The scope of habitats occupied by bacteria is unparalleled. They acquire energy in many different ways: some are photosynthetic (cyanobacteria), others are nutrient recyclers (saprophytes), and still others are symbiotic (living in or on other organisms). They have rapid reproduction—giving them the ability to evolve at lightning speed—which makes them one of the most respected (and perhaps feared) groups of organisms. Traditionally, all bacteria were placed in Kingdom Monera. Because of their complexity and morphological differences, bacteria have now been separated into two domains, Archaea and Bacteria **[Krogh sections 20.3–20.6]**.

The "extremophiles," such as those species inhabiting hot-water springs or frozen terrain, are now classified in **Domain Archaea**. Archaeans are probably much like the first life forms to exist on our planet. The chemoautotrophs that live in hydrothermal vents, such as the Galápagos Rift, are under 9,000 feet of water—where sunlight cannot reach them. They metabolize hydrogen sulfide and release methane as a by-product, and interestingly, some are large enough to be seen with the naked eye *(Ballard, 2000)*.

The most familiar to us, however, is **Domain Bacteria**. The word *bacteria* makes most people think of disease. Indeed, some of the most heinous diseases known to humans result from bacterial infections: anthrax, tuberculosis, cholera, botulism, diphtheria, Lyme disease, chlamydia, gonorrhea, bacterial meningitis, typhoid fever, and so on not to mention the nuisance of urinary tract infections, flatulence, and halitosis.

Most bacterial infections are treatable with antibiotics. But every time bacteria are exposed to antibiotics, they are given an opportunity to adapt to the presence of the drugs. This happens through natural selection favoring genes encoding for drug resistance.

Genes encoding for antibiotic resistance are usually found in plasmids (Figure 11.6).

Many bacterial infections that were once easily treated with antibiotics are now potentially life threatening, a result of the emergence of bacterial strains that are resistant to multiple drug treatments. A primary factor driving the emergence of antibiotic-resistant strains of bacteria is the unnecessary use of antibiotics. A recent survey conducted by the Centers for Disease Control and

153

Prevention (CDC) revealed that 86% of Georgia pediatricians routinely prescribed antibiotics for bronchitis (a use that is rarely justified), and that 42% prescribed antibiotics to treat common colds—which are caused by viruses and respond in no way to antibiotics *(Pichichero et al., 2000).* **[Krogh Chapter 17 Medialab Are Bacteria Winning the War? Natural Selection in Action]**

Recent developments have elevated the threat posed by some bacteria as agents of biological terrorism. Especially notable is *Bacillus anthracis* (Figure 14.1), which causes anthrax. This bacterium is normally found in wild and domestic animals. Although it still occurs in the "wild," *B. anthracis* is rare in the United States. Anthrax is still common in other parts of the world. The disease is typically spread by contact with spores through ingestion, inhalation, or through breaks in the skin (Figure 14.2).

Like some other bacteria, *B. anthracis* forms resistant endospores, or spores, under certain conditions. These spores can withstand up to 10 minutes in boiling water and may persist in contaminated soil for years. Spores have been shown to remain viable for up to 60 years in contaminated soil *(Wilson and Russell, 1964).*

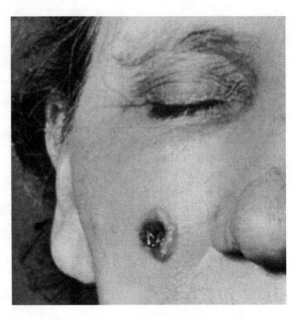

FIGURE 14.2   Lesion Associated with Cutaneous Anthrax, Which is the Most Common Form of the Anthrax Disease. It Occurs When Spores Enter the Body through Broken Skin.

Fortunately, when detected early, anthrax usually responds well to treatment with antibiotics, and a vaccine is available to prevent the disease. Transmission from human to human is unlikely. Under normal conditions, anthrax is acquired through exposure to spores associated with contaminated soil or animal products. Unfortunately, because of the resistance of spores, and because they can be prepared in an aerosol form, they may serve as a weapon of biological warfare.

Traditionally, penicillin was very effective in treating bacterial infections, but is less so now. There is currently some debate concerning the broad-scale distribution and prophylactic use of the antibiotic ciprofloxacin (Cipro®).

Prophylaxis refers to the prevention of or protection against disease. A drug used in prophylaxis is termed a *prophylactic.*

Should Cipro® be widely distributed and used by the general public to help prevent anthrax? Think about the implications of this practice. Do the risks of *not* taking ciprofloxacin outweigh the risks of creating an antibiotic-resistant *B. anthracis* strain? Is there a better option to prevent the spread of the disease?

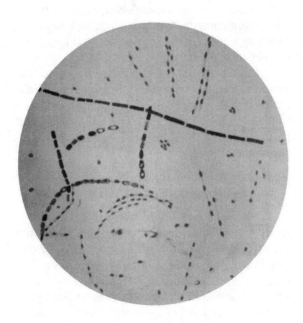

FIGURE 14.1   *Bacillus anthracis.*

While some bacteria are dangerous, most are completely harmless, and many bacteria are beneficial to humans. We use bacteria to make cheese and yogurt. They are used to degrade oil spills and to decompose our sewage waste. The bacteria in our guts live there and protect us, outcompeting harmful organisms that may find their way into our bodies. Bacteria "fix" nitrogen, making it usable for plants. They are very important decomposers, breaking down organic material and allowing it to recycle. And consider the contribution bacteria make in the food chain. They are the foundation of energy acquisition for other tiny organisms that rely on them for food supply; these organisms, in turn, are a food source for larger organisms, and so on.

As is the case with most organisms, so it is with bacteria. Sometimes, we can't live with them. But for the most part, we can't live without them.

Exercise 14.1   BACTERIAL TYPES

Bacteria may be divided into three main groups, based on shape. **Cocci** (singular, *coccus*) are spheres, **bacilli** (singular, *bacillus*) are rod-shaped, and **spirilli** (singular, *spirillum*) are corkscrew or spiral-shaped helixes (Figure 14.3). When the cell walls of coccus bacteria do not completely divide during fission, the resulting daughter cells may occur together in pairs (diplococcus), in chains (streptococcus), or in clumps (staphylococcus).

We will look at bacteria on prepared slides and in yogurt, which is made by adding certain strains of bacteria to milk. The tangy flavor of yogurt is from lactic acid, a by-product produced by the bacteria. *Lactobacillus acidophilus* is one species that enables yogurt to be labeled, "Contains live and active yogurt cultures." Look on the label of your next yogurt cup to see whether the company provides a genus or species for the bacteria involved. Why do you think the label doesn't read, "Contains live and active bacterial cultures"? Wouldn't that be more appropriate?

1. Observe the demonstration slide showing the various bacterial types. Locate each of the shapes described earlier. Note that even at high magnification, the bacterial cells still look quite small. Answer Question 1.

2. Now prepare your own wet mount to observe. Obtain a clean, dry slide. With a toothpick, smear a tiny amount of yogurt onto the center of the slide. The preparation should be thin enough so that you can see through it. Allow it to dry for a few minutes.

3. Place a drop of methylene blue onto the yogurt smear, and let it stain for 1 minute. Rinse the excess stain off with a few drops of water, and be careful not to rinse all the yogurt off the slide. Place a cover slip on the smear, and observe the specimen with your microscope. Answer Question 2.

4. Prokaryotic cells tend to be very small in comparison to eukaryotic cells. As outlined in Exercise 4.6, prepare a wet mount of cheek epithelial cells, stained with methylene blue, and examine it under high power. Answer Question 3.

Questions

1. **What type(s) of bacteria do you see on the prepared slide? Sketch and label them.**

2. **What type(s) of bacteria do you see on the yogurt slide? Sketch and label them.**

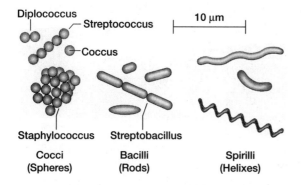

FIGURE 14.3   Types of Bacteria.

3. Do you also see bacteria on the cheek-cell slide? What kind are they? Sketch and label the cheek and bacterial cells.

Exercise 14.2   GRAM STAINING

Wear gloves and goggles.

Various groups of bacteria exhibit unique biochemical properties, and we take advantage of these to differentiate bacterial types. The morphology and biochemistry of the bacterial cell wall cause bacteria to respond differently to certain biological stains. In 1884, a Danish histologist, Christian Gram, developed a method now known as **Gram staining**. When the method of Gram staining is applied, Gram-positive bacteria will be deep purple or blue, and Gram-negative bacteria will be light pink.

The classification of disease-causing bacteria as Gram-negative or Gram-positive has traditionally been very important because Gram-positive bacteria respond to the antibacterial activity of penicillin, but Gram-negative bacteria usually do not. In this exercise, you will apply Gram staining to bacteria from *Bacillus megaterium*, *Escherichia coli*, and *Micrococcus luteus* cultures, and to bacteria from your nose and mouth.

1. With a wax pencil, label each of the three microscope slides BM, EC, and ML for each of the above-mentioned species. Your instructor may want you to stain all 5 of your slides at once; see Step 15.

2. Place a drop of distilled water onto the slide labeled BM.

3. Using a sterile bacteriological loop, transfer a small amount of *B. megaterium* bacteria to the drop of water and mix well.

4. Place the slide on the staining rack and allow to dry. Repeat steps 1 to 4 for *E. coli* and *M. luteus* using the other two slides you labeled.

5. Place a drop of absolute methanol onto the bacteria on each slide. This "fixes" the material. After 1 minute, turn the slide on its side to drain the methanol.

6. Cover the bacteria with crystal violet stain for about 90 seconds.

7. Rinse the slide gently under tap water.

8. Cover the bacteria with iodine-potassium iodide solution for 1 minute.

9. Rinse gently under tap water.

10. Cover the bacteria with acetone-alcohol solution for 30 seconds.

11. Rinse gently under tap water.

12. Cover the bacteria with safranin for 30 seconds.

13. Rinse gently under tap water, blot excess water off the bottom edge of the slide with a paper towel, and set aside until dry.

14. Examine each slide with the high-power objective and then the oil-immersion objective. Ask your instructor for help if necessary.

15. Next, obtain smears from around your teeth and inside your nose using two sterile cotton applicators. Roll the swabs onto labeled microscope slides. (Do not place a drop of water on the slide first.) Prepare Gram stain slides and observe under oil immersion, following steps 4 to 14.

Questions

1. Refer to Figure 14.3. What are the shapes of the bacteria that you see on each slide? Sketch them.

   BM:

   EC:

ML:

Nose:

Mouth:

2. **Indicate which bacteria are Gram-negative or Gram-positive in the preceding list.**

Exercise 14.3. COLIFORM BACTERIA

Extended two-week exercise

Having clean recreational and drinking water is something we tend to take for granted (see Figure 29.14 and Exercise 29.5). Many chemical and biological pollutants must be continually monitored to ensure the cleanliness and safety of our water supplies. One such measure of water quality is the number of bacteria present.

It is normal for bacteria to be found in water samples, but the presence of coliform bacteria (a specific type) is of special interest to us. Many types of coliform bacteria occur naturally

in soil and water, but fecal coliforms that normally live in the intestines of birds and mammals may also be found in water supplies. Because the presence of fecal coliform bacteria in water is an indication of fecal contamination, tests for fecal coliforms are often used as an important indicator of the suitability of water for recreation or human consumption. For instance, the bacterium *Escherichia coli* lives in the intestine of mammals, so the presence of *E. coli* in water is an indication of contamination with mammalian feces.

You are probably aware that some pathogenic strains of *E. coli* may cause serious human disease; however, remember that nonpathogenic *E. coli* strains are a normal part of our intestinal flora. Even if the *E.coli* is pathogentic, it is not the *E. coli* itself that poses the greatest health threat. Many other dangerous disease-causing agents may be transmitted through fecally contaminated water, including the organisms that cause cholera, typhoid fever, dysentery, many types of diarrhea, and worm infection. Because of these serious implications, if *E. coli* is found in drinking water, the water is usually declared unsafe for human consumption until the problem is resolved.

Water can be contaminated with bacteria in several ways. Quite often, *E. coli* levels will be elevated after periods of rainfall because of runoff from feedlots and other agricultural operations, and from the flooding of sewage systems.

In 1997, 649 beach closings and advisories resulted from sewage spills and overflows. In coastal Florida alone, 500 million gallons of untreated sewage were released annually through the 1980s (*Griffen et al, 2001*).

Because of greater ease in testing, total coliform count is sometimes the primary feature tested in large bodies of water, with the assumption that elevated total coliform counts are indicative of elevated *fecal* coliform counts. Total coliform levels typically are about 10 times those of fecal coliform levels; but this is not a very useful measure of water quality, because many coliform bacteria occur naturally in soil and water. Water that is actually safe and "clean" may test positive for the presence of coliform bacteria. Most decisions are now based on fecal coliform counts. For fecal coliform levels typically accepted, see Table 14.1. Each colony represents one bacterium in the original sample. When

TABLE 14.1 MAXIMUM FECAL COLIFORM LEVELS (*MICROLOGY LABORATORIES, LLC. [1997]; WASHINGTON VIRTUAL CLASSROOM*)	
Municipal drinking water (after treatment)	0 colonies/ 100 ml
Municipal drinking water (prior to treatment)	4 colonies/ 100 ml
Shellfishing water	70 colonies/ 100 ml
Recreational water	
Total Body Contact (swimming)	200 colonies/ 100 ml
Partial Body Contact (boating)	1,000 colonies/ 100 ml

levels exceed those given in the standards, the areas are closed to the activities listed (Figure 14.4).

In this exercise, you are going to test for the presence of coliform bacteria and fecal coliform bacteria using samples from local bodies of water. We will take advantage of the unique biochemistry of the organisms by using selective media. Selective media have nutrients that will support growth of only certain types of bacteria. If growth is successful, individual bacterial cells will reproduce and make colonies (plaques) on the media plates.

FIGURE 14.4   This Public Swimming Area Was Closed because of High Bacterial Levels in the Water.

We will sample for coliforms using **mEndo media**. This selective media will support the growth of coliform bacteria that produce the enzyme galactosidase, used in the break down of lactose. Most types of coliform bacteria will grow on mEndo media.

In order to test specifically for fecal coliforms (*E. coli*), we will use **EC + MUG media**. While this medium supports the growth of several coliforms, *E. coli* is the only coliform that produces an enzyme called glucuronidase. This enzyme reacts with MUG (4-methylumbelliferyl β-D-glucuronide) in the media, causing colonies of *E. coli* to exhibit fluorescence under long-wave ultraviolet (UV) light. The other colonies will not "glow" in this way.

Week 1

1. Obtain the following items: a water sample from a local source, one mEndo plate, and one EC + MUG plate.

2. Watch your instructor demonstrate the following technique, and then do it with your own plates. From the original, undiluted water sample, place 100 μl onto the center of the mEndo plate and spread with a glass spreader. Be sure to use proper sterile technique: Before touching the sample on the plate with the spreader, dip the spreader into alcohol and heat it briefly over the flame. Let the spreader cool for a few seconds before applying it to the plate (or cool it by touching it to the agar). Immediately after spreading the plate, close the petri dish, dip your spreader into alcohol, and flame it again. Fasten the petri dish together with short pieces of tape so that it will not open. Using a wax pencil or marker, label the perimeter of the plate with the *water sample location*.

3. Repeat the process by spreading the EC + MUG plate with the water sample. Be sure to use sterile technique!

4. The plates will be incubated for 48 hours at 37°C. After the incubation, your plates will be moved from the incubator to the refrigerator, where they will be held for you until next week. The colonies will essentially stop growing when placed in the refrigerator. If this isn't done, the colonies will "overgrow" and won't be countable.

Week 2

5. Count the number of colonies growing on the mEndo plate. On the basis of the number of colonies that you count on the plates, you can

estimate the number of bacteria/100 ml of water. Again, each colony represents one bacterium in the original sample.

6. To calculate the number of colonies growing/100 ml: Multiply the number of colonies that you count by 1000, since you placed 100 μl on the plate (100 μl = 0.1 ml, and 0.1 ml $\times$ 1000 = 100 ml.)

7. To determine the final sample concentration, take an average of those plates that had between 10 and 40 colonies per ml.

8. Repeat steps to 6 and 7 get the average number of *E. coli* colonies/100 ml on the EC + MUG plates.

9. Calculate the class average for each sample, and record this data in Table 14.2.

10. Compare your data to the standards given in Table 14.1. Complete Table 14.2 and answer the questions below.

Questions

1. **What is the total coliform count (per 100 ml) in the water sample you tested?**

2. **How does this compare to water samples from other local sources?**

3. **What is the fecal coliform count (*E. coli*) in the water sample you tested?**

4. **How does this compare to water samples from other local sources?**

TABLE 14.2   COLIFORM COUNTS

Water source:

	mEndo Total coliform	EC + MUG Fecal coliform
# colonies counted		
# bacteria per 100 ml (# colonies counted $\times$ 1000)		
Average coliform count for entire class		
Safe for recreation?		
Safe for drinking?		

5. **What might be the source(s) of the bacteria in each place?**

6. **What is the method of water-quality testing for your public water supply? What type of testing would reassure you?**

7. **Do you consider the source you tested as a safe drinking water source?**

8. **Is it safe for recreational purposes?**

Exercise 14.4   THE PLAGUE

Web exercise

Perhaps the most famous bacterial disease is bubonic plague, which is caused by *Yersinia pestis* (Figure 14.5). It is transferred from rodent to human or from human to human by the bite of a flea. During a phase of the human infection, however, it can become "airborne" when a person coughs, and droplets containing *Yersinia* are aerosolized into another person's breathing space. This disease killed 25 million Europeans during the Black Death of the fourteenth century—25 to 30 percent of the population. The name *Black Death* may have originated from the gangrenous, necrotic lesions associated with plague infection (Figure 14.6). Such lesions were probably more prevalent in the outbreaks of the Middle Ages than they are today. Plague is still common in many parts of the world and it occurs in the western United States, where 61 cases were reported from 1994 to 2001 *(Centers for Disease Control and Prevention, 2003)*.

It has been suggested that the nursery rhyme "Ring Around the Rosie" had its origins during one of the great plague episodes in Europe during the Middle Ages. Figure 14.7 shows one variation of the rhyme.

1. Research the origin of this rhyme, and provide a detailed explanation as to whether you think the nursery rhyme originated during the Black Death. This will require some research into the nursery rhyme and into the history and biology of the plague.

2. Your report should be a minimum of two double-spaced pages with at least three references. You may use books and Internet sources. See Appendix 1 for proper citation format.

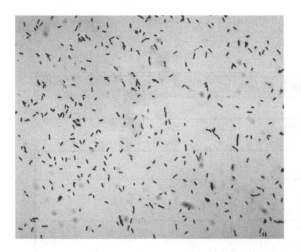

FIGURE 14.5   *Yersinia pestis*, the Bacterium that Causes Plague.

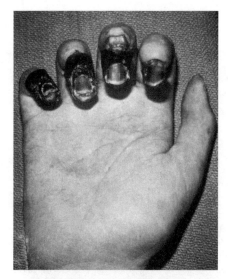

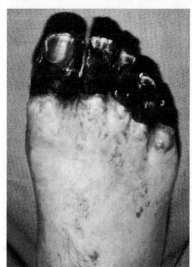

FIGURE 14.6   Gangrenous Lesions. This is a typical symptom of plague, giving the disease the name *Black Death*.

FIGURE 14.7   Kate Greenaway's *Mother Goose*, 1881.

15

Protists

What do you call a single-celled plant that acts like an animal? A protist. Some protists have plant-like or animal-like qualities, and some have both; still others have characteristics similar to fungi. Most protozoans are free living, but many are symbiotic, and protozoan parasites are responsible for some of the most serious diseases of humans and domestic animals.

Symbiosis refers to a relationship between two organisms in which at least one organism benefits from living in or on the other. A parasite is a symbiont that benefits from living on its host, and the host is harmed in some way.

Kingdom Protista gives us our first look at eukaryotic organisms. **Eukaryotes** are different from the prokaryotes because they have a true nucleus and membrane-bound organelles [**Krogh section 20.7**]. Like the bacteria, protists exhibit tremendous diversity in structure, nutrient acquisition, and distribution. They are found in freshwater, saltwater, soil, in other organisms—virtually anywhere that life is found.

Most protists are single-celled. Algae are often included in Kingdom Protista, but many algae are multicellular, as are the water molds. (Water molds are not "molds" like those that belong to Kingdom Fungi, however.) Take a look at

these protists, keeping in mind that they represent our very distant cousins, and perhaps our not-so-distant cousins.

Exercise 15.1   ANIMAL-LIKE PROTISTS

The animal-like protists (aka protozoans) are classified primarily on the nature of their locomotory structures, or lack of them. See Table 15.1.

AMOEBA

1. Examine *Amoeba proteus* (Figure 15.1) by preparing a wet mount from the culture containing living amoebae. For best results, take a drop

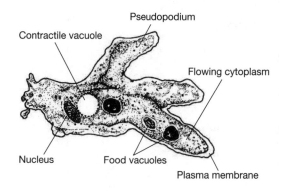

FIGURE 15.1   *Amoeba proteus.*

163

TABLE 15.1 PROTIST GROUPS		
Group	**Method of movement**	**Sketch (Label visible organelles)**
ANIMAL-LIKE PROTISTS (Exercise 15.1) **Amoeba** *Amoeba proteus* *Entamoeba histolytica*		
Plant-like Flagellates *Euglena gracilis* *Peridinium* *Volvox*		
Animal-like Flagellates *Giardia lamblia* *Trypanosoma*		
Ciliates *Paramecium caudatum*		
Apicomplexans *Plasmodium*		

TABLE 15.1 (*CONTINUED*)		
Group	**Method of movement**	**Sketch (Label visible organelles)**
FUNGUS-LIKE PROTISTS **(Exercise 15.2)** **Water Molds** *Phytophthora infestans* **Slime Molds** *Physarum polycephalum* *Dictyostelium discoideum*		
PLANT-LIKE PROTISTS **(Exercise 15.3)** **Algae** *Spirogyra* *Ulva* *Sargassum* **Diatoms** *Cyclotella* *Thalassiosira*		

FIGURE 15.2   Amoeba feeding by the process of phagocytosis, in which a small food particle is engulfed.

from the bottom of the culture. The amoebae achieve locomotion by **pseudopodia** ("false feet"), which are extensions of the cell's cytoplasm. Most amoebae are free living, but some are endosymbionts, living in other organisms. Amoebae feed by **phagocytosis** (Figure 15.2). Notice the amoeboid movement by pseudopodia. You may see contractile vacuoles, which are structures that enable the amoebae to get rid of excess water. You may also observe food vacuoles containing starch granules.

2. Obtain a slide of preserved specimens, and note the pseudopodia, cytoplasm, nucleus, and food vacuoles. Sketch them in Table 15.1.

3. One parasitic amoeba, *Entamoeba histolytica* (Figure 15.3), is the cause of a serious type of dysentery (bloody diarrhea) in humans. Close to 500 million people are infected with it at any given time, and it kills about 100,000 people each year *(Roberts and Janovy, 2000)*. Observe representatives of *E. histolytica* with an oil-immersion objective lens. The amoebae live in the large intestine, where they may invade intestinal tissue and cause ulcers. Before leaving with an infected human's feces, the amoebae change into cysts. The cyst is a stage that can live for several weeks in water or moist soil. Another person gets infected by ingesting cysts in fecally contaminated food or water.

PLANT-LIKE FLAGELLATES (CHLOROPLASTS PRESENT)

4. Observe living representatives of *Euglena gracilis* on a wet mount, using your compound microscope. (Place a drop of methyl cellulose on your wet mount slide before adding the cover slip. This will cause the specimens to move more slowly.) These organisms are unique flagellates because they are highly motile, like animals; but they also possess chloroplasts and carry out photosynthesis, like plants. Note the color of the organisms. Also note the jerky movement that is characteristic of flagellates, which move by means of their long, whip-like **flagellum**. The flagella of protists are similar to those of bacteria; but in bacteria, flagella rotate and *push* the organisms along, working like little propellors. In *Euglena*, however, the flagella rotate and *pull* the organism along in the direction of the flagellum. Sketch *Euglena* in Table 15.1, indicating the direction of their movement.

5. Dinoflagellates are another group of photosynthetic flagellates. Prepare a wet-mount slide of *Peridinium* spp., which is a large, nontoxic, freshwater species. Most dinoflagellates are marine, but some live in fresh water. Some exhibit bioluminescence, giving tropical seas a night-time "sparkle" or "glow." Some dinoflagellates release potent toxins that may kill aquatic organisms, especially fish. Some of these toxins may even cause serious illness in humans who consume tainted seafood.

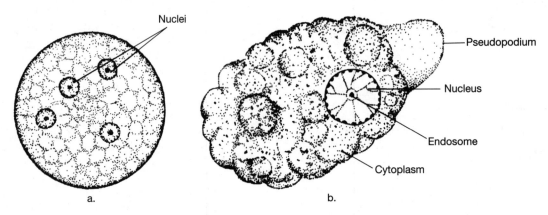

FIGURE 15.3   *Entamoeba histolytica.* a. Cyst. b. Trophozoite (feeding stage).

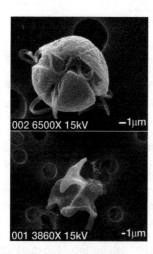

FIGURE 15.4 Scanning Electron Micrographs of *Pfiesteria piscicada*. Cyst stage (top) and Amoeboid form (bottom).

FIGURE 15.5 Fish Exposed to Toxic *Pfiesteria*.

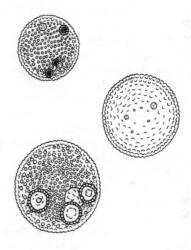

FIGURE 15.6 *Volvox* Colonies.

One group of important disease-causing dinoflagellates belongs to the genus *Pfiesteria* (Figure 15.4). This deadly dinoflagellate has been recognized as a serious threat to estuarine fisheries along much of the Atlantic coast. Aerosolized substances produced by *Pfiesteria* can affect the human nervous system as well. Exposure of laboratory workers led to skin and eye irritation, respiratory difficulties, gastrointestinal disturbance, and profound memory dysfunction in an Alzheimer-like condition (*Glasgow et al., 1995*). *Pfiesteria* may remain in a dormant cyst stage in sediment, but fish stimulate *Pfiesteria* to release toxins that cause the fish to become sluggish and unable to swim properly within minutes, and within hours skin lesions appear on the fish (Figure 15.5). Organic pollution, such as runoff from swine farms, leads to increased chances of *Pfiesteria* infection (as well as other dinoflagellates and algae) that often lead to the massive fish kills.

6. Make a wet mount of *Volvox* (Figure 15.6) and examine it with your microscope. Their green, hollow spheres may reach a diameter of 1 mm. These colonial organisms exhibit pronounced division of labor, with some individuals specializing in nutrition and locomotion, and other individuals specializing in reproduction. In addition to asexual reproduction by fission, *Volvox* exhibits sexual reproduction. Some biologists think that colonial protozoans are the ancestors of multicellular organisms. Sketch *Volvox* in Table 15.1.

ANIMAL-LIKE FLAGELLATES (NO CHLOROPLASTS)

7. The life cycle of *Giardia lamblia* (Figure 15.7) is very similar to that of *E. histolytica* except that *G. lamblia* lives in the small intestine of humans and other mammals. Infection with *G. lamblia* is common worldwide and can result in severe diarrhea, though it does not cause dysentery. Leeuwenhoek first observed *G. lamblia* microscopically in his own stool! *Giardia* is still readily available for study, but biologists get them from a company, not from their own homegrown variety. You might become a *Giardia* factory if you are not careful about your drinking water, especially during hiking and camping trips. Have you ever taken a refreshing drink from a running stream? Beaver

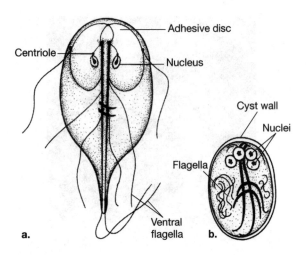

FIGURE 15.7   *Giardia lamblia.* a. Trophozoite (Feeding Stage). b. Cyst.

and other animals may contaminate the water with *Giardia lamblia* cysts. Obtain a slide of *Giardia* trophozoites and sketch them, labeling the nuclei and flagella.

8. Obtain a slide of *Trypanosoma rhodesiense* (or *T. gambiense*), and look at it with your compound microscope. Infection with *T. rhodesiense* (Figure 15.8) causes African trypanosomiasis. It is estimated that between 300,000 and 500,000 people have African trypanosomiasis (*Black and Seed, 2001*). In humans, these parasites reside in the bloodstream, then in the central nervous system in later stages of the infection, and this causes African sleeping sickness. This parasite is transmitted from host to host by the bite of a tsetse fly. Sketch *Trypanosoma* in Table 15.1.

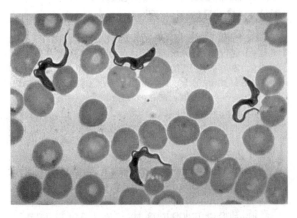

FIGURE 15.8   A Trypanosome Flagellate that Causes African Sleeping Sickness.

CILIATES

9. Representatives of phylum Ciliophora use **cilia** to move. Cilia are much shorter and more numerous than flagella and "beat" in rhythmic waves. Ciliary movement is important not only in locomotion but also in creating water currents that facilitate feeding and respiration. Prepare a *Paramecium caudatum* (Figure 15.9) wet mount from the culture vial. Use methyl cellulose to slow them down. Notice the smooth, gliding motion characteristic of ciliary movement.

10. Examine the preserved slide of *P. caudatum.* Pay particular attention to the two **nuclei** and **oral groove**. The oral groove is lined with cilia, which direct tiny food particles into the **cytostome** or cell mouth, where food vacuoles form. The contractile vacuoles function in osmoregulation by getting rid of excess water. Sketch these in Table 15.1.

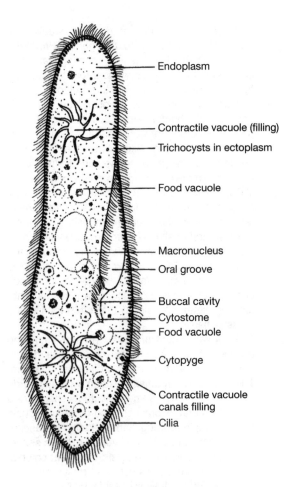

FIGURE 15.9   *Paramecium caudatum.*

APICOMPLEXANS: NO ORGANELLES FOR LOCOMOTION

11. Obtain a slide of *Plasmodium* (Figure 15.10) and look at it with an oil-immersion lens. Representatives of the phylum Apicomplexa are unique among the protozoans in that, for the most part, they have no locomotory structures. All apicomplexans are *endosymbionts*, living inside other organisms. *Plasmodium* species occur in the red blood cells of humans and cause malaria. Worldwide, about 800 million people are infected.

Approximately 1.7 million African children die every year—one child dies of malaria every 20 seconds (*Breman et al., 2001*).

This deadly disease is transmitted from person to person by the "bite" of mosquitoes of the genus *Anopheles* (Figure 15.11). Although this parasite was eradicated from the United States 50 years ago, malarial infections are on the rise again. In the summer of 2003, seven cases were reported in one Florida county (Centers for Disease Control, 2003). Malarial infections are also occurring at an alarming rate globally. As pointed out by Hotez (2002), malaria kills more children than any other disease, including AIDS. Sketch *Plasmodium* in Table 15.1.

Exercise 15.2   FUNGUS-LIKE PROTISTS

SLIME MOLDS AND WATER MOLDS

Slime molds and water molds are bizarre creatures, and include organisms that don't fit easily into any taxonomic group. Water molds are either free-living

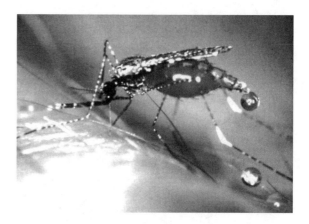

FIGURE 15.11   Adult *Anopheles dirus* taking a blood meal from a human. Female anophelid mosquitoes transmit malaria.

decomposers or they are parasites of plants and animals—just like fungi, but they have flagella during some stages of their life cycle. Yet, water molds have cell walls made of cellulose, like plants! Slime molds move around, and some also obtain their nutrients by phagocytosis, like amoeba. Yet, they reproduce in a manner similar to the fungi.

Aquarium fish sometimes develop white, fuzzy-looking patches on their scales, which is an infection with a water mold. The pathogen responsible for the nineteenth-century Irish Potato Famine, *Phytophthora infestans*, is also a water mold. In Ireland alone, approximately 1.5 million people died of starvation from 1845 to 1850. In addition, over 2 million people had left Ireland by 1855 to avoid starvation, cutting the population of Ireland by half—from 8 million to 4 million (*Quinnipiac University, 2000*).

Slime molds live in moist areas that are rich in organic matter. They are commonly found among decaying logs and leaf litter on the forest floor. There are two types of slime molds—true slime molds and cellular slime molds [**Krogh Figure 20.11**].

1. The **true slime mold** you are going to examine is *Physarum polycephalum* (Figure 15.12). In nature, this slime mold is found in cool, humid, dark places. Hold the petri dish containing the slime mold at an angle, and carefully look to find the specimen. The large, fan-shaped mass is called a **plasmodium**. It slowly creeps about over the forest floor like a giant amoeba, engulfing bacteria and other organic substances. The plasmodium contains many diploid nuclei, all

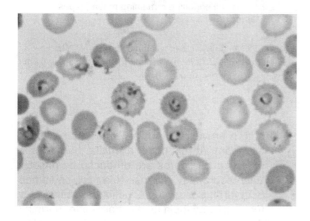

FIGURE 15.10   Malarial parasites (*Plasmodium* spp.) within red blood cells.

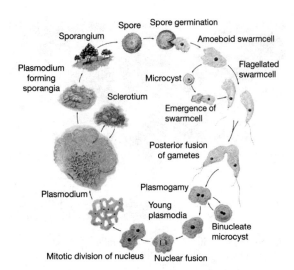

FIGURE 15.12 Life cycle of a true slime mold, *Physarum polycephalum*. © 1979, Carolina Biological Supply Company. Used by permission.

enclosed within a single plasma membrane. Now look at it with a dissecting microscope.

2. When food becomes scarce, the plasmodium sends up spore-producing **fruiting bodies**. If you look through the side of the petri dish, you will notice the stalks, and the fruiting body looks like a black golf ball on top of a tiny tee. The haploid spores produced by meiosis in the fruiting bodies are scattered by wind, rain, or other disturbances. Spores that land in favorable spots will germinate, giving rise to flagellated amoeba-like swarm cells. Two swarm cells fuse in a type of sexual reproduction to produce a diploid zygote. The zygote begins to grow and the diploid nucleus undergoes multiple mitotic divisions, giving rise to new single-celled, multinucleate plasmodia. Sketch a fruiting body and plasmodium in Table 15.1.

3. Obtain a plate of the **cellular slime mold** *Dictyostelium discoideum* (Figure 15.13), and examine it with a dissecting microscope. Cellular slime molds exist in nature as tiny, amoeba-like, single-celled organisms that creep about and feed by phagocytosis of bacteria. When bacteria become scarce, these amoeboid forms secrete chemical signals that are detected by other individuals, enabling them to merge together. Eventually, thousands of individuals unite, forming a mass called a **pseudoplasmodium** or **slug**. The slug begins to glide along

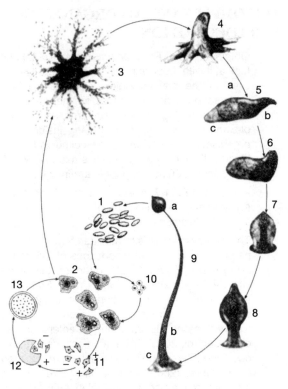

1. Spores
2. Myxamoebae (feeding stage)
3. Aggregation
4. Pseudoplasmodium
5. Migrating slug
6. Return to erect position
7. Beginning of sorophore formation
8. Elevation of sorogen
9. Sorocarp
   a. Sorus
   b. Sorophore
   c. Basal disc
10. Microcysts
11. Mating types (+ and –)
12. Diploid cell engulfing remaining mating types
13. Macrocyst

FIGURE 15.13 Life cycle of a cellular slime mold, *Dictyostelium discoideum*. © 1996, Carolina Biological Supply Company. Used by permission.

until it arrives at an area where nutrients are abundant. At this point, a fruiting body is sent up, in which spores are produced by meiosis. Each spore may give rise to a new, tiny amoeboid form. Sexual reproduction is facilitated by fusion of amoeboid forms shortly after germination of the spores. Sketch the cellular slime mold in Table 15.1.

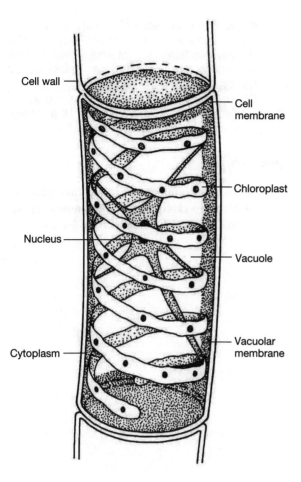

FIGURE 15.14   A *Spirogyra* cell.

Exercise 15.3   PLANT-LIKE PROTISTS

ALGAE

Algae are plant-like protists—they have chloroplasts and carry out photosynthesis. Algae may be unicellular, colonial, filamentous, or multicellular. Algae are named by the type of pigment they possess. There are green, golden brown, brown, and red algae. Regardless of the color of algae, it is important to remember that all have chlorophyll and carry out photosynthesis. In brown and red algae, other pigments mask the chlorophyll.

Algae are primarily aquatic, although some forms of green algae live on very moist rocks, trees, and soil. Green algae are believed to be the ancestors of all land plants. They grow in mats or masses on ponds, streams, and other bodies of fresh water. A common example of green algae is

Spirogyra (Figure 15.14), a **filamentous** green alga with spiral-shaped, ribbon-like chloroplasts. It is called a filamentous alga because the individual algal cells are joined together end to end in long filaments.

Seaweeds are multicellular algae. Some seaweeds are green algae, such as sea lettuce (*Ulva*), but most are red or brown algae. Although seaweeds come in various colors, all have chlorophyll and obtain their nutrients by photosynthesis. Seaweed has some important practical applications. It is an important food source, is rich in protein and offers much promise in helping to feed our growing population, and is an important source of fertilizer. Algin and carageenan are algal extracts that are used in dairy products such as ice cream, sherbet, and cream cheese to give them their smooth consistencies. Agar is also made from algal extracts **[Krogh Figure 20.13]**.

Algae play integral roles in the food chains of both freshwater and marine ecosystems. One type of seaweed, *Sargassum*, aggregates into large algal rafts that drift far out into the sea. These rafts of *Sargassum* serve as the basis for mini ecosystems, providing food and shelter for a host of organisms. For instance, juvenile sea turtles often live among *Sargassum* rafts, where they are protected from predators until they are large enough to survive in the open ocean *(Carr, 1995)*.

DIATOMS

Diatoms are single-celled photosynthetic protists—like *Euglena*, but without locomotory structures. Diatoms are common in the ocean, and a few occur in freshwater. These tiny organisms drifting freely in the water column are one of the most important components of phytoplankton, which serves as the main source of primary production in marine ecosystems. Diatoms have a cell wall containing silicon. The fossil shells of diatoms form thick deposits of *diatomaceous earth* that is mined for use as abrasive cleaners, including toothpaste and metal polish. These fossil cells are also added to paint to help make it sparkle.

Phytoplankton is comprised of microscopic photosynthesizing organisms that grow in abundance, drifting in the upper layers of oceans and other bodies of water. Phytoplankton is a major food resource in aquatic food chains.

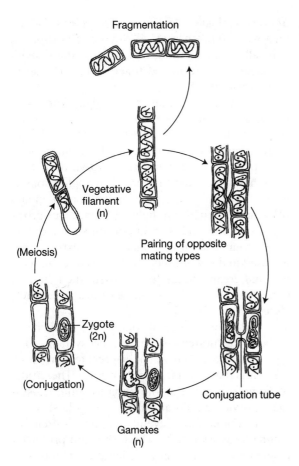

Fragmentation

Vegetative
filament
(n)

(Meiosis)

Pairing of opposite
mating types

Zygote
(2n)

(Conjugation)

Conjugation tube

Gametes
(n)

FIGURE 15.15   Life cycle of *Spirogyra*.

giving rise to new haploid filaments, and thus the cycle continues year after year. Sketch *Spirogyra* in Table 15.1, labeling the nuclei, haploid strand, conjugation tubes, diploid strand, and zygotes.

3. Observe the various preserved types of seaweed made available to you, such as *Ulva*. Sketch them and label their parts in Table 15.1.

Questions

1. **Give two examples of protists that might be difficult to classify into a taxonomic group. Provide a detailed explanation for each.**

1. Make a wet mount of the mixed-species culture of living diatoms (such as *Cyclotella* and *Thalassiosira*). Sketch several different types in Table 15.1.

2. Obtain a slide of *Spirogyra* and observe the various stages that occur in its life cycle (Figure 15.15). In late spring through early fall, *Spirogyra* haploid **filaments** occur in algal mats near the surface of a pond or gently flowing stream. Late in the fall, a **conjugation tube** forms between algal cells on adjacent strands. The cellular contents of one cell pass through the conjugation tube into a cell on the adjacent strand. The nuclei then fuse, forming the diploid **zygote**. At the onset of cold weather, the strands break apart and the diploid zygotes settle to the debris in the bottom. In the spring, when the climate warms, the zygotes undergo meiosis. The resultant haploid cells then divide by mitosis,

TABLE 15.2 PROTOZOAN RACES: RATE OF MOVEMENT				
	Locomotory structure	**Body length**	**Rate of movement**	**Displacement rate (body lengths/second)**
Amoeba				
Paramecium				
Euglena				
Human	Arms and legs		2 mph	
Car	Wheels	10 feet	50 mph	7.3 body lengths/sec

Exercise 15.4 PROTOZOAN RACES

(Adapted from Janovy, 1991)

In this exercise, you will compare the movement of various protozoans.

1. Using a stage micrometer (a ruler segment mounted on a slide) or an ocular micrometer, estimate the width of a field of view for your microscope on low power and on high power (1000 μm = 1 mm). Your instructor can help you with this.

   Low-power field = _____ mm = _____ μm

   High-power field = _____ mm = _____ μm

2. You will use 3 experimental protozoans in this exercise; *Amoeba proteus, Euglena gracilis*, and *Paramecium caudatum*. Make a temporary wet mount for each.

3. Estimate the length of each of your organisms. Record this in Table 15.2.

4. What are the locomotory structures utilized by members of each species? Take a few moments to observe and describe the movement of each.

5. Obtain a stopwatch, and estimate the speed of each of your experimental organisms, in body lengths/second. Use the timer to estimate how long it takes each organism to move across the field of view. (Don't wait for the amoeba to move across the stage. Just estimate how long they take to move a little bit, then extrapolate.) Do this three times for *Euglena* and *Paramecium*, and then record the average rate in Table 15.2.

6. Compare the rate of speed of each of your organisms to one another and to that of an automobile going 50 mph, also expressed in body lengths/second.

Conversions:
   1 mile = 5,280 feet
   1 inch = 25.4 mm
   1 car = 10 feet.

Example: Suppose a car that is 10 ft long is traveling at 50 mph. In one hour the car travels

50 miles/hour × 5,280 feet/mile =

$\qquad$ 264,000 feet/hour

264,000 feet/hour ÷ 10 feet/body length =

$\qquad$ 26,400 body lengths/hour

26,400 body lengths/hour ÷ 60 min/hour =

$\qquad$ 440 body lengths/minute

440 body lengths/minute ÷ 60 seconds/min =
$\qquad$ 7.3 body lengths/second

7. Assume that you can swim 2 mph. Determine your body length, and then calculate how fast you can swim and compare this rate to your experimental organisms. Record this in Table 15.2.

Questions

1. **What locomotory structure is most efficient? Explain your answer.**

TABLE 15.3 RELATIVE ABUNDANCE OF INFUSION CULTURE PROTOZOANS					
Protozoan					
First count: Avg. #/min					
Second count: Avg. #/min					
Third count: Avg. #/min					
Fourth count: Avg. #/min					
Fifth count: Avg. #/min					

Exercise 15.5   HAY INFUSION CULTURES

(Adapted from Janovy, 1991)

Extended two- to six-week exercise

This exercise is an introduction to a field of natural history called **ecology**, which investigates interactions between populations of organisms. **Populations** are groups of individuals of the same species that interact and live in a common environment. Different populations interact through competition and predator–prey interactions, live in a common environment, and constitute a **community**. Communities are not static things. They are dynamic, and the nature of interactions among the species in a community is continually changing. The changes that occur in a community, especially after a disturbance, are termed **ecological succession [Krogh section 31.7]**.

You will witness and document ecological succession in a community of protozoans maintained in the laboratory. Your instructor has created a *hay infusion culture* by placing a handful of vegetation in a jar of water, which was taken from an area of a pond that was "dried up." When a body of water starts to recede, many of the protozoans in it will encyst on plants and remain in an inactive state until water is once again available. In the encysted stage, the organisms are resistant to desiccation (drying out) and temperature extremes. The cysts

remain viable for months or even years—up to 40 years has been reported *(Pennak, 1953)*.

1. Prepare a wet mount from one of the infusion cultures. Observe any protozoans present on low power (10 × objective, total magnification = 100×). Sketch all kinds of protozoans found, and discuss with your classmates what to name each kind. All student groups will need to use the same names for the various organisms.

2. Using the low-power objective of a compound microscope, count the number of each kind of protozoan you see in 2 minutes. Be sure to move the slide so that you count protists in a different field of view each time to avoid counting the same individual twice. Each student in your group will do this, and then calculate the average number of individuals for each species present per minute. Record this in Table 15.3.

3. Observe the infusion cultures over a period of several weeks, and record how the population structure changes in Table 15.3.

4. Your instructor may ask you to design an experiment to examine the effects of an environmental variable on ecological succession in the infusion cultures. What might change in the natural environment? Consider testing the effects of temperature, pH, or light on your organisms. Or you may try to introduce a new species to your infusion culture, and observe how it changes the stability of the system. Record your experimental hypothesis in Question 1.

Questions

1. **Your hypothesis:**

2. **Prediction(s):**

3. **Independent variable:**

4. **Dependent variable:**

5. **Controlled variables:**

6. **Which species is predominant (represented by the greatest number of individuals) each time you count?**

7. **Dominant species, to a great extent, determine the conditions under which other species around them must exist. In your infusion cultures, which species is dominant? How do you know?**

Exercise 15.6   EMERGING DISEASES

Web exercise

Emerging diseases such as malaria and trypanoso-miasis are on the rise in various parts of the world. Investigate the occurrence of an emerging disease **caused by a protozoan**. If the disease you choose to study is caused by a parasite, refer to **Section 31.6 in your Krogh textbook.** Use several Internet sources to gather your information, starting with these Web sites:

http://www.fas.org/promed/about/index.html

http://www.who.int/health_topics/emerging_diseases/en/

http://www.paho.org/English/AD/DPC/CD/eid-eer-ew.htm

http://www.cdc.gov/ncidod

Questions

1. **What disease are you investigating?**

2. **Where does it occur?**

3. **List the Web sites you use for this assignment. See Appendix 1 for proper citation format.**

4. **What pathology does the disease cause?**

5. **Describe the organisms involved in the spread of this disease, including any intermediate hosts.**

6. **What does the word *prevalence* refer to?**

7. **What is causing this disease to increase in prevalence?**

8. **Is human activity directly or indirectly involved with its increase in prevalence? In what way(s)?**

9. **Explain whether any drug therapies have been useful in combating the disease.**

If you are investigating malaria, answer the following questions:

13. **What is preventing malaria from becoming reestablished in the United States?**

10. **Are the drug therapies readily available to the infected persons? Explain.**

14. **What disease control methods are being used today to prevent the spread of mosquito-borne diseases in the United States?**

11. **How does this disease affect the economy of the area where it is prevalent?**

15. **Will these strategies be effective in the long run?**

12. **How have humans worked to eradicate the disease, if at all?**

16. **If so, how do they work? If not, why are they still being used?**

16

Fungi

In His great wisdom, God gave us barley, which any fool can eat.
In His infinite mercy, He has shown us a better way.

R.E. Clopton (apocryphal)

The fungi are an extremely important group of organisms. Many fungi are involved in symbiotic relationships; they grow on plant roots and help with nutrient absorption, they join with algae to form lichens, and some fungi have adapted to a parasitic lifestyle, feeding on living tissue. But most fungi are saprophytic decomposers, obtaining their nutrients by secreting digestive enzymes and then absorbing the nutrients from their environment, feeding on decaying wood, leaves, and dead animals. We are thankful they decompose dead plants and animals, but when they also work to decompose our books, homes, and just about everything else, their presence suddenly becomes troublesome.

> *Saprophytic* organisms acquire their energy from dead or dying organic matter **[Krogh section 20.11]**.

There are probably more than 1.5 million species of fungi, but only 69,000 have been described. Fungi are multicellular eukaryotic organisms, except for yeasts, which are unicellular. The bodies of fungi are made up of filaments called **hyphae**, cylinders of cytoplasm, and many haploid nuclei. A collection of fungal hyphae is called a **mycelium** (Figure 16.1). Fungi may be divided into phyla on the basis of their structures that are used in sexual reproduction: the zygote-forming conjugation fungi (the bread mold *Rhizopus*, the dung

fungus *Pilobolus*), the sac fungi (yeasts, edible morels and truffles, blight), club fungi (mushrooms and shelf fungi, smuts, and rusts), and imperfect fungi, whose sexual reproduction structures have not yet been discovered. *Penicillium*, *Aspergillus*, and *Candida albicans* are usually considered imperfect fungi, but have recently been classified as sac fungi because their modes of sexual reproduction are now known.

Exercise 16.1 ZYGOTE-FORMING FUNGI

As the name implies, zygote-forming fungi have a diploid stage in part of their life cycle (Figure 16.2). Food spoilage happens in many ways: Bacterial contamination is common, but fungal contamination is also a problem from our kitchens to the laboratory to the crops in our farmlands. One reason we store food in our refrigerators is to inhibit bacterial and fungal growth. The black bread mold *Rhizopus* is a zygote-forming fungus, attacking breads and fresh fruits.

1. Obtain an agar plate with *Rhizopus* growing on it, and observe it with a dissecting microscope. Identify the sporangia, and sketch them in the following space.

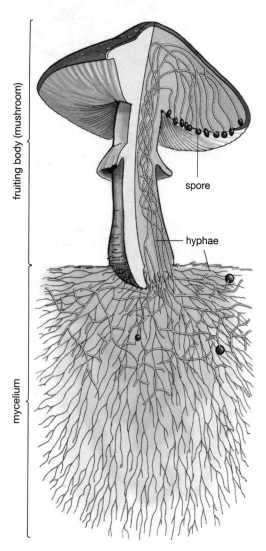

FIGURE 16.1   Structure of a Club Fungus.

2. Observe a prepared slide of *Rhizopus* and look at it with a compound microscope. Identify the sporangia, hyphae, and zygospores on your slide. Sketch them in the following space.

Sketches

1. Sketch the sporangia as they appear through the dissecting microscope.

2. Sketch and label the sporangia, hyphae, and zygospores, as they appear through the compound microscope.

Exercise 16.2   SAC FUNGI

In this exercise you will examine sac fungi, formerly known as imperfect fungi. The sac fungi are so named because some form elongate sacs with haploid spores. Many sac fungi resemble cups, and these are called cup fungi. *Penicillium* reproduces asexually by producing haploid spores on fruiting bodies called conidia. Most yeasts are also sac fungi, but they are unicellular. *Aspergillus flavus* is a mold that grows on stored grains and peanuts and produces aflatoxin, one of the most carcinogenic substances ever identified. The fungi that cause ringworm, athlete's foot, and jock itch also belong to this group; so does the yeast *Candida albicans*, which is responsible for vaginal yeast infections and an oral infection called thrush.

Yeasts are used by vintners as a source of ethanol, by bakers as a source of carbon dioxide, and by brewers as a source of both (*Raven et al., 1992*). Bakers take advantage of the carbon dioxide produced in the fermentation process. Used in this manner, yeast is a leavening agent like baking powder, making the dough rise. Most of the yeasts used in the production of wine, cider, and beer are various strains of a single species, *Saccharomyces cerevisiae*, or of *S. carlsbergensis*, in the making of lagers. Occasionally, yeasts of other species are used to give distinctive flavors to wine and beer. *Saccharomyces cerevisiae* is now virtually the only species used in making bread.

The bread mold *Penicillium* is the source of the antibiotic penicillin (Figure 16.3). Many antibiotics are derived from fungi, which naturally produce them. Bacteria and fungi have been competing—or engaging in warfare, if you will—for millions (or perhaps billions) of years. We take advantage of the weapons that they have developed

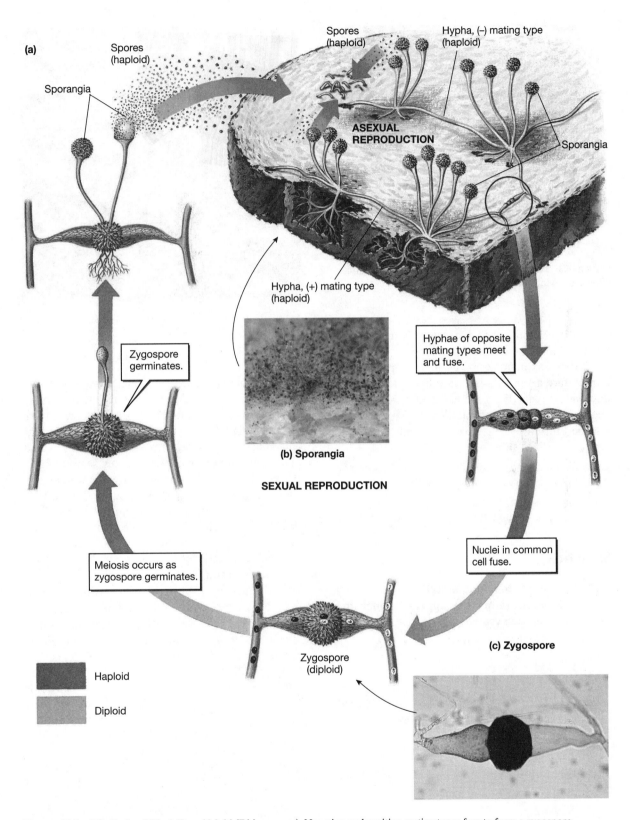

(a)

Spores (haploid)

Sporangia

Spores (haploid)

Hypha, (–) mating type (haploid)

ASEXUAL REPRODUCTION

Sporangia

Zygospore germinates.

Hypha, (+) mating type (haploid)

Hyphae of opposite mating types meet and fuse.

(b) Sporangia

SEXUAL REPRODUCTION

Nuclei in common cell fuse.

(c) Zygospore

Meiosis occurs as zygospore germinates.

Haploid

Diploid

Zygospore (diploid)

FIGURE 16.2   Life Cycle of Black Bread Mold (Rhizopus sp.). Negative and positive mating types fuse to form a zygospore.

FIGURE 16.3   Penicillin Kills Surrounding Bacteria as it Diffuses through the Agar in a Petri Dish.

FIGURE 16.4   Shelf Fungus Growing on a Tree.

against one another because nature is much more resourceful than our best technology. The upside to this is that we can utilize many "natural" products in our warfare against disease-causing organisms. The downside is that the genetic resourcefulness that leads to the production of these "weapons" may also be utilized by microbes to defend themselves against our "best shot" to fight them!

1. Obtain the petri dish with *Penicillium* growing on it, and observe it with a dissecting microscope.

2. Now observe the prepared slide of *Penicillium* with a compound microscope, noting its conidia.

Sketches

1. **Sketch and label the mycelium and conidia of *Penicillium* as they appear through the compound microscope.**

Exercise 16.3   CLUB FUNGI

The **club fungi** include those fungi you are probably most familiar with—the mushrooms and shelf fungi (Figure 16.4). The mushroom that extends above ground is actually a reproductive structure. The fungal hyphae associated with a mushroom may extend many feet underground, where they secrete digestive enzymes and absorb the liberated nutrients. Review the mushroom life cycle shown in Figure 16.5.

Occasionally, a circle of mushrooms will appear, called a **fairy ring** (see the figure at the end of the chapter). Fairy rings have been the source of much superstition for centuries. In Germany, they were traditionally called *hexen rings* and were thought to result from the dancing of witches on May Day eve. In France, it has been suggested that great toads with bulging eyes sometimes appear in the enchanted rings called *Rounds de Sorcieres*. The best-known European legend is that the fairy rings are little dance floors made by fairies dancing in a ring, with the mushrooms being stools upon which they sit. However, their occurrence may be explained by entirely natural phenomena. As the fungal mycelium expands, it begins to deplete nutrients in its center, and the hyphae associated with this region die. The hyphae on the outer margins of the mycelium send up fruiting bodies. This pattern of circular expansion can occur season after season. Some enormous fairy rings, estimated to be hundreds of

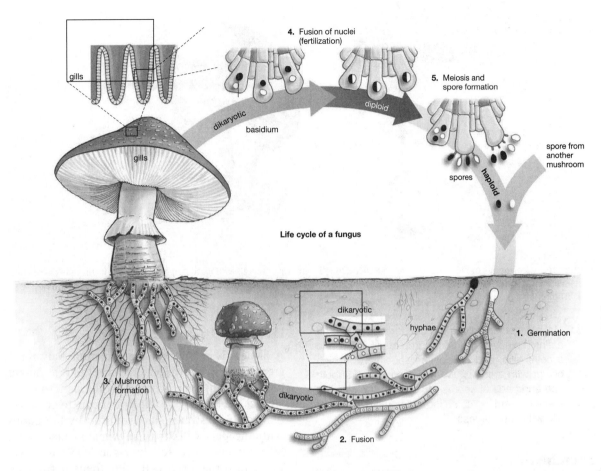

FIGURE 16.5   Life Cycle of a Club Fungus. 1. Spore formation: Haploid spores form by meiosis and are released from the gills of mushrooms. 2. Germination: Spores divide by mitosis and start growing haploid mycelia. 3. Fusion: The haploid cells of two different mycelia fuse, but their nuclei remain separate, hence the term *dikaryotic*. 4. Mushroom formation: The mushroom is a reproductive structure, initially made up entirely of dikaryotic cells. 5. Fertilization: Finally, the nuclei in the dikaryotic cells fuse, forming diploid cells. 6. Meiosis: The diploid cells divide by meiosis and form haploid spores.

years old have been found in the western United States *(Lincoff and Pacioni, 1981)*.

Fairy rings may be easily spotted because grass often grows in a lush circle on the periphery of the fairy ring and is very sparse in the center of the ring, where the fungi have depleted nutrients. Some fungi associated with fairy rings also secrete antibiotics that destroy bacteria essential to the normal growth of the grass. Fairy rings are considered a nuisance on lawns, golf courses, and playing fields.

Several mushroom species are toxic to humans. Great care should be taken when collecting mushrooms for consumption, because toxic species can resemble edible ones. Even one bite of the "death angel" *Amanita phalloides* can be fatal. On the other hand, consumption of the edible field mushroom *Agaricus campestris* is safe,

and supports a $14 billion worldwide market (Figure 16.6).

Another edible fungus is the prized truffle, which is actually a sac fungus that grows in symbiotic association with tree roots, particularly oaks and hazelnuts. Truffles usually grow below ground and are often hunted with the help of trained dogs or pigs.

Mushrooms also have cultural significance in Native American tribes, who use them in religious ceremonies *(Raven, et al, 1992)*.

1. Obtain an edible mushroom (a fruiting body) and note the stalk, cap, and gills under the cap.

2. With a knife or scalpel, slice the stalk lengthwise; with forceps or another sharp tool, separate the hyphae in the stalk—they will look like strands of cotton candy. Observe them with a dissecting scope.

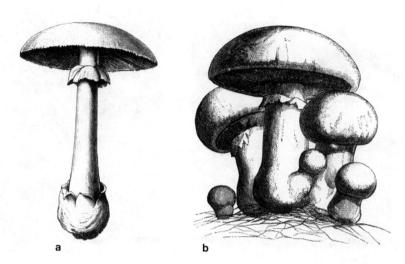

FIGURE 16.6   Poisonous or Not? a. *Amanita phalloides*, the "Destroying Angel" or "Death Angel." One bite of this may cause vomiting and hallucinations in a few hours, then kidney and liver failure, leading to death. b. *Agaricus campestris*, the Edible Meadow Mushroom.

3. Make a wet mount of a gill: Use forceps to tear off a single piece of gill, add a drop of tap water and a cover slip, and look at the specimen on high power with a compound microscope. You should see hundreds of spores attached to the gill, and some of them will come loose if you tap the cover slip with your forceps.

Sketches

1. **Sketch the gill and spores of your mushroom as they appear through the compound microscope.**

2. **Are the spores diploid or haploid?**

Exercise 16.4   LICHENS

A lichen is a symbiotic association between a green alga (or cyanobacterium) and a fungus. Lichens form when a fungal hypha penetrates an algal cell (Figure 16.7) and both begin to grow. The fungus absorbs nutrients produced by the photosynthetic algae cells. In time, fungal hyphae form a protective covering for the algae. While the algae suffer in terms of growth, they benefit from the protection afforded by the fungus. Reproduction occurs when a spore-producing piece of lichen breaks off and is dispersed **[Krogh section 21.5]**.

Because of the mutual benefits afforded by the symbiotic association, lichens occur where neither the algal nor fungal components could survive alone. Because the algal cells are photosynthetic, there is no need for external nutrients; because of the excellent shelter afforded by the tough, crusty fungal hyphae, the lichens can grow nearly anywhere. They are commonly found on rocks and trees. Lichens tend to grow slowly and are active only when moist. They take full advantage of moisture when it is available, absorbing up to 35 times their own weight in water (*Raven et al., 1992*).

Although lichens grow slowly, they are exceptionally long lived, with some believed to be as much as 4500 years old (*Raven et al., 1992*).

FIGURE 16.8 Three Lichen Types. Crustose lichens are the flat, light lichens in this photo. Foliose lichens have a leafy appearance, and fruticose lichens have a hairy appearance.

Sketches

1. **Sketch and label the three lichen types on display.**

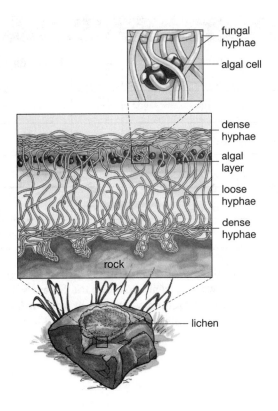

FIGURE 16.7 Structure of a Lichen.

Lichens are indicators of air quality. Lichens can easily absorb toxic materials of many sorts from the air; but with no means to get rid of these toxins, lichens are sensitive to pollution (*Starr, 1994*). Disappearance of lichens, as observed in London and New York at the height of their industrial revolutions, indicates polluted air. Lichens turned out to be a good means of monitoring contamination by radioactive fallout after the Chernobyl nuclear power plant disaster in 1986 (*Raven et al, 1992*).

1. Lichens may be roughly divided into three groups on the basis of their overall morphology (Figure 16.8). On your next walk outdoors, look for lichens on trees or rocks, or study the lichens on display in your lab. Crustose lichens have a flat, encrusting appearance. Foliose lichens are leaf shaped, hence the name. The fruticose lichens on display are bushy or branching lichens.

Exercise 16.5   ERGOTISM

Web exercise

Another mold that may be of profound historical significance is the ergot fungus, *Claviceps purpurea*. This mold infects rye, and causes a disease called ergotism. Some historians contend that the Salem witch hysteria in the seventeenth century was a result of this type of food poisoning.

1. Retype the questions and Table 16.1 in a new word processing document. Or, you may neatly fill in the table and answer the questions on these pages.

2. Research the notion that ergotism was responsible for the Salem witch episode. Using a search engine and two simple keywords, you will easily find numerous Web sites on the subject.

URL	Title	Author(s)	Date Web site Posted	Date Retrieved	Rank	Comments

TABLE 16.1   ERGOTISM WEB SITES

3. Evaluate five of these Web sites for accuracy and general reliability, completing Table 16.1 as you go. Rank the Web sites using a scale of 1–5, with 1 as the lowest score, and 5 as the highest score.

Questions

1. **What search engine(s) and keyword(s) did you use for this topic?**

2. **Describe the symptoms that ergotism causes in humans. Specifically describe what causes these symptoms. Cite the references that helped you answer this question, using Appendix 1 for proper format.**

3. **In your opinion, how likely is it that ergotism was the real cause for the Salem witch hysteria? Defend your answer with at least three examples. Again, cite your references.**

4. **Which Web site was the most reliable when conducting research on this topic? Give three reasons why you think it is most reliable.**

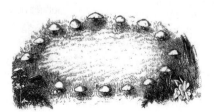

17

Plant Diversity

Little flower—but if I could understand what you are, root and all, and all in all, I should know what God and man is.
Alfred Lord Tennyson, 1869

Most botanists agree that land plants arose from a green alga ancestor. But how did plants adapt to life on land? Life on land presents unique challenges to organisms. Terrestrial organisms must possess mechanisms enabling them to do the following:

1. Prevent desiccation.
2. Stand upright in the absence of water's buoyant force. (e.g., how can trees grow so tall?)
3. Absorb and distribute water through the body.
4. Carry out sexual reproduction without the luxury of shedding gametes directly into water.

Before beginning these exercises, review the alternation of generations life cycle that is characteristic of plants (Figure 17.1) **[Krogh section 21.6]**. Knowing this life cycle is important in understanding how plants evolved. The first land plants appeared during the early Paleozoic era, about 500 million years ago (mya). These ancestral land plants first diverged into two major plant divisions, Bryophyta (mosses, liverworts, and hornworts) and Pterophyta (ferns and horsetails). Representatives of these groups live in moist environments. Their success as terrestrial plants depends on how well they have met the four challenges listed earlier. Pterophytes gave rise to

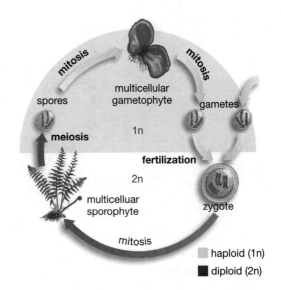

FIGURE 17.1   Alternation of Generations Life Cycle of Plants.

all other complex groups of land plants existing today (Figure 17.2).

Exercise 17.1   BRYOPHYTES: NONVASCULAR PLANTS

The bryophytes are small plants such as moss (Figure 17.3) and liverworts (Figure 17.4). All

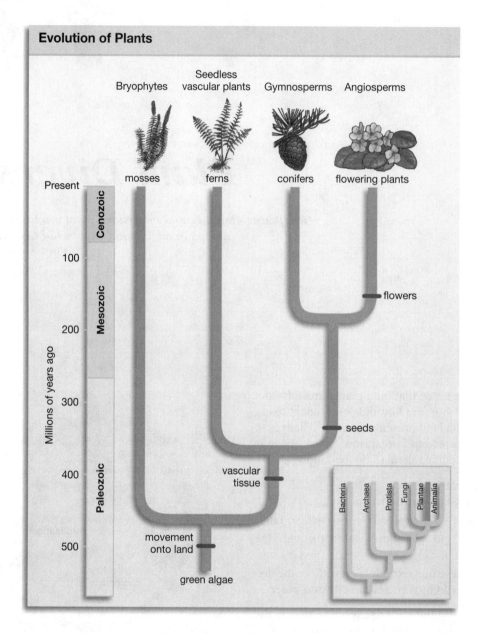

FIGURE 17.2   Evolution of Plants.

bryophytes lack vascular tissue, and they have no true roots, stems, or leaves. Because they have no roots, they cannot absorb water from the soil, but root-like **rhizoids** anchor the plant to the substrate. Simple diffusion is an adequate means of transporting water. The green, carpet-like portion of the "mat" of moss is comprised of the individual gametophytes, which are the predominant stage in the moss life cycle.

All of the male and female gametophytes comprising the moss "mat" are haploid. The male reproductive structure is called an **antheridium**, where sperm cells are produced by mitosis. (Remember, the gametophyte is already haploid; thus gametes are produced by mitosis, not meiosis.) The female reproductive structure, called the **archegonium**, is found at the top of the female gametophyte. Within the archegonium, the egg is produced by mitosis.

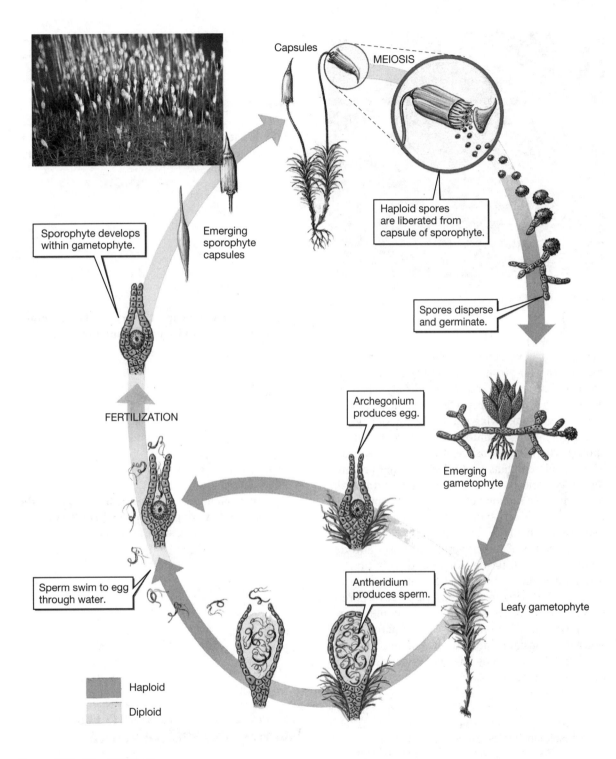

Capsules

MEIOSIS

Haploid spores
are liberated from
capsule of sporophyte.

Sporophyte develops
within gametophyte.

Emerging
sporophyte
capsules

Spores disperse
and germinate.

Archegonium
produces egg.

FERTILIZATION

Emerging
gametophyte

Sperm swim to egg
through water.

Antheridium
produces sperm.

Leafy gametophyte

Haploid

Diploid

FIGURE 17.3   Moss Life Cycle.

FIGURE 17.4   A Liverwort.

Bryophytes have retained swimming sperm, and sperm transfer from the antheridium to the archegonium can be accomplished only when the moss is very wet. The sperm fertilizes the egg within the archegonium, giving rise to the diploid zygote. The zygote grows by mitosis to produce the diploid **sporophyte**, which consists of a stalk and capsule. The capsule contains haploid spores that are produced by meiosis, which are released and distributed by the action of rain, wind, or other disturbances. Each spore may germinate to give rise to a new haploid gametophyte.

Because of their swimming sperm and dependence on diffusion for the distribution of water throughout the plant body, bryophytes are restricted to living in shady, moist places where it is very wet—at least seasonally. While many bryophytes have survived in such environments, botanists consider them to be an evolutionary dead end, because they have not given rise to any other plant groups.

1. Examine a clump of moss and remove a single gametophyte. It has a waxy coating to prevent desiccation, but because photosynthesis and respiration are occurring within the plant, oxygen and carbon dioxide must be able to pass into and out of the "leaves." There are microscopic pores in the waxy covering to permit the exchange of gases.

2. Find the sporophyte and identify its stalk and capsule.

3. Now observe a liverwort, named so for the lobed "leaves" resembling the lobes of the mammalian liver.

Questions

1. **What characteristics of mosses and liverworts restrict them to living in wet environments?**

2. **Sketch a moss and a liverwort. On the moss, label a sporophyte and gametophyte.**

3. **In a plant life cycle, which is haploid, the sporophyte or gametophyte? For bryophyes, which life-cycle stage is dominant?**

Exercise 17.2   PTEROPHYTES: SEEDLESS VASCULAR PLANTS

Ferns and horsetails represent the primitive seedless vascular plants, or **pterophytes**. All vascular plants have true roots, stems, and leaves.

Primitive means *unchanged*, like their relatives that lived long ago.

Vascular tissue is made up of long tube-like cells that grow end to end, forming little "pipelines" that transport water and nutrients throughout the plant. The vascular tissue responsible for carrying materials up from the roots throughout the plant is called **xylem**. The vascular tissue responsible for carrying the products of photosynthesis down the plant from the leaves is **phloem** (xylem up, phloem down). Vascular tissue is rigid and provides support for plants. It enables ferns to grow much larger than mosses and liverworts, and enables them to survive in drier environments than their nonvascular counterparts can withstand. Nevertheless, pterophytes have been restricted to wet habitats because of their swimming sperm.

Plants are composed of organic molecules that contain the element carbon. Plants are consumed by organisms that release this carbon in the form of carbon dioxide, a by-product of cellular respiration **[Krogh Figure 32.1]**. But some of this carbon is retained and buried in sediment in anaerobic conditions, where decay is only partial, forming a substance called **peat**. Over a *lot* of time, sedimentary rock may be deposited over the peat, exerting pressure on it and causing it to change into coal or other fossil fuels *(Raven et al., 1992)*.

Pterophytes were the dominant land plants during the Pennsylvanian period, when extensive fern forests were common (Figure 17.5). Pterophytes of the Pennsylvanian provided the material that we so heavily rely on as a source of energy to run our cars and heat our homes. The carbon in

FIGURE 17.6   A Fern Fossil.

fossil fuels you now use was once carbon dioxide captured by a fern-like plant about 300 million years ago (Figure 17.6).

Review the fern life cycle (Figure 17.7). The dominant stage of the pterophyte life cycle is the diploid sporophyte. The **fronds** are the leaves. The **roots** absorb water along with dissolved nutrients. You also notice long "runners" called **rhizomes**. Ferns may reproduce asexually by sending out rhizomes, which in turn may also send up new fronds. A young frond is called a **fiddlehead**.

1. Look at the specimen that has been "dug out" from the soil. Look at the underside of the mature fronds. Locate the small, nodular structures underneath some of the fronds. These are called **sporangia** (singular, *sporangium*), in which spores are produced by meiosis. When mature, the sporangia open and release the spores. When the spores germinate, they give rise to the haploid gametophyte generation. Also note the fiddleheads on the specimen provided.

2. Obtain a prepared slide of a fern gametophyte, called a **prothallus**, the tiny heart-shaped leaf.

FIGURE 17.5   Fern Forest of the Pennsylvanian Period.

Examine it with a dissecting microscope and notice that this type of fern has a prothallus that bears both the male antheridia and female archegonia, in which the sperm and eggs are produced by mitosis. (Some ferns have male and female sexual structures on separate prothalli.) Sperm must swim from the antheridium to an archegonium, where fertilization takes place. Fertilization of the egg results in the production of a diploid zygote, which matures into a diploid sporophyte. The plant growing up from the prothallus on your specimen is a young sporophyte.

3. Observe the fossil fern specimens on demonstration.

Questions

1. **Why are ferns restricted to growing in wet environments?**

2. **What advantage do pterophytes have that bryophytes do not, enabling them to inhabit drier environments?**

3. **Sketch and label a fern, a frond, rhizome, fiddlehead, sporangia, roots, and gametophyte with male and female reproductive organs.**

4. **What is the dominant life-cycle stage of a fern?**

5. **During what geological era and geological period did the first land plants (mosses and ferns) arise? Refer to the geological time scale on the inside back cover.**

6. **What else was occurring during that geological period?**

7. **During what geological era were the ferns the dominant plant group?**

8. **What else was occurring during that era?**

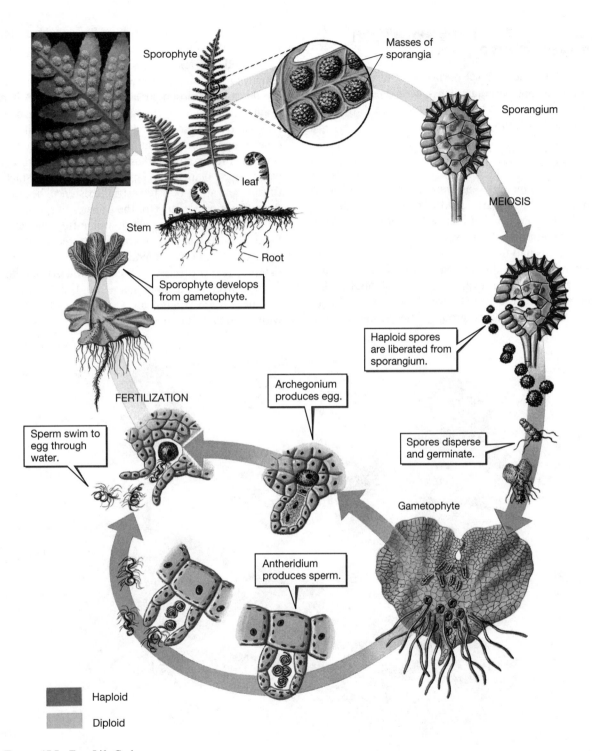

Sporophyte

Masses of sporangia

Sporangium

leaf

Stem

Root

MEIOSIS

Sporophyte develops from gametophyte.

Haploid spores are liberated from sporangium.

FERTILIZATION

Archegonium produces egg.

Spores disperse and germinate.

Sperm swim to egg through water.

Gametophyte

Antheridium produces sperm.

Haploid

Diploid

FIGURE 17.7   Fern Life Cycle.

Exercise 17.3   GYMNOSPERMS: NAKED-SEED PLANTS

The **gymnosperms** (also known as evergreens or conifers) are vascular plants with seeds that are not protected by a fruit. They appeared during the Mississippian era, about 350 mya.

The word *gymnosperm* means "naked seed."

Continental seas were retreating and new masses of higher, drier land were appearing during the late Paleozoic and early Mesozoic eras. With the evolution of the seed, the primitive gymnosperms had the evolutionary innovation needed to fully exploit this terrestrial frontier. The gymnosperms include several groups of plants, but the conifers are the best known. Conifers represent the largest remaining group of the gymnosperms,

and include trees such as pines, firs, sequoias, and redwoods.

Pollen grains represent the immature male gametophyte stage and are composed of two non-motile cells enclosed in a dry outer coat. Pollen grains are produced in huge numbers by the male cones and are disseminated by the wind. Thus, in the case of gymnosperms, the sperm move by floating through the air rather than by swimming through water. These plants overcame the last major obstacle to full utilization of a terrestrial existence, because they do not need water for the transfer of sperm.

Review the life cycle of a pine tree, shown in Figure 17.8. When you look at a pine tree, what you see is the diploid sporophyte. Because of drought-resistant leaves, protective seed coats, reduced gametophytes, airborne pollen grains, and a well-developed vascular system, conifers are a very successful group, with more than 700 species existing today.

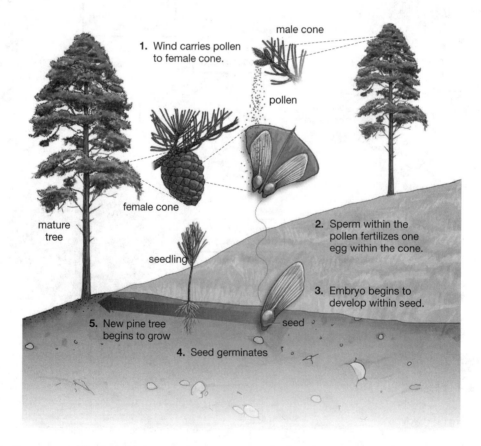

FIGURE 17.8   Gymnosperm Life Cycle. Pine trees have two kinds of cones. The small male cones produce pollen, and the larger female cones produce eggs. When the wind carries pollen onto the female cone, the sperm within the pollen fertilizes one of the eggs within the cone. An embryo then begins to develop inside a seed, which falls to the ground. Once conditions are suitable, the seed germinates and a whole new pine tree begins to grow.

1. If you have pine trees near your campus, study their branches and needles. The waxy surface and needle shape of the leaf minimizes **transpiration**, the loss of water through leaf surfaces. By having less area exposed to the air, desiccation is minimized.

2. Now turn your attention to the male and female pinecones. Carefully separate the scales of a female cone to reveal the **ovule**, or seed. Remove the seed with a pair of forceps. Notice the winglike structure attached to the ovule. Toss two or three seeds into the air and see what happens.

3. Observe a prepared slide of pine pollen grains with a compound microscope.

Questions

1. **Sketch and label a needle, a female pinecone, a male pinecone, and an ovule.**

2. **What is the function of the ovule? What is the function of the wing attached to the ovule?**

3. **Sketch a pollen grain. What is the function of the structures that look like ears?**

4. **Name two major differences between gymnosperms and pterophytes.**

Exercise 17.4 ANGIOSPERMS: FLOWERING PLANTS

Flowering plants exhibit greater diversity than all other plants combined, with 272,000 described species. They exhibit marvellous variety, and their adaptations for succeeding on land are remarkable.

Angiosperms belong to Division Anthophyta. *Anthos* is the Greek term for "flower."

The flowering plants are relative newcomers, first appearing on the scene about 127 mya. When angiosperms were undergoing adaptive radiation, the global climate was becoming cooler, inland seas and swamps were expanding, and large mountain ranges such as the Andes, Himalayas, and Rockies were forming. In 20 to 30 million years (a very short time, geologically speaking), the angiosperms became the dominant terrestrial life-form on the planet, and have remained so ever since. Let's take a look at the adaptations that have enabled them to be so successful.

1. Review the angiosperm life cycle in Figure 17.9. Angiosperms developed a new reproductive structure, the **ovary**, which houses and protects the **ovules** and which become seeds after fertilization. They also developed **flowers** and **fruit**, which facilitated more efficient pollination and seed dispersal. Broad leaves for efficient light collection also enabled the angiosperms to radiate into more species than all other land plants combined **[Krogh Chapters 23 & 24]**.

2. Observe the whole flower and identify the structures listed in Figure 17.10, then sketch them in the space provided in Question 1.

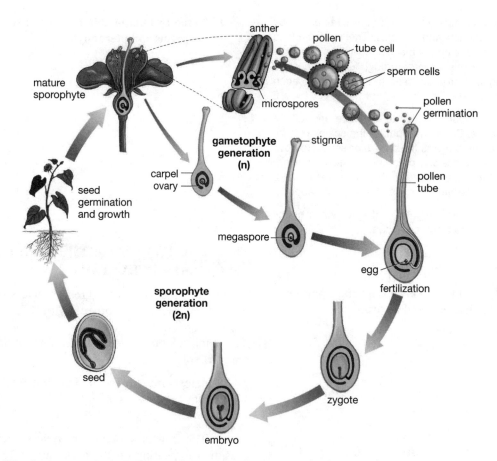

FIGURE 17.9   Angiosperm Life Cycle. The mature sporophyte produces pollen grains on the anther, and a female egg within the ovary. Pollen grains, with their sperm inside, move to the stigma of the plant. There, the tube cell of the pollen grain sprouts a pollen tube that grows down toward the egg. The sperm cells then move down through the pollen tube, and one of them fertilizes the egg. This results in a zygote that develops into an embryo, which is protected by a seed. If the seed ends up in a favorable spot, it will grow into the sporophyte, starting the cycle over again.

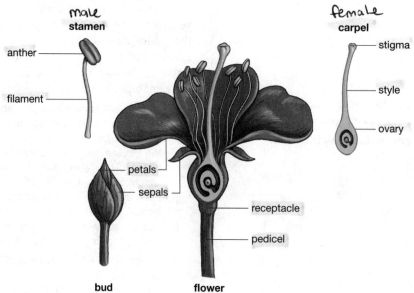

FIGURE 17.10   Parts of a Flower.

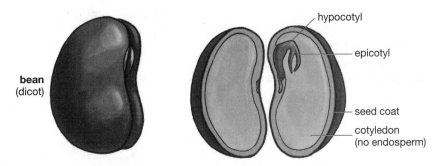

bean
(dicot)

hypocotyl

epicotyl

seed coat

cotyledon
(no endosperm)

FIGURE 17.11   Bean Seed.

3. On most flowers, the sepals are green, but on some flowers, the sepals and petals look alike. The sepals cover the petals before the bud blooms. The petals are nearer the stamen and carpel. Remove the sepals, and then the petals.

4. Remove a stamen, and tap the anther on a white sheet of paper. The powdery substance you see is a collection of pollen grains.

5. Remove the remainder of the stamen. Touch the stigma and note its texture.

6. Using a knife or scalpel, cut the ovary lengthwise, and use the blade to slightly lift the ovules out of the ovary. Look at the specimen with a dissecting microscope.

7. Observe the various food items on display. You may use a knife or scalpel to dissect them if necessary.

8. Examine a bean seed (Figure 17.11). The endosperm provides nutrients to sustain the developing embryo, and it is used by the cotyledons (seed leaves) early in development (Figure 17.12). Then, the cotyledons store nutrients for the young plant, which cannot yet support itself by photosynthesis. The cotyledons are the first (lowest) leaves to appear at germination, but quickly degenerate during the early stage of growth of the plant. The bean seedlings on demonstration illustrate various stages of development **[Krogh Figure 24.39]**.

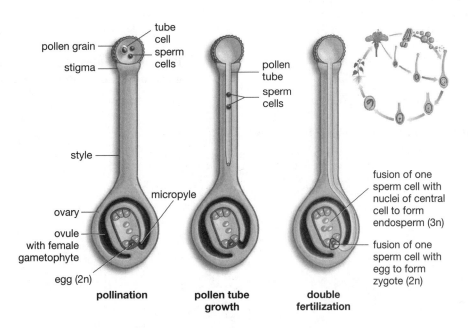

pollen grain

tube cell
sperm cells

stigma

pollen tube
sperm cells

style

micropyle

ovary

ovule with female gametophyte

egg (2n)

fusion of one sperm cell with nuclei of central cell to form endosperm (3n)

fusion of one sperm cell with egg to form zygote (2n)

pollination

pollen tube growth

double fertilization

FIGURE 17.12   Double Fertilization in Angiosperms. When the male pollen lands on the stigma, the tube cell within it sprouts a pollen tube, through which the two sperm pass in moving to the ovule. One sperm cell fertilizes the egg, forming a zygote—the new sporophyte plant. The other sperm cell fuses with the two nuclei of the central cell, forming the endosperm that will later nourish the embryo.

Questions

1. Sketch your flower and label the sepal(s), petals, anthers, and carpel(s) with their stigma, style, and ovary.

2. What does the stigma feel like? Why is this important?

3. What are the round structures that you observe within the ovary? What will they become as they mature? Sketch and label them.

4. What does an ovary become as it matures?

5. What is the function of the sepals and the petals?

6. Are the foods you see fruits or vegetables? How do you know?

7. List the reasons why living on land can be difficult for an organism.

8. Why do you think the angiosperms are the most successful of all the plant groups?

18

Flowering Plants: Structure and Function

Without the gift of flowers and the infinite diversity of their fruits...
man might still be a nocturnal insectivore
gnawing a roach in the dark.

Loren Eiseley, 1957

In this chapter, you will investigate the principal characteristics of flowering plants that enable them to function in their role as primary producers. What enables a plant to absorb and move water? What does a plant use to keep itself from drying out? How do plants react to environmental stimuli?

The sporophyte of flowering plants (Figure 18.1) consists of an aboveground axis called the **shoot** (composed of the stem and leaves) and a belowground axis called the **root**. The shoot functions as a support, transports nutrients, and bears the leaves. The root anchors the plant in soil and absorbs, stores, and transports water and nutrients **[Krogh section 23.2]**.

Angiosperms are divided into two basic groups—the monocotyledons (monocots) and dicotyledons (dicots). The **monocots** have one seed leaf during development; the group includes grasses, grains, palms, lilies, orchids, and others. The **dicots** have two seed leaves during development, and include the broadleaf plants—legumes, herbs, and most deciduous trees and shrubs. Figure 18.2

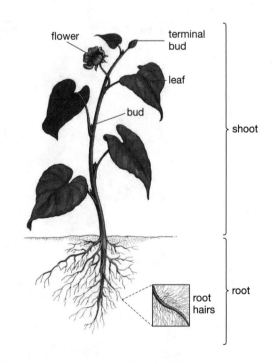

FIGURE 18.1 Anatomy of a Flowering Plant.

199

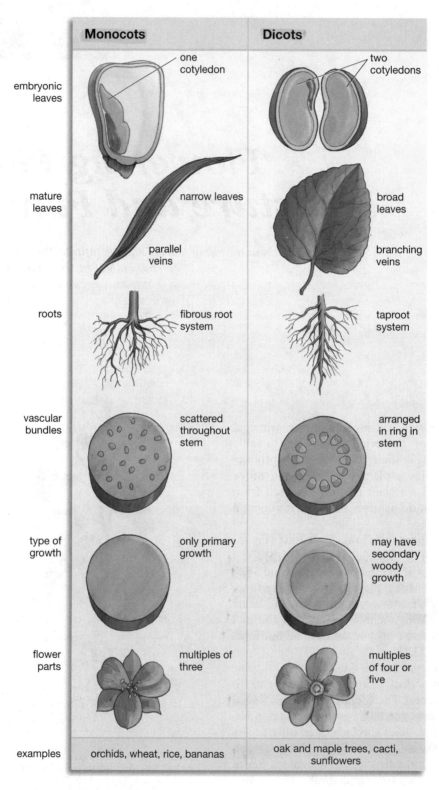

FIGURE 18.2   Characteristics of Monocots and Dicots.

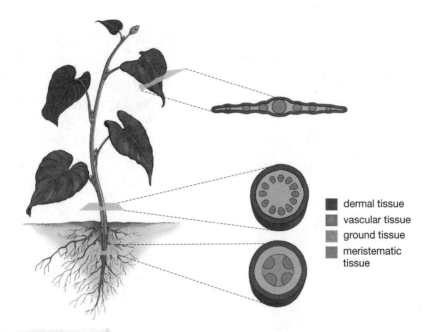

FIGURE 18.3 Four Types of Plant Tissues.

summarizes the fundamental differences between the monocots and dicots.

There are four types of plant tissue—dermal, vascular, meristematic, and ground tissue (Figure 18.3). **Dermal tissue** consists of a single layer of cells that covers the plant and protects it from entry of pathogens and from desiccation. The dermal tissue of plants is analogous to our skin. **Vascular tissue** functions in the transport of water, nutrients, and photosynthetic products. **Ground tissue** essentially fills up space between vascular and dermal tissue. It plays an important role in nutrient storage and makes up the bulk of young plants. **Meristematic tissue** is the site of growth. Recall the plant mitosis slides? Those specimens were onion root tips—meristematic tissue with many dividing cells **[Krogh section 24.3]**.

Exercise 18.1 ROOT ANATOMY

1. Obtain a prepared slide of a cross section of a buttercup *(Ranunculus)* root. Using Figure 18.4 as a guide, locate the vascular tissue in the center of the root.

2. Locate the epidermis, the single layer surrounding the root.

3. All of the cells between the vascular tissue and the epidermis constitute ground tissue (cortex). Starch granules are stored in the ground tissue.

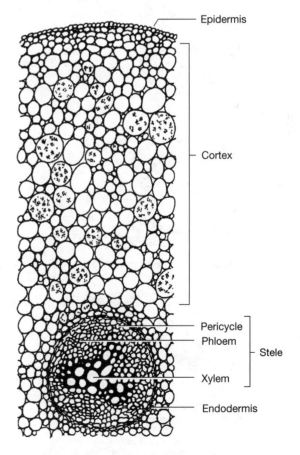

FIGURE 18.4 Herbaceous Dicot *(Ranunculus)* Root Cross Section.

Sketches

1. **Sketch the vascular tissue, ground tissue, epidermis, and starch granules in the buttercup root.**

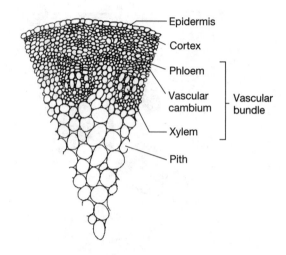

FIGURE 18.6   Dicot Stem Cross Section.

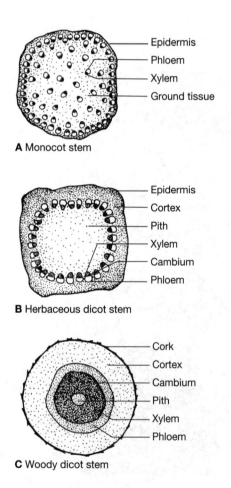

FIGURE 18.5   Arrangement of Vascular Tissue in Various Types of Plant Stems.

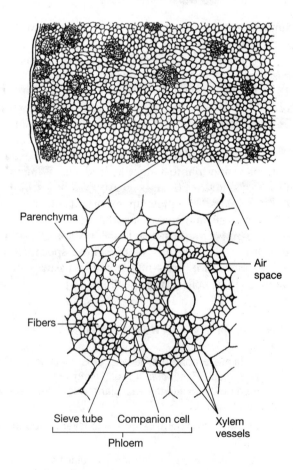

FIGURE 18.7   Monocot Stem Cross Section.

Exercise 18.2   HERBACEOUS STEM ANATOMY

1. See Figures 18.5,18.6, and 18.7. Obtain a prepared slide of a cross section of a sunflower stem *(Helianthus)*, which is a herbaceous plant. Using Figure 18.6, locate the epidermis, vascular bundles containing xylem and phloem, and ground tissues in the stem. Sketch them (see Question 1).

2. Obtain a prepared slide of a cross section of a corn *(Zea)* stem, which is also an herbaceous plant. Using Figure 18.7, locate the epidermis, vascular bundles containing xylem and phloem, and ground tissues in the stem. Sketch them (see Question 2).

Questions

1. Sketch and label the epidermis, vascular bundles containing xylem and phloem, and ground tissues of the sunflower stem. Is the sunflower a monocot or a dicot?

2. Sketch and label the epidermis, vascular bundles containing xylem and phloem, and ground tissues of the corn stem. Is corn a monocot or a dicot?

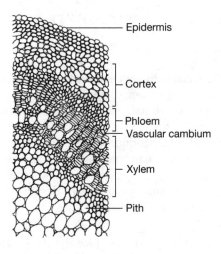

FIGURE 18.8   Young Woody Stem Cross Section.

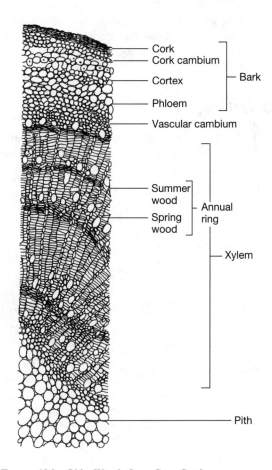

FIGURE 18.9   Older Woody Stem Cross Section.

Exercise 18.3 WOODY STEM ANATOMY

See Figures 18.5, 18.8, and 18.9. In the woody dicots, the stem is divided into **bark, vascular cambium**, and **wood (xylem)**. The bark consists of the **cork, cork cambium, phelloderm**, and **phloem**. During each growing season, new layers of xylem and phloem are produced at the **vascular cambium**. A ring of xylem makes an annual ring in a tree. Each ring is composed of two bands—a lighter colored spring wood and a darker summer wood. As the tree grows, it adds new rings outside the old ones.

Read *A Tree's History Can Be Seen in Its Wood* on **page 515 of your Krogh textbook**.

1. Obtain a cross section of a basswood *(Tilia)* stem, which is a tree. Using Figure 18.8, identify the epidermis, ground tissue, and vascular bundles comprised of xylem and phloem. Sketch them (see Question 1).

2. Examine one of the cross sections of a mature tree. Using Figure 18.9, identify the bark, cork cambium, cortex, phloem, vascular cambium, xylem, and pith.

Questions

1. **Sketch the epidermis, vascular bundles containing xylem and phloem, and ground tissues of the basswood stem. Is basswood a monocot or a dicot?**

2. **How old is your basswood specimen? What part of the stem do you observe to find out?**

3. **How old is the mature tree specimen?**

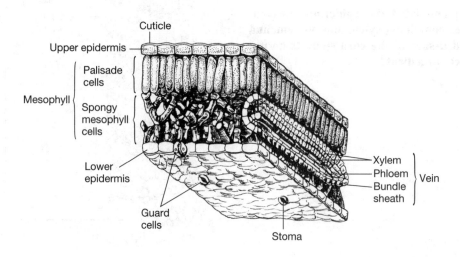

FIGURE 18.10 Leaf Cross Section.

Exercise 18.4   LEAF ANATOMY

[Krogh Figure 23.7]

The function of a leaf is to carry out photosynthesis, but how is a leaf adapted to do so? The leaf is covered on the top and bottom by epidermal tissue (Figure 18.10). The interior portion of the leaf, between the two dermal layers, is referred to as **mesophyll**. Chloroplasts that capture solar energy are packed in structures near the top of the leaf. In the lower part of the leaf, the cells are loosely organized. Water is transported in by vascular tissue, but is lost by **transpiration** when it evaporates, a process similar to perspiration in animals. The waxy cuticle works to keep water in the leaf.

Next, the leaf needs carbon dioxide; but the waxy cuticle restricts the flow of gases. This is resolved by small openings called **stomata** (singular, *stoma*) in the lower side of the leaf. Having stomata in the lower side of the leaf minimizes transpiration. When the stomata are open, carbon dioxide gas enters the leaf, but it cannot be used in its gaseous state. It must first be dissolved in the water that keeps the membranes of the plant cells moist. Oxygen is released across the cell membranes and leaves the plant through the stomata.

Stomata open and close so as to optimize the amount of carbon dioxide entering, while minimizing the amount of water lost to transpiration. This is accomplished by the activity of **guard cells**. When the plant is well hydrated, the guard cells are swollen due to turgor pressure, which keeps the stoma open. When the plant becomes dehydrated, the guard cells become flaccid and the stomata close, conserving water. The plant maintains a delicate balance, obtaining enough carbon dioxide to perform photosynthesis while restricting water loss. Even so, about 90% of the water taken in by roots is released by the plant as water vapor **[Krogh section 23.2]**.

In a growing season, one tomato plant will lose about 125 liters of water to transpiration. A single adult maple tree can lose 200 liters of water in one hour (*Ferry, 1959; Hopkins, 1995*).

1. Obtain a prepared slide of a leaf cross-section and observe it with a compound microscope. Note the cuticle, upper and lower epidermis, mesophyll, and vascular bundles. Depending on where the leaf was cut, you may be able to see the guard cells of the stomata. Sketch these characteristics (see Question 1).

Questions

1. Sketch and label the cuticle, upper and lower epidermis, mesophyll, vascular bundles, and stomata in your leaf.

2. Why is it an advantage for a plant to have chloroplasts in the top layer of a leaf, rather than in the bottom layer?

3. Look again at the upper epidermis and the lower epidermis of your leaf. Which of these seems to have more stomata, and why might this be an advantage for the plant?

Exercise 18.5   A VISIT FROM DARWIN

Extended four-week exercise

When you think about studying the behavior of organisms, the first thing that probably comes to mind is how insects are attracted to light, or how lab mice run though a maze. But plants also exhibit specific behaviors in response to certain stimuli. In this exercise, you are going to examine the behavior of plants in response to light [Krogh section 23.4].

One of the defining characteristics of life is *irritability*, which is the ability to respond to external stimuli. Plants exhibit certain behavior in response to touch, light, and gravity. Such a movement of plants in response to a stimulus is called a **tropism**. The equivalent directional movement of animals in response to a stimulus is called **taxis** (see Chapter 30). A tropism associated with touch is called **thigmotropism**, a tropism associated with gravity is called a geotropism or **gravitropism**, and a tropism associated with light is called a **phototropism**. If the plant moves or grows in the direction of the stimulus, it is said to be a **positive tropism**. If it moves or grows away from the stimulus, it is said to be a **negative tropism**. For instance, a seedling growing toward a light constitutes a positive phototropism.

Charles Darwin is most often associated with his book *On the Origin of Species*, but he also conducted a lot of other scientific research with various groups of organisms, including corals, barnacles, worms, and plants. The lab experiments that you will be conducting in this exercise are based directly on Darwin's study of the movement of plants.

Darwin gave the term *heliotropism* to the phenomenon of plants bending toward a source of light. At the time of Darwin's writing, there was a lot of speculation about the mechanisms responsible for the differential growth leading to the movements of plants. We now know that in most instances, such movement (called differential growth) results from unequal distribution of the plant growth hormone **auxin**. Auxin moves to the shady side of the plant in a positive phototropism. This behavior may also occur because cells on opposite sides of the same plant respond differently, even if they contain the same level of auxin.

1. Read excerpts from Darwin's book, *The Power of Movement in Plants* (1880), in Appendix 2.

2. Work in groups to investigate the effects of light on the growth of sunflower seedlings (*Helianthus*) or another species.

   You will have four weeks to complete your experiments. Write a hypothesis, prediction(s), and design an experiment to test your prediction(s). Don't forget to include control groups and to identify your dependent and independent variables. Be creative in your experimental design—you are not restricted to merely repeating one of Darwin's experiments. You will have the following materials at your disposal, and you may request other materials if you need them:

   young seedlings, light sources of various intensities, a photometer (an instrument used to measure light intensity), boxes and paper punches of various diameters, string, weights, paper, pencils, rulers, and protractors.

3. Set up your experiment. You may be asked to present a report to your classmates to discuss your findings in light of what you learn by reading Darwin; refer to Table 1.1 for guidance in writing a scientific paper. Decide who is responsible for writing each part of the paper.

Experimental Design

1. **What is the effect of sunlight on plant behavior, according to Darwin?**

2. **Hypothesis:**

3. Prediction(s):

4. Independent variable:

5. Dependent variable:

6. Controlled variables:

7. Experimental group:

8. Control group:

9. Do your data support your hypothesis?

10. If your results are different from what you expected, why do you think this is so?

Exercise 18.6   APPLIED BOTANY AND MYCOLOGY

Web exercise

For this Web exercise, you may choose option A, B, or C listed below. Answer only the questions corresponding to the option you choose. Your instructor may allow you to write all your answers on these pages, or may ask you to type and double-space your answers.. Cite your references, using Appendix 1 for proper format.

Option A: Investigate and report on the geographic origin of a plant or fungus used for human consumption, such as wheat, corn (maize), rice, legumes, tomatoes, apples, pears, pineapples, bananas, grapes, citrus fruits, coffee, tea, peppers, various seasonings and spices, cocoa, rubber, cotton, flax, or mushrooms.

Questions for Option A

1. **What is the plant you are investigating? Give its scientific and common name.**

2. **Where did this plant originate?**

3. **Describe how and when these plants "migrated" around the globe.**

4. **How are these plants used by people, and how has their use changed over time?**

5. **Cite your references here:**

Option B: Investigate the origin of a drug, including those that are consumed illegally in the United States. These plants and fungi include—but are not limited to—coca, marijuana, opium poppy, tobacco, mushrooms, and peyote cactus.

Questions for Option B

1. **What drug are you investigating? What is the scientific and common name of the plant this drug is derived from?**

2. **Describe the effects of this drug on the human body.**

3. **Is the use of this drug illegal in its country of origin? Why do you think this is so?**

4. **Is the use of this drug regulated in the United States, and why or why not?**

5. Cite your references here:

Option C: Investigate a plant from which a nonprescription medicinal herb (drug) is derived, such as Jesuit bush, St. John's wort, echinacea, rose hips, wild "Mexican" yams, periwinkle, willow, and others.

Questions for Option C

1. What is the name of the plant you are investigating? Give its scientific and common names.

2. Describe the effects of this herbal medicine on the human body.

3. Is the sale and use of this drug regulated by the U.S. Food and Drug Administration? Why or why not? What risks does it pose?

4. Cite your references here:

19

Invertebrates

The truth is that we need invertebrates but they don't need us.
E.O. Wilson, 1987

In the following chapters, you will examine animal diversity. You will get a tour of the animal kingdom—one that you will not see in any zoo or museum. The study of animals will be approached within an evolutionary context, examining the tremendous diversity of animal life that exists on our planet. This approach allows us to discover what makes an insect different from (and similar to) a worm or a starfish or a human. In this chapter, we will focus on invertebrates, which are animals without backbones.

As we survey some of the better-known of the 36 animal phyla, bear in mind that although animals exist in many forms, we all have a common origin. This is the history of a journey from that common origin to the wonderful diversity of animal forms that we see now. Figure 19.1 is an evolutionary tree showing the relationships of the major animal phyla. It indicates when major evolutionary innovations have arisen.

An evolutionary tree has one serious limitation. As pointed out by Stephen J. Gould (1989), the tree is problematic because it "conveys the impression of continuously increasing diversity and range." There is no more diversity of life on the planet now than has existed during many periods throughout the long history of life. The tree

has always been bushy (more so at some times than others), but as growth proceeds, most of the branches are lost along the way. The lost branches represent the myriad extinct forms. Some are known only from the fossil record; most can only be imagined because they are lost to history forever **[Krogh section 19.1]**.

Exercise 19.1 EMBRYOLOGICAL DEVELOPMENT

Many characteristics separating animal phyla are associated with early embryological development, so you will study embryology in order to understand how phyla differ **[Krogh section 29.1]**.

Shortly after fertilization, an embryo begins dividing by mitosis. The first mitotic divisions are referred to as **cleavage** divisions. The embryo passes through the 2-, 4-, 8-, 16-, and 32-cell stages. A mass of 16 or more cells with no large cavity at its center is called a **morula**. During these early divisions, little growth is occurring. The embryo is not increasing in size, but the number of cells is increasing, doubling at each division. For instance, the 32-cell stage is no larger in diameter than the unfertilized egg.

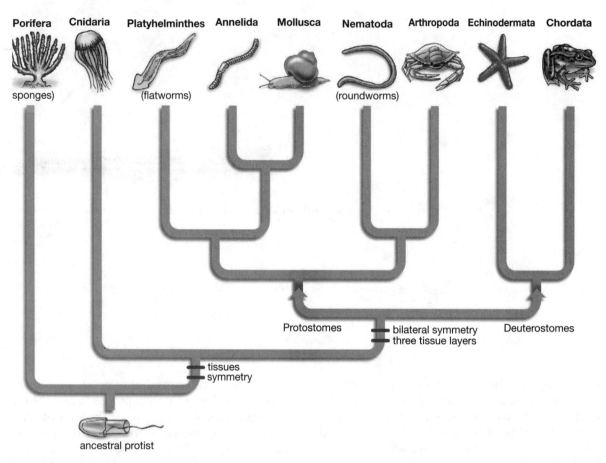

FIGURE 19.1   Evolutionary Tree of the Animal Kingdom. Each branch leads to a different phylum.

As division continues, the cells form a single layer around a hollow cavity. It looks like a hollow ball and is called a **blastula**. The single layer of cells forming the "cover" of this ball constitutes the first tissue layer (germinal layer) known as the **ectoderm**. The hollow portion of the ball is the **blastocoel**. As development proceeds, one side of the blastula begins to invaginate, forming a new cavity called an **archenteron**. This process is called **gastrulation** and the resultant stage is called a **gastrula**. The cells lining the newly formed archenteron constitute the second tissue layer, called the **endoderm**. The opening to the archenteron is the **blastopore**.

In most animals, a third germinal tissue layer arises, called a **mesoderm**. Animals exhibiting only two tissue layers (ectoderm and endoderm) are said to be **diploblastic** (the cnidarians); those exhibiting three layers during development

(ectoderm, endoderm, and mesoderm) are referred to as **triploblastic** (all animals except poriferans and cnidarians).

Triploblastic animals may be divided into two basic groups, based on the developmental fate of the blastopore. In some animals, the blastopore becomes the mouth. These are known as the **protostomes**. In other animals, the anus forms from the blastopore and the mouth forms elsewhere. These are known as the **deuterostomes [Krogh section 22.2]**.

The embryological features that separate most phyla appear at or before the gastrula stage of embryological development. While there are many differences in general appearance and in specific aspects of early embryological development, all animals exhibit these stages—including you, a few (?) years ago.

1. Obtain a microscope slide labeled "starfish developmental stages" or "sea urchin developmental

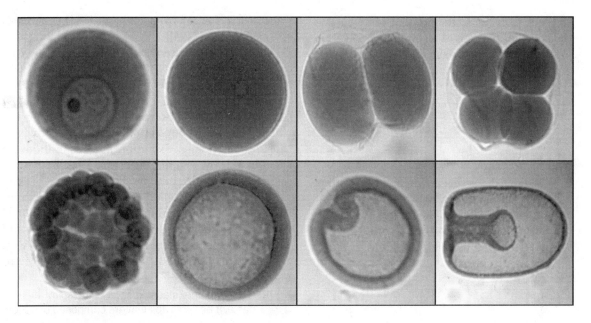

FIGURE 19.2   Stages in the Early Embryological Development of a Starfish.

stages." Study Figure 19.2 as you identify the stages of development.

2. It makes sense to start looking for unfertilized eggs, but they may be difficult to find. An unfertilized egg may be distinguished from a fertilized egg because it has a very distinct nucleus (with a very dark nucleolus in it). These circular structures are lacking in the fertilized egg.

3. Next look for the two-, four-, and eight-cell stages, and a morula. Sketch these and all the following stages as you study them (see Question 1 below).

4. Identify a blastula, and remember that it is a hollow ball, with tiny cells surrounding a space, like a ping-pong ball.

5. Identify a gastrula with its archenteron and blastopore, which is like a ping-pong ball that has an indentation on one side. The archenteron may resemble a mushroom if you happen to be looking at the embryo from its side. A gastrula may also look like a light purple ball with a dark ring inside it; the middle section of the ring is lighter. In this case, you are looking up through the blastopore of the embryo, or down from the top of the embryo.

6. Find and identify an older embryo. The embryos continue to develop into larvae with billowing margins, and you may even notice some specialization of the tissues within it.

Questions

1. **Sketch and label each stage of embryological development, including an unfertilized egg.**

2. **What happens to the size of an embryo as development proceeds through all the stages?**

3. **What happens to the amount of cytoplasm in each of the cells of an embryo as development proceeds through all the stages?**

4. **Why does the archenteron look darker than the rest of the gastrula? Why does it appear as a ring (lighter in the center, not like a solid-color circle) if you are viewing the gastrula from its top or bottom side?**

5. **Most "big" differences separating phyla have to do with embryological features. Why might a "genetic change" or mutation that is manifested in an embryo have a greater impact on the organism than one that is expressed at a later stage of development?**

Exercise 19.2   PHYLUM PORIFERA: SPONGES

(Adapted from Janovy, 1991)

Use caution with bleach. Keep it off your skin and clothes.

Sponges belong to phylum Porifera. Sponges exhibit a cellular level of organization, have no true tissues or organs, and are asymmetrical—with varying size and shape.

For a discussion about symmetry, see **Krogh section 22.2**.

Sponges are unique animals, and zoologists believe that they may have arisen from a protozoan ancestor different from that of all other animals **[Krogh section 22.3]**.

Bodies of sponges are covered with tiny pores opening into a series of canals (Figure 19.3). Water moves into the sponge through **ostia**, into the **spongocoel**, and out through the **osculum**. Food particles are filtered out of the water by the flagellated **choanocytes**, and are digested by cells inside the sponge. Excretion and respiration are facilitated by simple diffusion. The bodies of sponges are stiffened by the presence of tiny skeletal elements called **spicules**, which are composed of calcium carbonate or silicon, stabilized by protein fibers (spongin). The composition and shape of spicules are used to classify sponges. Spicules may be needlelike or may have three to six rays.

1. Examine the glass sponge on display at your table. It has spicules that are composed of silicon.

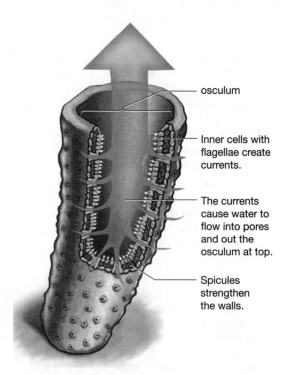

osculum

Inner cells with flagellae create currents.

The currents cause water to flow into pores and out the osculum at top.

Spicules strengthen the walls.

FIGURE 19.3   Anatomy of a Sponge. Sponges filter water through themselves in order to live. Cells on their interior surface have flagella, whose rapid back-and-forth movement draws water in through pores in the sponge's side walls. After filtering this water for nutrients, the sponge then expels it through the osculum. Sponges lack true tissue layers, but possess many specialized cell types. The cells are held together in a stable arrangement by tiny, hard structures called spicules and by the flexible protein called collagen.

2. Examine representatives of the sponge genus *Sycon* with a dissecting microscope. Note the tiny ostia covering the body. Sketch them (see Question 1).

3. Before you can examine the spicules individually, the other tissues must be removed by dissolving them in bleach. Using forceps, tear off a 5-mm piece of sponge belonging to the genus *Leucosolenia* and place it on a microscope slide. Place a drop or two of bleach on the specimen, place it on a hot plate, and allow it to dry. After the specimen is thoroughly dried, place a drop of water on the slide, add a cover glass, and examine it with a compound microscope. Sketch the spicules (Question 3).

4. Repeat Step 3 with a piece of *Scypha* sponge.

Questions

1. **Sketch the ostia of *Sycon* as they appear through your dissecting microscope.**

2. **How does the sponge keep water flowing through its ostia and out through the osculum?**

3. **Sketch and label the different types of spicules in *Scypha* and *Leucosolenia*.**

4. **How are sponges similar to plants? How are they different from plants?**

Exercise 19.3   PHYLUM CNIDARIA: TISSUE TYPES AND *OBELIA* LIFE CYCLE

The phylum **Cnidaria** includes jellyfish, hydroids, sea anemones, and corals. Only the cnidarians are diploblastic, with an ectoderm and an endoderm with several specialized cell types. The phylum gets its name from stinging cells called **cnidocytes**. More than 9000 species of cnidarians are known. They are most commonly found in shallow marine waters, especially in warm temperate and tropical regions **[Krogh section 22.4]**.

Cnidarians exhibit radial symmetry and a tissue level of organization, but do not have organs. They have a gastrovascular cavity, analogous to our stomach, with a single opening that serves both as mouth and anus. Surrounding the opening to the gastrovascular cavity are many tentacles, usually covered with cnidocytes. Each cnidocyte contains a barbed **nematocyst**, a "weapon" that is discharged when a trigger on the cnidocyte is touched (Figure 19.4). Depending on the type, the tentacles may bear barbs, inject poison, or both. The tentacles provide protection

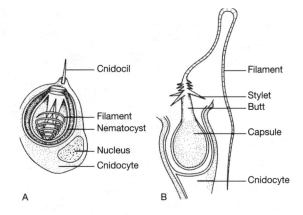

FIGURE 19.4   Cnidarian Stinging Cells. a. Nematocyst within cnidocyte. b. Discharged nematocyst.

and capture food for the organism. Cnidarians feed on a wide range of prey, depending on size, from tiny crustaceans to fish. Humans are refractory to the "sting" of most jellyfish; however, the stings of a few, such as the Portuguese man-o-war (Figure 19.5), can be quite painful and even dangerous.

Cnidarians are responsible for the construction of *coral reefs*, massive mounds of the skeletal remains of billions of individual cnidarians. After a cnidarian colony dies, its soft tissue decays, but its calcium carbonate skeleton remains cemented to its substrate. A new larva may find its way to the old skeleton and use it as its substrate, and so the process of reef building continues for millions of years, and will continue in this manner as long as the substrate remains at a shallow depth. As the

ocean floors shift positions, these coral reefs can gradually sink, allowing new cnidarians to build on top of the old ones, and so great coral walls can slowly form. Alternatively, the ocean floor may be upheaved above the water's surface, exposing the coral formations to the atmosphere. Some U.S. State Parks are built on expanses of fossil coral beds. Charles Darwin wrote about his coral reef observations during his voyage on the *Beagle*.

1. If your instructor has living *Hydra* available, place the specimen in a watch glass and observe it with a dissecting microscope. It may take a few minutes for the animal to relax. When it does, note its tentacles and watch to see if it captures anything to eat. Sketch the specimen (see Question 1 below).

2. Obtain a prepared slide of the longitudinal section of *Hydra* and examine it with a compound microscope. Using Figure 19.6, identify the cell types in the tissue layers.

3. Using Figure 19.7 as your guide, examine a prepared slide of *Obelia* polyps, and a prepared slide of *Obelia* medusae.

4. Pick up the coral specimens on your lab bench. The hard calcium carbonate is secreted by the corals and is all that's left after the animals die. In living coral, polyps anchor themselves in the cup-like **theca**, and share digested nutrients with other polyps in the colony. Note the differences, if any, in the size and shape of the theca among the different colonies and within each colony. The brain coral has a unique arrangement of theca in rows, with space between each row, so that the colony has ridges and valleys that take on the appearance of a human brain.

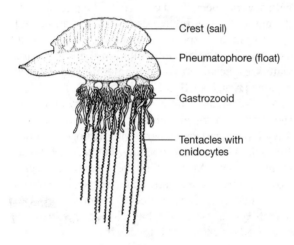

FIGURE 19.5   Portuguese Man-o-war (*Physalia*).

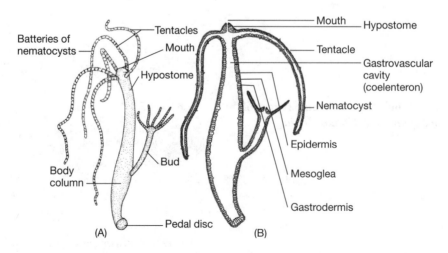

FIGURE 19.6   *Hydra*. a. Whole specimen. b. Longitudinal section.

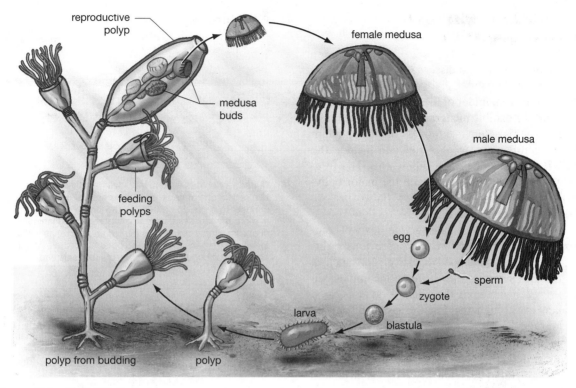

FIGURE 19.7 *Obelia* Life Cycle.

Sketches

1. Sketch the living *Hydra* and label its stalk, mouth, and tentacles.

2. Sketch 2 different coral specimens, emphasizing the arrangement and size of theca in the colonies.

3. How is *Obelia* like a plant? How is it different?

4. How are corals like sponges? How are they different?

Exercise 19.4 PHYLUM PLATYHELMINTHES

The sponges exhibit no discernable symmetry, and the cnidarians exhibit radial symmetry. Representatives of the remainder of the animal kingdom (all triploblastic) exhibit **bilateral symmetry** and have organ systems. Representatives of the phylum Platyhelminthes are acoelomates and include the planarians, trematodes, and tapeworms. The free-living planarians are aquatic, with the majority living in freshwater, but all other flatworms are parasitic **[Krogh section 22.5]**.

Flatworms have a much greater level of complexity than organisms we have previously discussed, with well-developed digestive, reproductive, and nervous systems. They have an incomplete digestive system, meaning that there is no anus. Food is taken in through the mouth, passes through a short esophagus, and is digested in the branched intestine (also called ceca). Undigested food material is regurgitated through the mouth. Respiration is facilitated by simple diffusion. Excretion is facilitated either by simple diffusion, or with excretory cells called *protonephridia*. The flatworms are **hermaphroditic**, meaning that each individual worm has both a complete male and a complete female reproductive system.

It is important for you to remember that development of sensory structures and nervous tissue in the head region (cephalization) are general characteristics of organisms with bilateral symmetry.

Cephalization provides a way for organisms to sense stimuli in a region as they are entering it. Have you ever seen an animal with eyes on its posterior end?

Flatworms have a double ventral nerve cord (Figure 19.8). Ganglia lie in the anterior end. They are analogous to the brain, although it would be overly generous to imply that a flatworm has a brain, let alone two. Note the sensory structures that look like ears, called **auricles**. Also note the **ocelli**, sensory structures that do not form images but serve as photoreceptors, enabling the worms to detect and respond to various levels of light intensity.

Tapeworms are unique enough among the flatworms to warrant separate consideration. First of all, tapeworms are exclusively parasitic,

Flatworm anatomy

Dugesia's nervous system includes primitive eyes and two collections of nerve cells, the cerebral ganglia, that connect to nerve cords that run the length of the animal. In reproduction, *Dugesia* is hermaphroditic, meaning it possesses male sex organs (testes and penis) as well as female sex organs (ovaries and other structures). When two *Dugesia* copulate, each projects its penis and inserts it in the genital pore of the other.

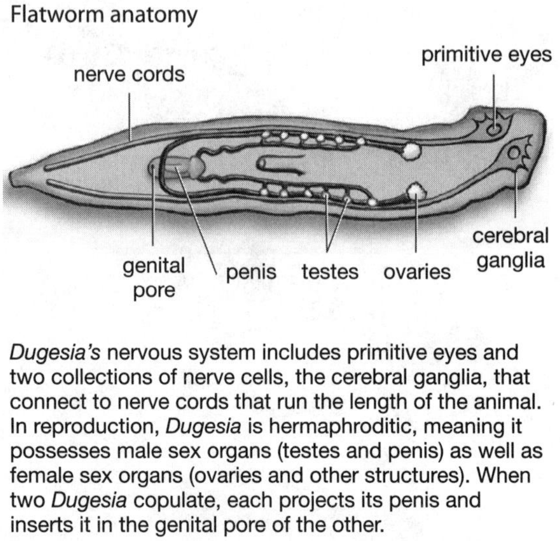

FIGURE 19.8   Body Plan of a Flatworm. *Dugesia's* nervous system includes primitive eyes and two collections of nerve cells, the cerebral ganglia, that connect to nerve cords that run the length of the animal. *Dugesia* is hermaphroditic (monoecious), meaning it possesses male and female sex organs.

so well adapted to their parasitic lifestyle that they have completely lost their mouth and digestive system. They absorb all of their nutrients across their body wall—like an intestine turned inside out. The loss of unused structures is a common theme in natural history. It takes energy to produce and maintain any structure, and if that structure is lost, this energy may be reallocated. This is a general rule in nature—use it or lose it!

Basic tapeworm anatomy is shown in Figure 19.9. Tapeworms have a complex life cycle, and the odds of any given individual surviving are low. To compensate for the overwhelming odds, tapeworms produce thousands of eggs in each **proglottid**. The anterior end of the tapeworm is called a **scolex**; it is armed with hooks, suckers, or both, and is used to attach to the intestinal wall of the host. Proglottids are produced by budding from the neck region. Those proglottids nearest the neck are immature. As they are pushed further back by new proglottids, they mate with proglottids from another worm and eventually fill with eggs. In some species, the egg-filled proglottids break off and disintegrate, releasing eggs that are passed in the feces of the host. An individual tapeworm may

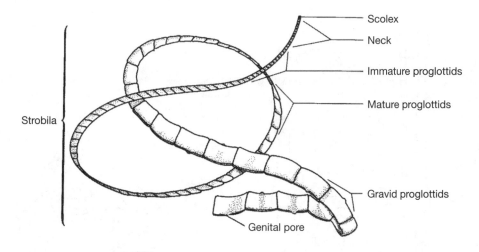

FIGURE 19.9 Tapeworm Anatomy.

live for decades, continually producing proglottids. In other species, whole proglottids are passed in the host's feces. Do you have a cat? See Figure 19.10.

1. Observe movement of living *Dugesia* in a container of water on a white background. Note the trail it leaves behind. The worm secretes mucus, and then uses cilia to glide smoothly along the trail.

2. Obtain a prepared slide of *Dugesia* and look at it with a compound microscope. Locate its pharynx, intestine, ocelli, auricles, and ganglia.

3. Obtain a prepared slide of *Taenia pisiformis*, a dog tapeworm (Figure 19.11). Locate its scolex

with hooks, immature proglottids, mature proglottids, gravid proglottids (filled with eggs), ovaries, uterus, seminal vesicles, and testes.

Questions

1. **Describe the direction and movement of the *Dugesia* specimen.**

2. **Is *Dugesia's* movement most similar to *Euglena* or to *Paramecium* (both are protists)? Why?**

FIGURE 19.10 Proglottids of the Tapeworm *Dipylidium caninum* in Cat Feces. Cat litter pebbles are on the left, and there are five small, white proglottids "crawling" on the feces.

3. **Note that *Dugesia* moves the head first. Why is it an advantage for an organism to**

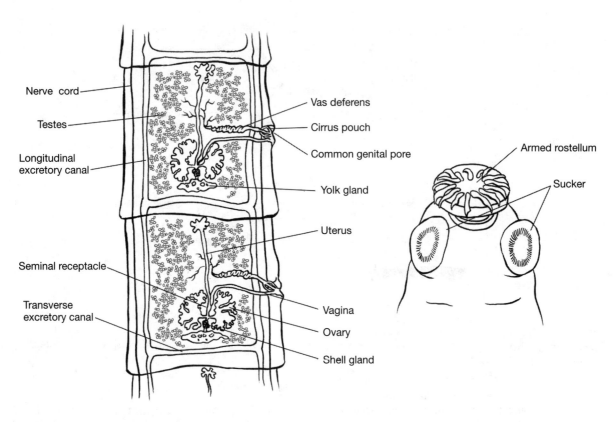

FIGURE 19.11   Mature Proglottid and Scolex of *Taenia pisiformis.* Note that both male and female reproductive organs are there in each proglottid.

have sensory structures on the anterior end, rather than on the posterior end?

4. **What is the potential advantage of losing an organ or organ system that is not being used?**

5. **Few parasitic worms that live inside their hosts have eyespots or pigmentation, yet many in the free-living stages (outside the host) in the life cycle do. Why might this be an advantage for an organism?**

Exercise 19.5 PSEUDOCOELOMATES

(Adapted from Hickman & Hickman, 1993)

The pseudocoelomates comprise several phyla whose evolutionary relationships are not well understood, including Nematoda, Acanthocephala, Nematomorpha, Gastrotricha, Rotifera, and others. Most are small, although a few may exceed a meter in length. Most have a complete digestive system, meaning that it is composed of a mouth, intestine, and anus. Respiration is facilitated by simple diffusion. Excretion is facilitated by simple diffusion or by specialized cells in a canal system. These organisms have a complex nervous system with numerous sensory structures. All have separate sexes and exhibit sexual dimorphism, meaning that you can differentiate the sexes by looking at them. They all have a **hydrostatic skeleton,** in which hydrostatic pressure exerted by the fluid in the pseudocoel gives the body shape and support.

Nematodes are the most abundant pseudocoelomates. These prolific creatures occur in nearly every conceivable environment on the planet. Nematodes are found in both terrestrial and aquatic settings, both freshwater and marine.

It has been suggested that if Earth were to completely disappear leaving only the nematodes, the general contour of the planet would still be visible from outer space *(Cobb, 1915).*

Although most are free living, nematodes parasitize virtually every kind of plant and animal, leading to incredible amounts of crop destruction and serious human disease. Free-living nematodes feed on bacteria, fungi, algae, or small invertebrates, depending on the species **[Krogh section 22.8].**

Representatives of the phylum Nematomorpha (Figure 19.12) are nematode-like worms that are occasionally seen in streams, ponds, puddles, and even in swimming pools and campsite showers. The common name *horsehair worm* came from a belief that these worms arose by spontaneous generation from horsehairs that had fallen into water. Nematomorphs, which are parasites of invertebrates, exhibit a most bizarre life cycle. Most of the worm's life is spent within the invertebrate host, which may be a cricket, grasshopper, crab, or beetle. When the parasite is mature, it fills virtually the entire body cavity of its host. When the host is near water, the worm breaks free, killing its host, leaving

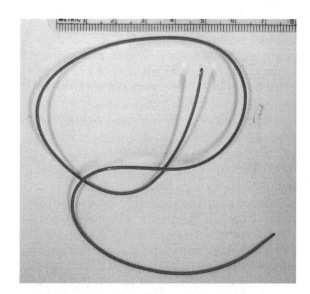

FIGURE 19.12 An Adult Nematomorph.

little more than an empty exoskeleton of the insect behind (Figure 19.13). Nematomorphs are rather unusual in this regard, as most parasites do not kill their hosts. It is not clear whether infection due to the parasite causes the host to seek out water, or whether the parasite "chooses" to burst forth while the insect is near water. The nonfeeding adults mate outside the host, continuing the life cycle.

One common and widely distributed genus, *Gordius,* gets its name from the tendency of two or more worms (one female and several males) to congregate in hopelessly tangled masses.

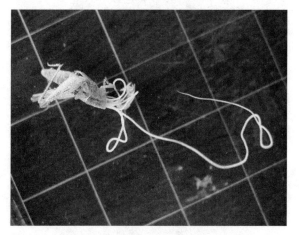

FIGURE 19.13 Adult Nematomorph Emerging from the Body Cavity of a Grasshopper.

The Gordian knot refers to the myth about a rope that was tied by Gordius, King of Phrygia, and could supposedly be untied only by the future lord of Asia. The knot was cut by the sword of Alexander the Great.

After mating, the male dies and the female deposits large numbers of eggs in a mass that swells to a greater size than the worm that laid them. The larval parasites hatch and then are thought to encyst in vegetation. The life cycle is completed when the vegetation and its embedded larval parasite is ingested by an appropriate host.

Another interesting group of pseudocoelomates is phylum Acanthocephala, the thorny-headed worms. About 1,100 species of acanthocephala have been described. As adults, acanthocephalans are parasitic in the vertebrate small intestine. Like the tapeworms, acanthocephalans have completely lost their mouth and digestive system, as an adaptation to a parasitic lifestyle. The hallmark characteristic of the phylum is the **proboscis** covered with hooks and spines, which enables worms to attach to the host intestine.

The acanthocephalan shown in Figure 19.14 belongs to the species *Centrorhynchus robustus* and is from the intestine of a northern spotted owl. You may be familiar with the northern spotted owl because of the controversy about logging of old-growth forests of the Pacific Northwest. Without old-growth forests to provide nesting sites for owls, some ornithologists (biologists who study birds) predict that the northern spotted owl will become extinct. If the owls disappear, their worms will disappear too. When we lose any species on this planet, we are losing more than what meets the eye.

The loss of one species may signal the loss of many others. Often, when a species dies out, symbiotic species that depend on it die out as well. The disappearance of forest habitat has a devastating effect on biodiversity. If an animal as big as the

(a)
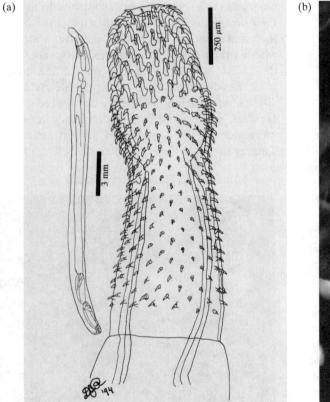

(b)

FIGURE 19.14   a. Proboscis (Spiny Head) of *Centrorhynchus robustus*, an Acanthocephalan that Occurs in the Northern Spotted Owl and the Great Horned Owl. b. Northern Spotted Owl.

spotted owl is dying, how many more species are we losing—ones not so easily recognizable?

Incidentally, the worm drawn in Figure 19.14 was obtained from the intestine of an owl that was found dead under logging debris near Corvallis, Oregon.

1. Make a wet-mount slide from a culture of vinegar eels, *Turbatrix aceti*, and look at the slide with a compound microscope. These nematodes live in vinegar and other fermented fruit juices (wines), and if the vinegar is unpasteurized, it will contain enough yeast and bacteria to support countless worms.

2. Notice the nematodes' serpentine movement, facilitated by longitudinal muscles—they lack circular muscles and so can only whip from side to side, and are not capable of directional movement. Take the cover slip off your slide, add a pinch of sand to the solution, replace the cover slip, and look at them again. Note how they are moving now, using the sand as a substrate for "crawling."

3. If you warm your wet-mount slide, the worms will slow down enough for you to see their internal anatomy. Heat the slide by holding it close to an incandescent bulb, or over a gentle flame for a few seconds.

4. Using Figure 19.15, find the following organs on a worm: mouth, pharyngeal bulb, intestine, and anus. Females are larger than males and give birth to live young. When you find a female, you may see 2–5 larvae coiled up within her uterus. Sketch a female worm.

5. Observe the nematomorph on demonstration. Answer questions 2–4.

6. Now obtain a prepared slide of an acanthocephalan, and observe it with a compound microscope. Note the proboscis with its rows of hooks at the anterior end. The shape and number of hooks are specific for each type of acanthocephalan.

7. Figure 4.7 shows members of two other pseudocoelomate phyla that are abundant in freshwater, Gastrotricha and Rotifera. Observe the living gastrotrichs (*Lepidodermella* sp.) and rotifers (*Philodina* sp.) with a compound microscope.

Questions

1. **Sketch a female *Turbatrix* specimen and label the mouth, pharyngeal bulb, intestine, anus, and uterus with larvae.**

2. **Estimate the length of the nematomorph specimen on display. Give your answer in cm. How much of the host's body cavity volume was probably filled by this worm?**

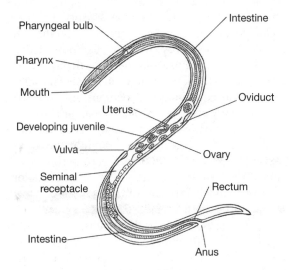

FIGURE 19.15   Female Vinegar Eel, *Turbatrix aceti*.

3. Parasites don't usually kill their hosts. Why is not killing a host an advantage for a parasite?

4. What mechanisms have nematomorphs used in order to overcome their need for a host at that particular stage in their life cycle?

5. Why might the acanthocephalan's hooks be useful? How do you think this worm moves within the intestine of its host?

6. How are the gastrotrich and rotifer similar to the protozoa that you looked at in Chapter 15?

7. What evidence do you see that these are complex multicellular animals?

8. What is the function of the ciliated organ on the rotifer?

Exercise 19.6   PHYLUM MOLLUSCA

Phylum Mollusca includes snails, slugs, clams, squids, and octopuses. There are nearly 50,000 species of molluscs that exhibit tremendous diversity in form, ranging in size from a few millimeters to 20 meters. Molluscs are found in a wide range of aquatic and semiaquatic habitats worldwide [Krogh section 22.7].

Molluscs are complex organisms—they have a complete digestive system, a circulatory system containing blood, a three-chambered heart, kidneys, and a nervous system with many sensory structures, including a complex image-forming eye in the octopus. Advanced sensory structures like this are common in marine animals that hunt for their prey. It is interesting to think that when you are viewing an octopus in an aquarium, that octopus also perceives you in a very real way. Molluscs exhibit a tremendous amount of diversity in feeding habits, and most feed by scraping food into their mouth with a rasping organ called a **radula**. Snails are herbivores or carnivores. Squids and octopus are keen predators, feeding on much larger prey. Bivalves use filter feeding, capturing particles from the water column.

Molluscs interact with humans in many ways. Consider the different types we eat—escargot, oysters on the half-shell, clams, and calamari. Certain snails serve as intermediate hosts for devastating parasites of humans and domestic animals. Recently, tiny zebra mussels have become established in the Great Lakes and the Mississippi and Ohio River drainage systems, where they were introduced by European ships. In the absence of their natural predators, the tiny mussels are flourishing and causing millions of dollars in damage by blocking filter screens of water and sewer systems.

Introduced species such as the zebra mussel tend to out-compete native species, often driving them to extinction.

Molluscs are also important in that the collection of "seashells" offers hours of enjoyment for many people. This activity will provide you with knowledge of basic molluscan shell anatomy. Take a closer look the next time you go to the beach.

1. Following the diagram shown in Figure 19.16, study the major parts on each of the gastropod shells provided. The whelk shell is a gastropod. Locate the **aperture** or large opening: this is the opening through which the body of the mollusc extends. The **siphonal canal** contains the incurrent and excurrent siphons, through which water passes. Molluscs do not shed their shell; it grows with them throughout their life.

2. All gastropod shells are coiled. To determine whether your shell is right-coiled **(dextral)** or left-coiled **(sinistral)**, hold the **spire** (the point) up, with

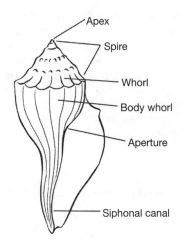

FIGURE 19.16 Gastropod Shell Anatomy.

the aperture facing you. If the aperture is on the right side, your shell is dextral or "right-handed." If the aperture is on the left side, the shell is sinistral or "left-handed."

3. Pick up a bivalve shell and note that the two halves are capable of closing tightly. Using Figure 19.17, find the **umbo**, the oldest part of the shell. Radiating out from the umbo you can find the growth rings or **annuli**, which were secreted by the animal's soft body as it grew. The inside surface of the

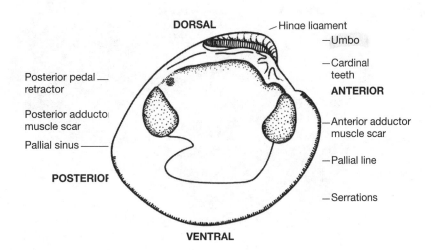

FIGURE 19.17 Bivalve Shell Anatomy.

shell has scars where the soft parts once were. **Adductor muscles** were used to pull the shell halves closed. The circular scars on both ends of the shell indicate where the adductor muscles were attached. The **pallial line** shows where the soft body (mantle) was attached to the shell, and the **pallial sinus** shows where the siphons once were. The **hinge ligament** allowed the shell halves to open, and left a black line near the umbo.

Sketches

1. **Sketch and label the parts of your gastropod shell.**

2. **Sketch and label the parts of the bivalve shell.**

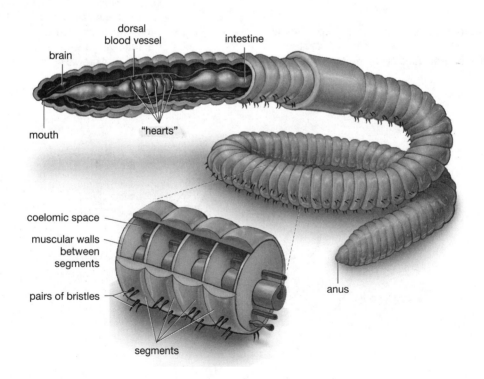

FIGURE 19.18   Earthworm Anatomy.

Exercise 19.7   PHYLUM ANNELIDA

Phylum Annelida includes the earthworms (oligochaetes), leeches (hirudineans), and polychaetes. The annelids exhibit **body segmentation** (Figure 19.18). That is, their body is divided into a series of identical repeated segments. This body plan is considered a major evolutionary innovation, because it provides the evolutionary process "material to work with." One or several segments can become modified for a specific function, and the remaining segments are available to carry out other functions. This makes a pronounced division of labor possible among the various body parts. If the body were composed of only two or three segments, the organisms would not have the luxury of such "evolutionary experimentation." **[Krogh section 22.6]**

Most annelids are marine polychaetes, but you are probably more familiar with their terrestrial counterparts, the earthworms. Earthworms play an important role in aerating and "revitalizing" soil. They eat the soil and then defecate on the surface, bringing up essential plant nutrients from lower levels. An earthworm can ingest up to its own weight in soil every 24 hours *(Hickman and Roberts, 1994)*. Darwin (1881) estimated that every year 10 to 18 tons of dry earth per acre passes through the intestines of earthworms.

Another interesting way in which annelids play an important role for humans is through the use of the medicinal leech *(Hirudo medicinalis)*. For centuries it was believed that "bad blood" was responsible for diseases, and leeches were used in the practice of blood letting. The practice became so popular in Europe during the nineteenth century that the medicinal leech was nearly driven to extinction. The emperor Napoleon imported nearly 6 million leeches in one year to treat his soldiers *(Halton, 1989)*. Today, leeches are commonly used for very legitimate medical applications. Two aspects about the saliva of leeches make them useful. First, they secrete an anesthetic, so that feeding is essentially painless. Their three sharp cutting plates or jaws produce a Y-shaped cut from which they suck blood (Figure 19.19). Typically, a single medicinal leech will ingest 10 to 15 cc (ml) of blood at each feeding. Second, they secrete an anticoagulant called **hirudin** that prohibits the clotting of blood. Wounds continue to ooze blood for up to 24 hours after application. Leeches are used to stimulate circulation and to drain off unwanted blood in an injured or reattached body part *(Halton, 1989)*. See Figure 19.20.

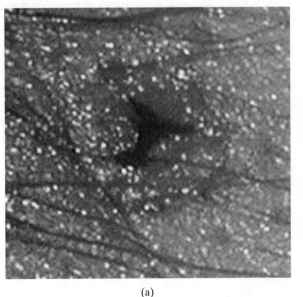

(a)

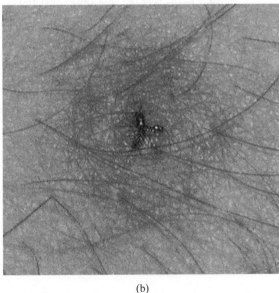

(b)

FIGURE 19.19   a. Y-shaped Wound Produced by the Cutting Plates of a Medicinal Leech. The leech that produced this lesion fed on the author's arm for approximately one hour. Note the circular impression left by anterior sucker. b. Leech Wound 48 Hours after Feeding. Note bruising.

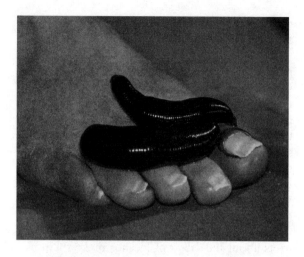

FIGURE 19.20   Leeches removing excess blood from injured tissue. Leeches are commonly used for such medical applications.

1. Obtain a living or preserved specimen of each of the three major groups of annelids—oligochaetes, polychaetes, and leeches. Observe each with a dissecting microscope and note their general differences, such as external structures found in one but not in the others.

Questions

1. **Sketch the three different annelid specimens, and label the external anatomy of the earthworm.**

2. **How is the external morphology of the three annelids adapted to their habitats?**

Exercise 19.8   PHYLUM ECHINODERMATA

The phylum Echinodermata includes starfish and brittle stars, sea urchins, sea cucumbers, sand dollars, sea lilies, and feather stars. These organisms are unique because the adults exhibit radial pentamerous symmetry, a type of radial symmetry in which the rays or arms occur in multiples of five. Adults exhibit radial symmetry, but the larvae are bilaterally symmetrical. There is strong evidence suggesting that echinoderms share a common ancestor with the chordates. We will investigate this in Chapter 21 **[Krogh section 22.10]**.

Echinoderms possess a unique water vascular system (Figure 19.21). Water enters the water vascular system through a sieve-like opening called the **madreporite**, passes through a canal system, and then moves down the **rays**. The canal system is connected to **tube feet**, which are equipped with internal bulb-like structures that work like the bulb on a pipette, creating suction that allows the tube feet to adhere to substrates.

1. Obtain a preserved starfish, and locate the rays, madreporite, mouth, and tube feet. Sketch them (see Question 2).

Questions

1. **How many rays does the starfish have?**

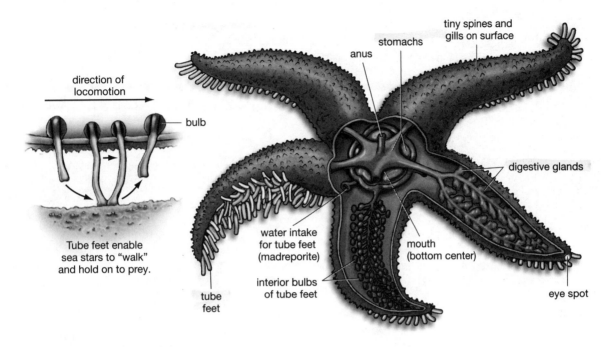

direction of
locomotion

bulb

Tube feet enable
sea stars to "walk"
and hold on to prey.

tube
feet

anus

stomachs

tiny spines and
gills on surface

digestive glands

water intake
for tube feet
(madreporite)

interior bulbs
of tube feet

mouth
(bottom center)

eye spot

FIGURE 19.21   Starfish Anatomy.

2. **Sketch the starfish and label the rays, madreporite, mouth, and tube feet.**

3. **How are echinoderms similar to and different from sponges?**

20

Arthropods

Watch where you step. Be careful of little lives.
E.O. Wilson, 1991

Phylum Arthropoda is so impressive that its members are being addressed in their own chapter. These invertebrates are the most extensive in the animal kingdom both in the numbers of species and in the sheer numbers of individuals represented. More than 75% of all known animal types belong to phylum Arthropoda—shrimp, crabs, spiders, ticks, scorpions, centipedes, millipedes, insects, and many others. More than 900,000 species have been described and many, many more remain undescribed. In addition, because of their hard exoskeleton, the ancestral arthropods left a rich fossil record. Table 20.1 provides the classification of some arthropods. As you complete the exercises in this chapter, write the characteristics of each animal group in the table **[Krogh section 22.9]**.

Arthropods are of considerable importance to humans. Certainly shrimp, crayfish, crabs, and lobsters are considered by many to be some of the finest cuisine available. In addition to enriching our lives as food, arthropods are essential in the pollination of most of our food crops. They also provide useful medicines, dyes, and silk. On the other hand, arthropods compete with humans for food, and we spend millions of dollars per year to kill them. Arthropods also serve as vectors (carriers) for the most serious diseases known to humans—plague, malaria, yellow fever, African sleeping sickness, river blindness, and Lyme disease, to name a very

few. Today, Americans are fighting the spread of diseases such as West Nile virus and eastern equine encephalitis, both vectored by mosquitos.

Like annelids, arthropods exhibit a segmented body plan and conspicuous segmentation. See Figure 20.1. On the prototypic (primitive) body plan, each segment bears a pair of **jointed appendages**. Arthropods have a hard **exoskeleton** composed primarily of **chitin**. This hard exoskeleton is beneficial because it protects the soft body underneath. Unlike the shell of a mollusc, the

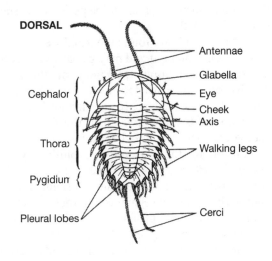

FIGURE 20.1   A Trilobite.

231

TABLE 20.1 ARTHROPOD CLASSIFICATION		
Taxon	**Representatives**	**Characteristics**
Subphylum Trilobita	Trilobites	
Subphylum Chelicerata		
Class Merostomata	Horseshoe crabs	
Class Arachnida	Ticks, mites, spiders, scorpions	
Subphylum Crustacea	Shrimp, crabs, lobster, barnacles	
Subphylum Uniramia		
Class Chilopoda	Centipedes	
Class Diplopoda	Millipedes	
Class Insecta		
Order Coleoptera	Beetles	
Order Hymenoptera	Bees, wasps, ants	
Order Orthoptera	Grasshoppers, crickets, roaches	
Order Lepidoptera	Butterflies, moths	
Order Hemiptera	True bugs	
Order Isoptera	Termites	

exoskeleton of an arthropod does not grow with the animal; therefore, it has to be periodically shed to permit growth.

Arthropods are found in virtually every conceivable environment and may be carnivores, omnivores, or symbiotic. But most of them are herbivorous—terrestrial forms feed on plants, and aquatic forms eat algae. In the animal kingdom, the arthropods are unrivaled in the diversity of ecological distribution. Two features are primarily responsible for the tremendous evolutionary success of the arthropods—the hard exoskeleton providing protection, and the fusion of segments into **tagma**. The fusion of segments into these functional units facilitated a pronounced division of labor among different regions of the body. Through **tagmatization**, the arthropods have capitalized on the basic segmented body plan.

Exercise 20.1   SUBPHYLUM TRILOBITA: TRILOBITES

Obtain the fossil remains of a trilobite from your instructor. Please be careful with the fragile specimen. You are holding a piece of *natural history* in your hand. The trilobites (Figure 20.1) first appeared in the late Precambrian Era, and flourished during the Cambrian and Ordovician periods. Trilobites disappeared at about the same time that the first dinosaurs appeared, 200 mya in the Triassic period.

Stephen Jay Gould *(1989)* said of these early invertebrates, "They are grubby little creatures of a sea floor 530 million years old, but we greet them with awe because they are the Old Ones, and they are trying to tell us something."

Aside from the sheer wonder of trilobites, we examine them because they are a good example of the basic arthropod body plan, with their repeated segments. If you have an exceptionally good specimen you will notice a pair of appendages on the ventral side of each segment.

Sketches

1. **Sketch and label the eyes, segments, and appendages of your fossil trilobite.**

2. **What else was happening while the trilobites lived? Refer to the geologic time scale on the inside back cover. Imagine a trilobite as a living organism on a Cambrian sea floor, 500 mya.**

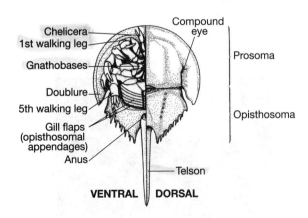

FIGURE 20.2 Ventral and Dorsal Views of a Horseshoe Crab (*Limulus*).

Exercise 20.2 SUBPHYLUM CHELICERATA: HORSESHOE CRABS AND SPIDERS

There are only five species of horseshoe crabs (Class Merostomata) in existence today. The horseshoe crab is said to be a *primitive* animal (Figure 20.2). What does a primitive animal look like, anyway? The word **primitive** simply means "unchanged." The fossil record suggests that living horseshoe crabs are very similar to those that lived during the Triassic period, more than 200 mya.

The class Arachnida includes the mites, ticks, scorpions, and spiders. The body is divided into a cephalothorax (fused head and thorax) and abdomen (Figure 20.3). Among the public at large, the arachnids are not a terribly popular group of organisms. It is difficult to find a redeeming quality in a tick. Spiders, on the other hand, often get a bad rap, and the popular portrayal of spiders in movies like *Arachnophobia* hasn't done much for their reputation. Innate fear of spiders and other animals that "are most readily evoked by sources of peril that have existed in the natural world throughout humanity's evolutionary past" is part of the human makeup (Wilson, 2002). This *biophobia* is the antithesis of biophilia discussed in Chapter 2. But spiders are actually beautiful and fascinating creatures. Most are very beneficial to humans because they feed on many less desirable critters. So, instead of smashing spiders that you find in your house, admire them and then let them go about their business.

There are a couple of exceptions, however; two species of spiders found in the continental United States are considered dangerous. The brown recluse (*Loxosceles reclusa*; Figure 20.4) occurs throughout the southeastern United States, and into the Midwest. These spiders are common in human dwellings and are easily recognized by the violin-shaped marking on the dorsal side of the cephalothorax. The venom of the brown recluse and related spiders is **hemolytic**, which causes tissue necrosis (rotting) at the site of the bite. A bite from the brown recluse can be disfiguring and even fatal, if the necrosis isn't treated properly. Occasionally, claims are made of brown recluse bites occurring outside the spiders known geographic range. It is likely that such cases are

 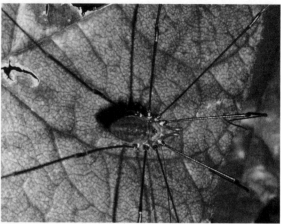

FIGURE 20.3   a. The Cellar Spider (*Pholcus phalangioides*) is one of the most common spiders of human dwellings. Its common name is daddy long legs, a name often applied to other organisms, including the harvestman b. which is not a spider. An urban legend sometimes associated with both these organisms is that they have the most deadly venom known, yet there is no scientific evidence to support this claim. They are completely harmless.

bites of the "yellow sac spider" (*Chiracanthium mildei*; Figure 20.5) that cause similar lesions as the brown recluse but are much less serious. Spielman and Levi (1970) found that *C. mildei* was the most common spider in houses in the Boston area. It is important to note that bites by the yellow sac spider are very rare.

The other dangerous spider that occurs in the United States is the black widow (*Latrodectus mactans*). The female black widow is usually easy to spot because of the bright red hourglass-shaped marking on the ventral side of her abdomen (Figure 20.6). The black widow is distributed widely throughout the United States, but it is much more common in the western, southern, and southeastern states. The bite

of the black widow is usually much more serious than that of *Loxosceles*, because the venom is a **neurotoxin** (affecting nerves) and has a systemic effect.

1. Examine a horseshoe crab on display, a representative of the genus *Limulus*. Note its eyes, carapace, four pairs of walking legs, pedipalps, chelicerae, gills, and telson.

2. Using Figure 20.7, examine some of the spiders on display. If the specimen is small, use a dissecting microscope to view its appendages. Note the four pairs of walking legs on the spiders' thorax, one pair of pedipalps, and one pair of chelicerae on their heads, which help to stuff food in their mouths. The chelicerae bear fangs that inject poison into the spider's prey.

 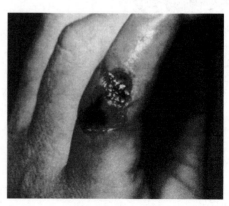

FIGURE 20.4   a. A Brown Recluse (*Loxosceles reclusa*) with its Violin-shaped Marking. b. Necrosis resulting from a brown recluse bite.

3. Examine the brown recluse specimen, and note the "fiddle" shape on the dorsal side of its cephalothorax.

4. Now look at the black widow, and look for the "hourglass" shape on the ventral side of its abdomen. In live specimens, the hourglass is red, but on preserved specimens it often appears white. The black widow's fangs are much larger than those of the brown recluse.

Sketches

1. **Sketch and label the cephalothorax, abdomen, eyes, walking legs, pedipalps, and chelicerae of a spider on display.**

FIGURE 20.5   A Yellow Sac Spider (*Chiracanthium mildei*)

FIGURE 20.6   A Female Black Widow (*Latrodectus mactans*) with her Hourglass-shaped Marking.

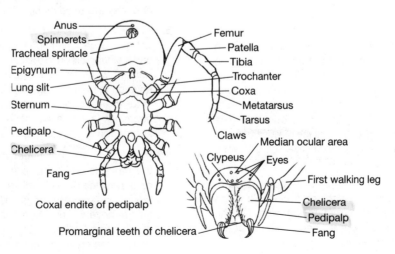

FIGURE 20.7   Spider Anatomy.

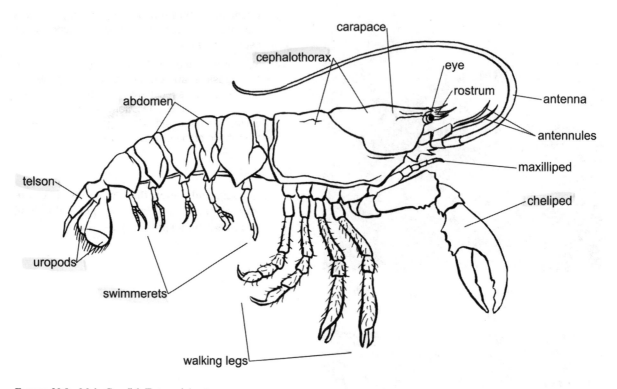

FIGURE 20.8   Male Crayfish External Anatomy.

Exercise 20.3   SUBPHYLUM CRUSTACEA: CRAYFISH DISSECTION

Wear gloves and goggles.

Subphylum Crustacea includes arthropods that are commonly considered to be good to eat (lobster, shrimp, crab, crayfish), as well as many others. A crayfish illustrates the concept of tagmatization very well and provides a good representative crustacean for dissection.

1. Using Figure 20.8 as your guide, study the external anatomy of the crayfish. Before proceeding with the dissection, spend a few moments observing live crayfish, if they are available. Note the hard exoskeleton covering the crayfish. The body is divided into two major regions—the **cephalothorax** (fused head and thorax) and **abdomen**. The **carapace** covers the gills that the crayfish uses for respiration. Note the **rostrum** extending beyond the eyes.

2. Now we will look at the various appendages of the crayfish. Think about the pronounced division of labor the various structures provide. On the head you will find the eyes and two pairs of antennae. These are sensory structures.

3. Behind the antennae, you will see the three pair of **maxillipeds** surrounding the mouth. These aid in feeding. The first pair of walking legs bears large pincers and are called **chelipeds**. These are used in food handling and defense.

4. Behind the chelipeds are four more pairs of walking legs. If you turn the crayfish over and view the ventral side, you will see a pair of **swimmerets (gonopods)** on each abdominal segment. In the male, the first pair of swimmerets is enlarged, and they are used as copulatory organs.

5. At the very posterior portion of the abdomen, you will see large fan-shaped **uropods** on each side and a single **telson** in the center. These function as rudders when swimming and in the protection of eggs on the female. You will note the anus just anterior to the telson. Using forceps, remove all appendages from one side of the crayfish, place them in order on a piece of paper, and label each part.

6. Insert your scissors under the posterior end of the carapace, and make a longitudinal incision on either side of the "midline" of the carapace extending nearly to the eyes. Using forceps, carefully remove the pieces of the carapace. Expose the "feathery" **gills**, which are the respiratory organs of the crayfish. Using Figure 20.9, examine

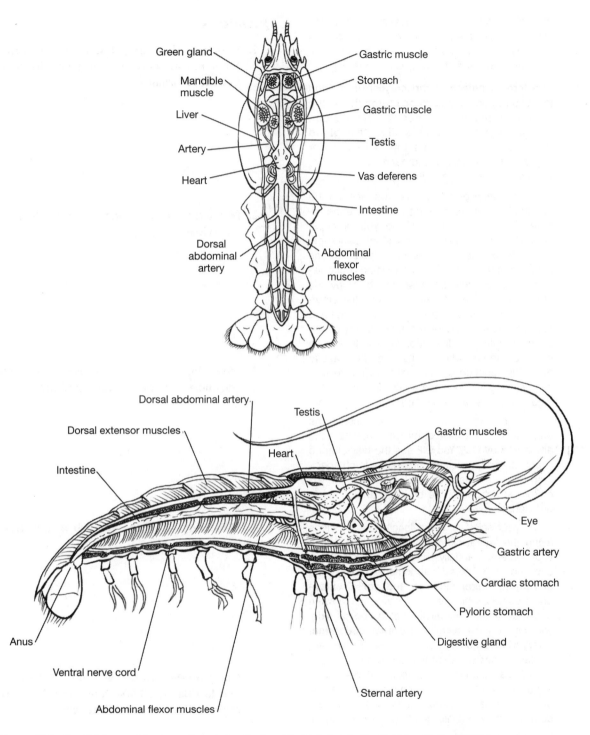

FIGURE 20.9   Crayfish Internal Anatomy. Dorsal view (top) and Lateral view (bottom). (After Huxley, 1880.)

the internal anatomy of the crayfish. Remove the gills to expose the internal thoracic anatomy of the crayfish. Note the large muscles that move the mandibles when the crayfish is "chewing."

7. The large stomach of the crayfish is divided into two regions, the **cardiac stomach** and the **pyloric stomach**. On either side of the stomach you will see the large **digestive gland** (hepatopancreas). This secretes digestive enzymes into the pyloric stomach. It also functions in the storage of nutrients.

8. In the posterior portion of the carapace, on the dorsal side, you will see the **heart**. Note the arteries leaving the heart. Using your forceps, carefully remove the heart and hepatopancreas. This will expose the intestine, which extends back out of the pyloric stomach. Using your scissors, sever the intestine just behind the pyloric stomach. Now very carefully remove the stomach. Using your scissors, open up the gastric stomach. Carefully feel the inner lining of the stomach with your fingertip. You will find three chitinous teeth, which constitute the **gastric mill**. The function of the gastric mill is to mechanically grind food materials, facilitating the digestive process, just like our chewing of food.

9. After removing the stomach, look inside the body cavity at the base of the antennae near the anterior body wall. You will see the disc-shaped **green glands**, which are green in a fresh specimen (not in a preserved one). These are the excretory structures of the crayfish, analogous to our kidneys.

10. Using your scissors, make longitudinal incisions along the lateral sides of the abdomen. Using your forceps, carefully remove the exoskeleton covering the dorsal abdomen. Notice the **abdominal flexor muscles and dorsal extensor muscles**, which enable the crayfish to swim backward. Along the "midline" of the dorsal side, you will see the prominent **dorsal abdominal artery** lying just on top of the intestine. Shrimp are similar to crayfish; when you eat shrimp you are actually eating abdominal flexor and dorsal extensor muscles. When the shrimp are advertised as being "deveined," this means that the dorsal abdominal artery and intestine have been removed.

11. Using your forceps, very carefully remove the pieces of exoskeleton from the ventral portion of the abdomen. This will reveal the **ventral nerve cord** that is typical of an invertebrate nervous system. Follow the nerve cord up through the body as far as you can.

Questions

1. How does the aggregation of sensory structures in the head region (cephalization) benefit an organism?

2. Name the sensory structures on the head of a crayfish.

3. How does the fusion of body segments (tagmatization) benefit an organism?

4. What regions of the crayfish demonstrate tagmatization?

5. Compare the external anatomy of the crayfish to that of a horseshoe crab. What are their similarities and differences?

Subphylum Myriapoda

Exercise 20.4   SUBPHYLUM UNIRAMIA: THE MILLIPEDE

Millipedes (Figure 20.10) and centipedes (Figure 20.11) belong to a group of arthropods know as the myriapods (Class Diplopoda and Class Chilopoda, respectively). They have only two tagmata, a head and a trunk. Centipedes have only one pair of legs per segment, and millipedes typically have two pair per segment. **[Krogh Figure 22.21]** Centipedes are predators, feeding most commonly on other small invertebrates, but millipedes are herbivores. The venom of most centipedes does not affect humans, but by now you have learned that generalizations are almost always false—the bites of a few centipedes can be dangerous.

1. Observe the millipedes available in your lab, or if you don't have one available, look for a smaller, native specimen in moist leaf litter outside your building. Notice the millipede's hard, segmented exoskeleton. Handle it carefully, because a fall to the floor could seriously injure it.

2. Notice that the millipede has two pairs of legs per segment. The name **millipede** literally means "thousand legs." Estimate the number of legs on its body. Notice the graceful, rhythmic movement of the legs as it crawls. Be sure to wash your hands after handling the millipede, because it can secrete a chemical that irritates your eyes if you rub them.

Chilopoda

Diplopoda

(a)

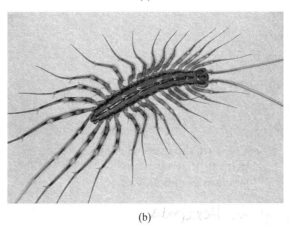

(b)

FIGURE 20.11   a. A Garden Centipede (Lithobius forficatus). Centipedes have one apir of legs per segment. This type is found among leaf litter or other decaying vegetation. b. A House Centipede (Scutigera coleoptrata). This species is commonly found in human dwellings.

Questions

1. **How do centipedes and millipedes provide good examples of the basic arthropod body plan?**

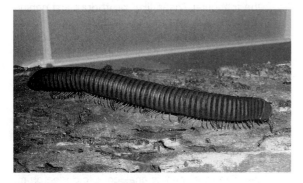

FIGURE 20.10   A Millipede Commonly Found in the Southeastern United States. Millipedes have two pairs of legs attached to each segemnt.

2. **Sketch your millipede and label the antennae, eyes, head, trunk, and legs.**

3. **Compare the external anatomy of the millipede to the external anatomy of an earthworm. What are their similarities and differences?**

4. **Compare the external anatomy of the millipede to that of a crayfish. What are their similarities and differences?**

Subphylum Hexapoda
- insects (Not centipedes : millipedes)
- 200 mil | per each human present
- Only invertebrate that flies
- Class Insecta
    - Metamorphis

Exercise 20.5   SUBPHYLUM UNIRAMIA: INSECT ANATOMY

No longer used

(After Duncanson, 1895)

The insects constitute, by far, the most extensive class in the phylum Arthropoda. More species of insects have been described (800,000) than species of all other animals combined. In sheer numbers of individuals, the mighty nematodes or planktonic crustaceans are their only potential rivals, although these rivals do not exhibit a diversity that is anywhere near that found in the insects. Among the insects, the beetles are by far the most successful. Of all animal species discovered so far, one in five is a beetle!

Insects are characterized by having a distinct head, thorax, and abdomen. The head consists of 5 fused segments, and they typically have 3 pairs of legs and 2 pairs of wings attached to the thorax; however, some insects have one pair of wings or no wings at all. Spend a few moments observing the behavior of living grasshoppers or other insects in your classroom terrarium, and then begin examining the specimens provided.

Wear gloves and goggles.
---

1. Using Figure 20.12 as your guide, study the external anatomy of the grasshopper. Notice the general outline of the body, both from a dorsal and lateral aspect. Note the division of the body into a **head, thorax**, and **abdomen**.

2. Note the antennae of the grasshopper. How many segments comprise the antenna? Note the position and shape of the **compound eyes**. Examine one of the compound eyes at the highest magnification possible under a stereoscopic microscope. Estimate the number of facets (divisions) in one compound eye.

3. Search for a pair of **simple eyes**, or ocelli, located in front of the compound eyes. Can you find other simple eyes?

4. Insects have evolved to feed on virtually everything imaginable. Observe the mouthparts of the grasshopper, which are modified for chewing. Note the moveable flap in front, called the **labrum**. Remove the labrum by cutting along the hinge. Observe the hard, jagged jaws called **mandibles** underlying the labrum. Manipulate the mandibles with your forceps to observe their "operation." Remove the mouthparts and try to reconstruct them in the proper order, on white paper.

5. Note the general form of the thorax. You will find the triangular, "cape-like" **prothorax** just behind the head. Carefully examine the legs. With your forceps, remove the legs on one side. Identify the parts of the legs, including the **tarsal hooks** (claws).

6. Spread the wings on the right side. Note the venation pattern of the wings.

7. Observe the general form of the abdomen, and count the abdominal segments. On the side of the first abdominal segment, you will find the **tympanum**. This is the auditory organ of the grasshopper. The tympanum is analogous to our eardrum. Observe the series of **spiracles** lining the lateral side of the abdomen. This allows air to enter the respiratory system of the grasshopper. They are

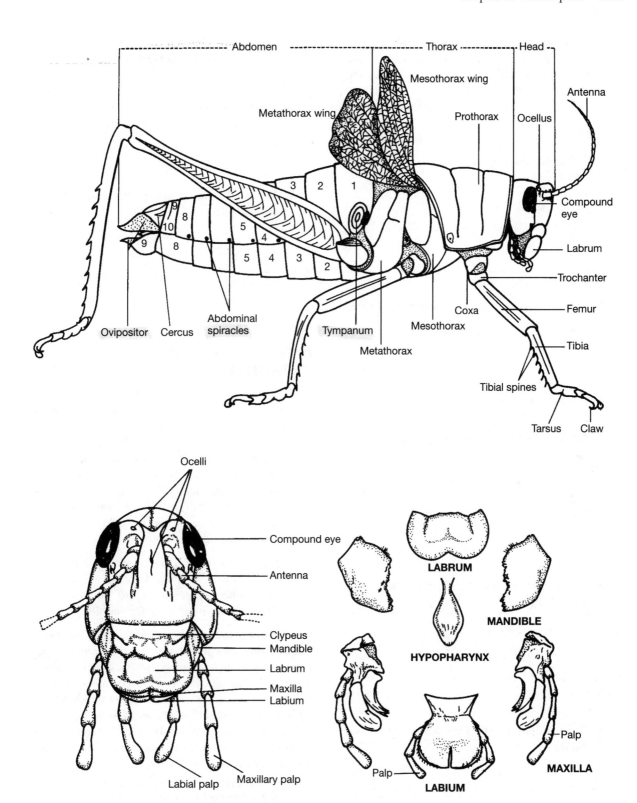

FIGURE 20.12 External Anatomy of the Grasshopper. Lateral view (top). Head and mouthparts (bottom).

connected to a complex system of air tubes called **tracheae**. The abdomen of the female ends in two pairs of appendages comprising the **ovipositor**, which is an egg-laying apparatus.

Questions

1. **What are the primary sensory structures of the grasshopper?**

2. **Where are the sensory structures located?**

3. **What can you conclude about the grasshopper's range of vision?**

4. **What might be the specific function of each pair of legs?**

5. **What might be the function of the tarsal hooks?**

6. **Of the two pairs of wings, which would be most useful in flight? Why?**

7. **What is the primary function of the forewings?**

Exercise 20.6   INSECT METAMORPHOSIS

Most insects undergo **complete metamorphosis**, which has egg, larva (with several substages called instars), pupa, and adult stages. The hard exoskeleton of insects must be periodically shed to allow for growth, and the substages between molts are the instars. In Chapter 10, you may have studied genetics using the fruit fly *Drosophila melanogaster*. Review the fruit fly life cycle, which is an example of complete metamorphosis, in Figure 10.1. Butterflies also undergo complete metamorphosis.

Larvae have structures that are not found in the adult, such as prolegs (Figure 20.13), whereas juveniles are immature stages that are like "miniature adults."

Most of the remainder of insects undergo **incomplete metamorphosis**, which lacks a pupal stage. In these terrestrial insects the larvae are called nymphs, and aquatic larvae are called naiads. A very few primitive insects undergo **direct development** with an egg, juvenile stages, and adult. A juvenile possesses no unique larval structures. An example is a silverfish, whose juvenile stages look just like the adults, except for their size (Figure 20.14).

1. The life cycle of the grain beetle *Tenebrio molitor* and a dragonfly are on display. Compare the specimens and note the differences between the stages of complete and incomplete metamorphosis.

FIGURE 20.13 Life Cycle of the Luna Moth: Pupa, Adult, and Larval Stages. (Egg Not Shown.) Note the prolegs on the caterpillar, which are called "larval structures" because they are lacking in the adult months.

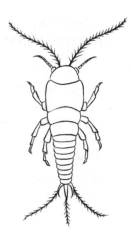

FIGURE 20.14 A Silverfish, a Wingless Primitive Insect Commonly Found in Homes. It undergoes direct development.

Questions

1. Sketch and label the stages of each of the insect life cycles on demonstration. Which exhibits complete metamorphosis, and which exhibits incomplete metamorphosis?

2. Label your sketches with the type of food each stage eats, and where each stage lives.

3. What is a larva? What is the difference between a larva and a juvenile?

4. What is the common name for a larval butterfly? What does the larval butterfly eat? What does the adult butterfly eat?

5. Why might it be an advantage for insect species to have larvae that utilize different food sources than adults?

Exercise 20.7   HUMAN-INVERTEBRATE RELATIONSHIPS

Web exercise

Use the Internet to investigate how humans use an invertebrate (you do not have to restrict your search to arthropods) for consumption as food products, jewelry, medicines, clothing, and other uses. This link might be helpful:

http://www.food-insects.com/links_to_ food_ insects_sites.htm

1. Answer the following questions: What is the common name of the organism? What is it used for? Is it sold in the worldwide market, or is its use local only? Where is the organism found? Is the organism threatened with extinction because of overharvesting or other factors? Are any regulations in place to prevent overharvesting? Is this organism farmed commercially?

2. Type your answers in a two-page, double-spaced report, and cite three references. Be sure to cite your Web references properly (Appendix 1).

21

Chordates

... man, with all his noble qualities ... still bears in his bodily
frame the indelible stamp of his lowly origins.
Charles Darwin, 1880

There is strong evidence indicating that representatives of phylum Chordata share a common ancestry with the echinoderms, including shared radial cleavage and similarities in patterns of coelom formation. The first chordates probably appeared late in the Precambrian era, about 570 mya. The common ancestor between the chordates and echinoderms is thought to have been a sessile filter feeder with free-swimming larvae, much like modern-day tunicates (*Hickman & Roberts, 1994*).

At first, it seems a foreign idea that our (chordates) closest ancestors are the echinoderms, but the two groups diverged half a billion years ago and have been evolving along separate evolutionary pathways ever since.

Figure 21.1 shows the evolutionary relationships of the chordates. From the tunicate-like ancestor, one lineage gave rise to the present-day echinoderms while another gave rise to a free-swimming protochordate that was probably like *Amphioxus*, which you will examine soon. This protochordate is thought to have given rise to three lineages: one leading to the present-day tunicates (Subphylum Urochordata), one to the present-day lancets (Subphylum Cephalochordata), and another to the fish (Subphylum Vertebrata).

The first fish appeared early in the Paleozoic era. Note that not all chordates are vertebrates—two of the three chordate subphyla are comprised entirely of invertebrates.

Exercise 21.1   FOUR CHORDATE CHARACTERISTICS

Subphylum Cephalochordata contains 25 species of lancets. The lancet that we will examine belongs to the genus *Amphioxus*. *Amphioxus* provides an excellent specimen for examination, because it clearly shows all four hallmark (distinctive) chordate characteristics, which are unique to the chordates and are found in all chordates.

1. Using Figure 21.2, learn these characteristics and observe them by looking at a prepared slide of *Amphioxus* with a compound microscope on low power. All chordates possess a **notochord** during some stage of development. This flexible, rod-shaped cord is the first part of the endoskeleton to appear during embryological development. The notochord serves as a stable axis for muscle attachment, and the flexible nature of the notochord permits undulating swimming movements.

2. All chordates possess a **single, dorsal, tubular nerve cord**. This is different from the typical ventral nerve cord of invertebrates, which is often

245

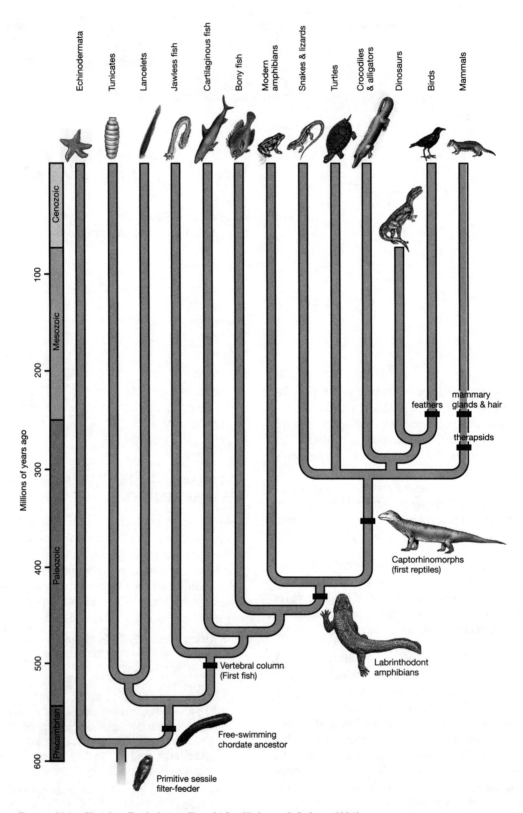

FIGURE 21.1   Chordate Evolutionary Tree. (*After Hickman & Roberts, 1994.*)

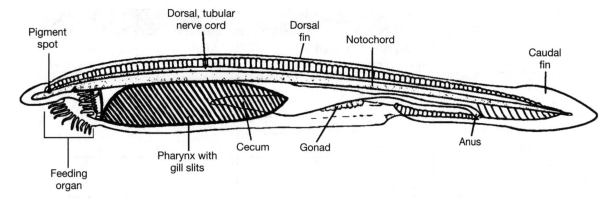

Pigment spot
Dorsal, tubular nerve cord
Dorsal fin
Notochord
Caudal fin
Feeding organ
Pharynx with gill slits
Cecum
Gonad
Anus

FIGURE 21.2 Lateral View of a Lancet.

double, and on the ventral side. The anterior end of the nerve cord is enlarged to form the brain.

3. All chordates possess **pharyngeal gill slits**, at least in some stage during development. In the urochordates, cephalochordates, fish, and some amphibians, the gill slits mature into functional gills used for respiration. In the other chordates, the gill slits give rise to other structures, including the eustachian tubes, middle ear cavity, tonsils, and parathyroid glands in humans.

4. All chordates possess a **postanal tail**, at least in some stage during development.

Sketches

1. **Sketch and label the four chordate characteristics on your lancet.**

2. **What chordate characteristic(s) is found in adult humans?**

Exercise 21.2   VERTEBRATE SURVEY

Subphylum Vertebrata is divided into seven classes (Table 21.1). The vertebrates are defined by the possession of a vertebral column. The vertebral column replaces the notochord as the chief mechanical axis (axis for muscle attachment), and the notochord disappears. The stronger axis provided by the vertebral column laid the foundation for the first vertebrates, which were fast-moving predators.

As you complete this exercise, read the following text and study the specimens on your lab benches. Fill in the "Appeared in fossil record" column in Table 21.1 as you proceed.

1. The fish are a very successful and ancient group. The fish were the first vertebrates to appear on the scene, about 500 mya near the beginning of the Ordovician period. Representatives of Class **Agnatha** are the hagfishes and lampreys, the jawless fishes. The hagfishes are marine scavengers. Lampreys occur in both marine and freshwater environments, and about half of them are parasitic on other fish. The parasitic forms attach to fish with their sucker-like mouth. Using their sharp teeth, they create a wound and suck fluids from underlying tissues [**Krogh Figure 22.29**].

2. Representatives of Class **Chondrichthyes**, the cartilaginous fish, have a skeleton composed entirely of cartilage. These vertebrates appeared about 400 mya and include the sharks, skates, and rays. Interestingly, the cartilage skeleton is secondarily derived, meaning that the cartilaginous fishes arose from an ancestor that had a bony skeleton like the other vertebrates [**Krogh Figure 22.30**].

3. Class **Osteichthyes** comprises the bony fish, which also appeared about 400 mya. This is the most extensive group of vertebrates both in

TABLE 21.1 VERTEBRATE CLASSES			
Class	**Characteristics and some representatives**	**Extant species**	**Appeared in fossil record (mya)**
Agnatha Jawless fishes	Parasitic or scavengers hagfish, lampreys	60	
Chondrichthyes Cartilaginous fishes	Skeleton of cartilage sharks, skates, rays	750	
Osteichthyes Bony fishes	Skeleton of bone trout, sunfish, sea horses	24,000	
Amphibia Amphibians	Breathe with lungs and skin frogs, toads, salamanders	5300	
Reptilia Reptiles	Scaly skin turtles, snakes, alligators	7000	
Aves Birds	Feathers penguin, eagle, ostrich, owl	10,000	
Mammalia Mammals	Hair, mammary glands platypus, human, bat, opossum	5000	

numbers of species and numbers of individuals. There are more species of fish than of all other vertebrates combined **[Krogh Figure 22.31]**.

4. The first vertebrates to emerge onto land were the amphibians. The initial transition to land was also made during the Devonian period, around 400 mya. (For reference, the first dinosaurs appeared around 200–230 mya, and the last dinosaurs disappeared about 70 mya.) It is thought that these early amphibians arose from a lobe-finned fish **[Krogh Figure 22.32]**. This entire group of fishes was thought to have become extinct during the Mesozoic era some 70 mya. Then, in 1938 a living lobe-finned fish (a coelacanth) was discovered off the coast of South Africa near Madagascar. The fins of these fish "predisposed" them for use in land travel, and over time limbs evolved for terrestrial movement (Figure 21.3). Thus began the rise of a group of terrestrial vertebrates called **tetrapods**, which are four-limbed vertebrates. The terrestrial environment of the Devonian period, with its blossoming new plant life, allowed the tetrapods to colonize many new niches. Figure 21.1 summarizes the evolution of the terrestrial vertebrates.

5. During the Paleozoic era, just over 300 mya, the early amphibians gave rise to the first reptiles. These lizard-like insectivores were a very successful group of animals, and gave rise to all other major groups of terrestrial vertebrates. The early reptiles underwent a pronounced period of adaptive radiation, with various lineages giving rise to turtles, dinosaurs, crocodiles, alligators, birds, snakes and lizards, and therapsids, which later gave rise to mammals.

Snakes are mostly harmless, but like spiders, they have become the victims of media hype. There are, however, four types of venomous snakes in the continental United States. Their bites are dangerous—even fatal (Figure 21.4).

6. About 150 mya, a flying animal, the size of a crow, drowned in a tropical lake in what is now Bavaria, Germany. Its fossil remains were found in 1861 in a limestone quarry **[Krogh Figure 22.36]**. At first glance, the animal appeared to be a reptile because of its teeth, claws, bony tall, and abdominal ribs (only reptiles have abdominal ribs). There was, however, the unmistakable imprint of feathers (Hickman & Roberts, 1994). This animal, Archaeopteryx lithographica, was undoubtedly the nearest thing to a missing link that had ever been found, exhibiting both distinctly reptile-like and birdlike characteristics. Figure 21.5 shows the reptilian structures associated with A. lithographica. More recent fossil finds, particularly some from China, have further solidified the connection between birds and the reptiles (Ackerman, 1998). The Chinese fossils also push the origin of birds back to about 70 million years. Examine a bird skeleton on display, and compare its features to those in Figure 21.5.

(a) Ray-finned fish   **(b)** Lobe-finned fish   **(c)** Amphibian

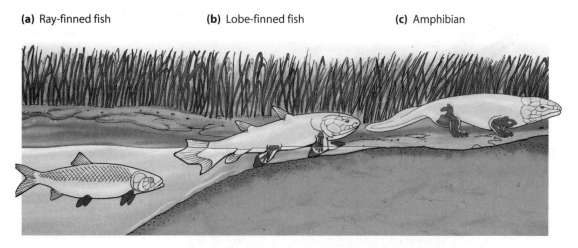

FIGURE 21.3   Vertebrates onto Land. a. Our ray-finned fish ancestors lacked bones in their fins. b. The lobe-finned fish that evolved from ray-finned fish had small bones in their fins that enabled them to pull out of the water and support their weight on tidal mudflats and sandbars. c. Finally, the early amphibians lost ray fins altogether, and had only bones in their limbs.

FIGURE 21.4   Poisonous snakes in the United States. a. Copperhead (*Agkistrodon contortrix*) b. Cottonmouth, or water moccasin (*Agkistrodon piscivorus*) c. Timber rattlesnake (*Crotalus horridus*), one of several rattlesnake species. d. Coral snake (*Micrurus* sp.)

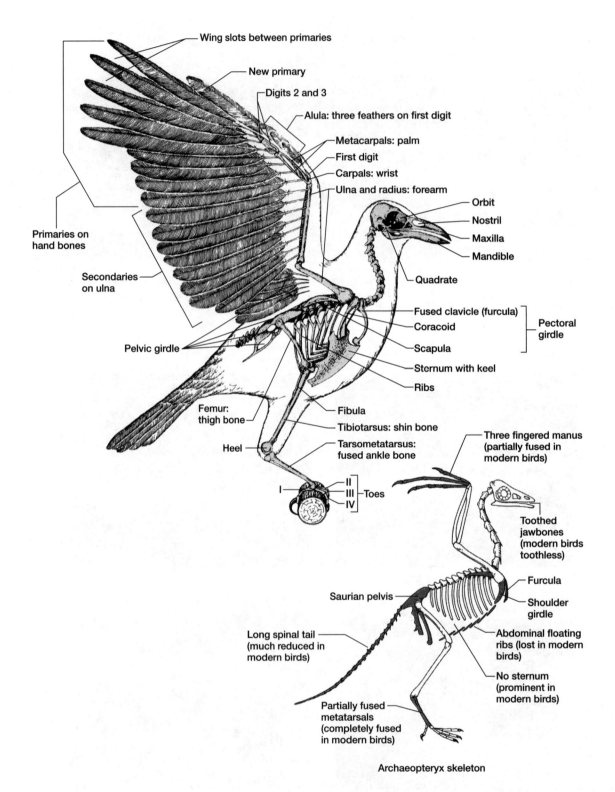

FIGURE 21.5   Comparison of Fossil and Modern Birds. *Archaeopteryx lithographica* is shown on the bottom. Reptilian structures are shaded. The skeleton of a crow is at the top. From Hickman & Roberts, *Biology of Animals* 6th ed. © 1994 William C. Brown. Reproduced with permission of the McGraw-Hill Companies.

7. The two hallmark characteristics of mammals are the presence of *hair* and *mammary glands*. The **therapsids** were insectivorous mammal-like reptiles that probably gave rise to all mammals (Figure 21.6). The first mammals appeared late in the Triassic period—about 180 to 200 mya—and were insectivores, probably similar to the present-day tree shrews (Figure 21.7). Like the reptiles from which they arose, the first mammals were egg layers. One primitive group that diverged from the "main line" of mammalian evolution is the monotremes, which lay eggs and are represented by the modern-day duck-billed platypus.

Just how lucky are we to be here? All mammals arose from therapsids, which were just one group of many reptiles called synapsids that were abundant 300 mya. But the therapsids were the only synapsids to survive the Permian Extinction, in which 96% of all species on the planet became extinct.

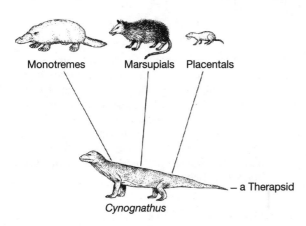

Monotremes   Marsupials   Placentals

Cynognathus

— a Therapsid

FIGURE 21.6   Mammalian Evolution.

FIGURE 21.7   A Tree Shrew.

After the divergence of the lineage leading to the monotremes, the next group to arise was the marsupials. These mammals give birth to poorly developed offspring that make their way into a pouch on the mother's ventral surface, where they attach to a nipple and continue development. Near the beginning of the Jurassic period, the lineage leading to the marsupials diverged from that leading to the placental mammals, which give birth to well-developed offspring. Near the beginning of the Cenozoic Era, the placental mammals exhibited pronounced radiation, giving rise to most mammalian orders that exist today. The mammals were agile, intelligent, adaptable, gave birth to well-developed young, and had a built-in means of feeding their young *(Hickman et al., 1993)*, predisposing them to the great success that they would exhibit. The success of the mammals was reinforced as new niches appeared, after the last dinosaurs vanished 70 mya.

Questions

1. **How does the distribution of plants affect the distribution of animals?**

2. **Relatively speaking, how common is life on land when compared to the number of aquatic species?**

3. **What adaptations enable terrestrial vertebrates to thrive on land?**

4. List the vertebrate classes that possess each of the four chordate characteristics in their adult stages:

Notochord:

Dorsal nerve cord:

Pharyngeal gill slits:

Postanal tail:

Exercise 21.3   MAMMALIAN DENTITION

Evolutionary relationships among the mammals can be deciphered by examining their patterns of dentition. Their varied dentition patterns will allow us to predict differences in their feeding habits. Using the skull of the modern human and Figure 21.8, learn the human dentition pattern. There are four types of teeth. The bladelike **incisors** are used for clipping or cutting plants or meat. The pointed **cuspids** (canines) are for tearing and slashing meat. **Premolars** (bicuspids) and **molars** with flattened crowns and prominent ridges are for crushing and grinding food. Herbivores usually have prominent molars for grinding plant parts.

For any given species, the dentition may be expressed as a **dental formula** that gives the number and kinds of teeth on one side for both the upper and lower jaws (*Janovy, 1991*).

The dental formula is not just a way to count the teeth, but shows the proportions of each kind

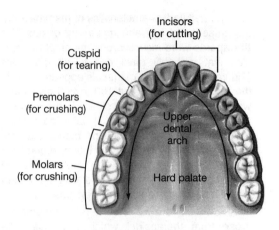

FIGURE 21.8   Human Dentition.

of tooth. For example, a human has two incisors, one cuspid, two premolars, and three molars per side on both the upper and lower jaws; so the dental formula for a human is

$$I + C + P + M$$
$$2 + 1 + 2 + 3$$

1. After you have become familiar with the names, arrangements, and functions of human teeth, carefully study the teeth of the mammal skulls on display. Describe the appearance of the teeth of at least three of the skulls in Table 21.2. Your instructor will provide you with a key to the skulls. As a group, make a list of the various mammals represented and divide them based on feeding habits.

Questions

1. **What animal might the skull be from? What are the specific types of teeth adapted for in each case? Fill in Table 21.1 for each species.**

2. **On the basis of dentition patterns, sort the animals into different groups, assuming that their teeth can be used to predict their feeding habits. Record these in Table 21.2.**

TABLE 21.2 DENTITION OF MAMMALS		
Skull dentition formula I + C + P + M	**Species**	**Feeding habits** **(Herbivore, Carnivore, or Omnivore)**
2 + 1 + 2 + 3	Human (*Homo sapiens*)	

Exercise 21.4   CONSTRUCTING AN ANIMAL KEY

Using the specimens from many phyla that are available to you, construct a dichotomous key like the one in Exercise 13.2.

1. Working in groups of three or four, construct a key with couplets that a reader will use to identify organisms. You will need to discuss the features of animals you have selected, and divide them into groups. Divide your main groups further, and then start to write the characteristics in opposing pairs. Do not use taxa as key features (for example, "Animal is an arthropod" v. "Animal is an annelid"), but describe the characteristics of each group ("Animal has hard exoskeleton" v. "Animal has many repeating segments").

2. When you are finished, exchange your key with that of another group. Can you identify their organisms?

Questions

1. What types of characters did you find the most helpful when identifying organisms? And what characteristics were the least helpful?

2. Did the other group's key reflect the evolutionary relationship of the organisms that you have learned about in class? If not, what could they do to make it that way?

UNIT 6
Mammalian Anatomy and Physiology

22

Tissues and the Skeletal System

The exercises in these next four chapters address basic mammalian anatomy (structure) and physiology (function). Aside from being interesting in its own right, there is great practical value in acquiring an understanding of anatomy. At the very least, these exercises will give you a better understanding of your own body, but they should also help you understand how the parts of the body work together to make a living organism.

One of the characteristics of life is *organization*. Living organisms are made of atoms, which make up molecules, which make up cells, which make up tissues, which make up organs, and so on **[Krogh section 25.3]**. A tissue is a group of cells working together to carry out a common function. Four basic tissue types comprise the body of animals—epithelial, connective, muscle, and nervous tissue.

Exercise 22.1   EPITHELIAL TISSUE

Epithelial tissue covers all surfaces of your body. This includes the skin and the inner linings of organs. They are the surfaces that control what enters or leaves your body, and they protect you by preventing the entry of pathogens. There are four basic types of epithelial tissue, categorized by the shape and arrangement of the cells (Figure 22.1)

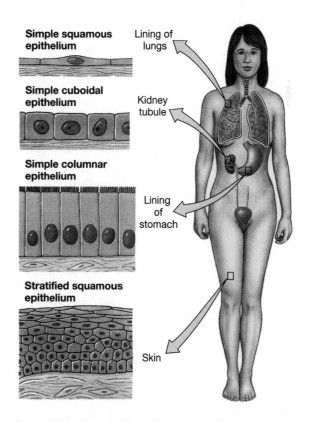

FIGURE 22.1   Some Epithelial Tissues in the Human Body.

[Krogh section 25.4]. Using the prepared slides, examine each epithelial tissue type on low and high power with a compound microscope. Space is provided for sketches of each type.

1. A good example of **simple squamous epithelial tissue** is the lining of the air sacs (alveoli) of your lungs, and the walls of the blood vessels. Obtain a prepared slide of human lung tissue, and examine the flat, single-layered, simple squamous epithelial cells.

2. The cells of **stratified squamous epithelial tissue** are like simple squamous epithelial cells, but instead of occurring in a single layer, they are stacked in layers. This adds strength to the tissue. Examples of stratified epithelial tissue are found in the lining of the mouth (cheek cells) and in the skin. Examine the prepared slide of stratified epithelium of frog skin.

3. **Simple columnar epithelial tissue** gets its name from the single-cell layer of column-shaped cells. The inner linings of the uterus and intestine are composed of simple columnar epithelial tissue. Examine these in the prepared slide of a cross section of human small intestine (Figure 22.2).

4. **Cuboidal epithelial tissue** lines ducts throughout the body, such as those of kidneys and glands. Examine the prepared slide of cuboidal epithelial tissue.

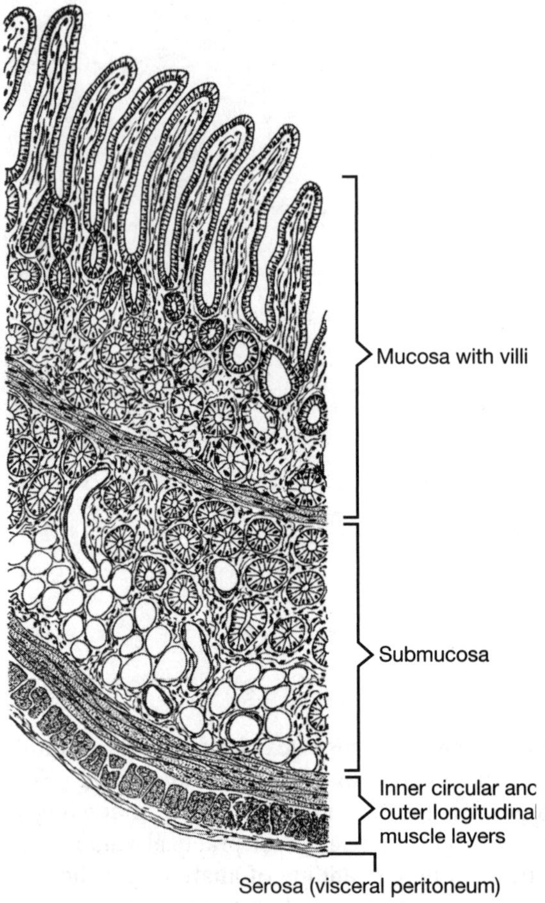

FIGURE 22.2   Cross Section of the Small Intestine.

Sketches

1. **Sketch and label the four epithelial tissue types you observed.**

Simple squamous:

Simple columnar:

Stratified squamous:

Cuboidal:

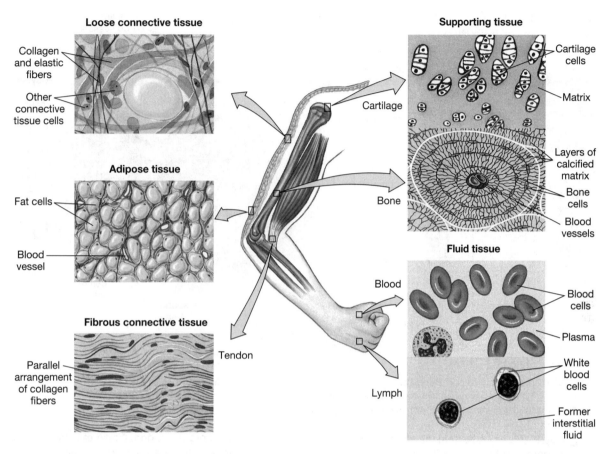

FIGURE 22.3   Connective Tissues in the Human Body.

Exercise 22.2   CONNECTIVE TISSUE

Connective tissue has four functions—supporting and protecting the body and its organs, transporting materials within the body, storing energy, and defending the body against pathogens. The cells of connective tissue are always embedded in a **ground tissue** that forms the "matrix" of the connective tissue (Figure 22.3). Observe each connective tissue type on low and high power with a compound microscope. Sketch each tissue type in the space provided.

1. **Adipose tissue**, or fat, is one example of loose connective tissue. Adipose tissue provides cushioning and insulation and is a source of stored energy. Observe the adipose tissue associated with the serosa (outer lining) of the human small intestine (Figure 22.2).

2. Examine the prepared slide of ground **bone tissue** (Figure 22.4). Bone tissue is highly organized into what is known as an **osteon** or Haversian system.

Bone cells called **osteocytes** are found in cavities called **lacunae** (singular, *lacuna*) arranged in concentric circles within the calcium matrix of the osteon. The center of the osteon is the central canal (Haversian canal), which contains blood vessels and a nerve. Bone is a living tissue that requires oxygen and nutrients, and it must get rid of metabolic wastes. The blood vessels entering bone through the Haversian canals facilitate this exchange of materials. Osteocytes within lacunae communicate with one another through pseudopod-like extensions lying in little canals that radiate outward.

3. **Cartilage** is also a supporting connective tissue (Figure 22.3). Like bone cells, cartilage cells called **chondrocytes** are contained within lacunae embedded in a matrix; however, cartilage tissue is not arranged into osteons. Examine the prepared slide of hyaline cartilage. Note the irregular arrangement of the lacunae. Hyaline cartilage is found on the ends of long bones and in the nose, ears, trachea, and other parts of the

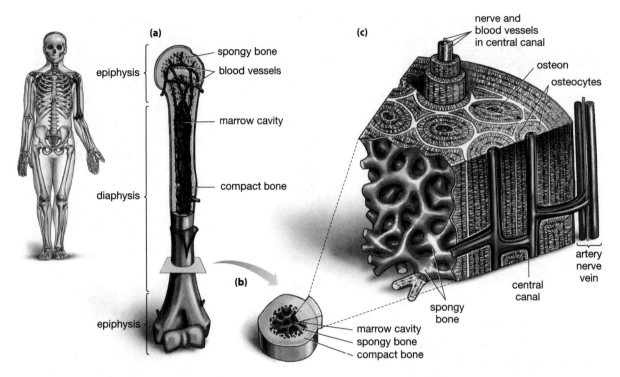

FIGURE 22.4   Features of a Typical Bone. The periosteum is the outer layer covering the bone, which is depicted as a peeled layer in the figure. The building units of compact bone, the osteons, run parallel to the long axis of the bone. Spongy bone does not contain osteons, but instead is composed of an open network of calcified rods or plates.

body. It gets its name from the smooth, glass-like appearance of the matrix. The matrix of cartilage tends to hold a lot of water, making cartilage relatively pliable. That is why cartilage tends to tear, but bone breaks. Also, because cartilage does not have as rich a blood supply, injured cartilage tends to heal more slowly than injured bone.

4. **Fluid connective tissue** includes blood and lymph. Observe a prepared slide of human blood. Blood is a tissue made up of cells in a fluid matrix called plasma (Figure 22.5). Blood cells are divided into two basic types—red blood cells called **erythrocytes**, and white blood cells called **leukocytes** (Figure 22.6). Mature human red blood cells are biconcave (indented on both sides) and have no nucleus. Leukocytes are part of the immune system that protects our bodies from various pathogens, and are much less common than red blood cells. They do most of their work in the tissues of the body; most seen in the bloodstream are on their way to sites of infections from the bone marrow, spleen, or lymphatics where they are produced.

White blood cell types include lymphocytes, monocytes, eosinophils, basophils, and neutrophils, each of which plays different roles in the immune

system. There is a "normal" ratio of these cells in your blood at any given time, and elevations of certain types of white blood cells are indicative of various diseases. For instance, high macrophage counts are indicative of chronic viral or bacterial infections, elevated neutrophil numbers suggest bacterial infection, and elevated numbers of eosinophils are indicative of worm infections.

Can you identify any of the white blood cell types using Figure 22.6? Sketch a red blood cell and a white blood cell.

Sketches

1. **Sketch and label the components of the four connective tissue types.**

   Adipose:

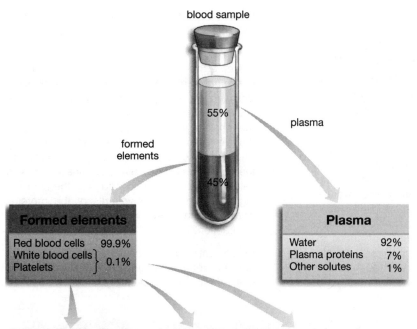

blood sample

formed
elements

plasma

55%

45%

Formed elements

Red blood cells	99.9%
White blood cells	
Platelets	0.1%

Plasma

Water	92%
Plasma proteins	7%
Other solutes	1%

Red blood cells

White blood cells

Platelets

FIGURE 22.5   The Composition of Blood. Blood that is spun in a centrifuge separates into its two primary constituent parts, formed elements and plasma. Formed elements are cells and cell fragments that help make up the blood—red blood cells, white blood cells, and the cell fragments called platelets. Blood plasma, the liquid in which the formed elements are suspended, is composed mostly of water, but also contains proteins and other materials.

Bone:

Cartilage:

Red and white blood cells:

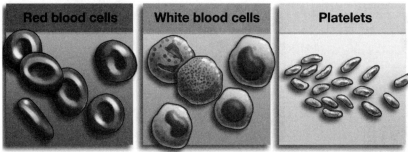

Agranular Leukocytes

Thrombocytes

Lymphocytes

Monocyte
—10μm—

Erythrocytes

Eosinophil

Neutrophil

Basophil

Granular Leukocytes

FIGURE 22.6   Human Blood Cells.

Exercise 22.3   MUSCLE TISSUE

Mammals have three basic types of muscle tissue—smooth, skeletal, and cardiac (Figure 22.7) **[Krogh section 25.4]**. Observe each type of muscle tissue on low and high power with a compound microscope, and then sketch them in the space provided.

1. Observe the **smooth muscle** in the circular layer of the human small intestine cross section (Figure 22.2). Smooth muscle is found in the walls of hollow organs. It is also called involuntary muscle, because it is responsible for contractions controlled by automatic nerve impulses, not the ones that we "command" our brains to send. These involuntary contractions include peristalsis, which moves material through hollow organs such as the small intestine and urinary bladder. Smooth muscle cells have a single, centrally located nucleus, and they are long and pointed at each end.

2. Observe the prepared slide of **skeletal muscle** tissue under high power. This muscle was taken from the leg of a grasshopper. Skeletal muscle is also known as *voluntary muscle* because it is responsible for voluntary movement. Skeletal muscle cells are cylindrical with blunt ends. They have several nuclei that are located near the cell membrane. Skeletal muscle is also called *striated muscle* because of the alternating arrangement of myofibrils of the proteins **actin** and **myosin**, which are responsible for contraction.

    Smooth muscles also contain myofibrils, but they are not organized in such a way as to give the striated appearance seen in skeletal muscle.

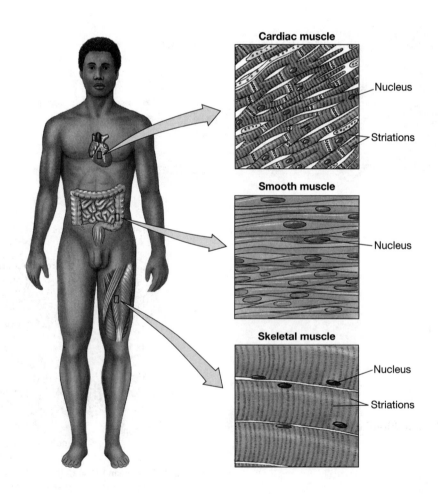

FIGURE 22.7   Muscle Tissue in the Human Body. The three types of muscle tissue are cardiac, smooth, and skeletal.

3. Observe the prepared slide of **cardiac muscle** from a human heart. Cardiac muscle is also striated like skeletal muscle, but it functions more like smooth muscle in that it is involuntary. Cardiac muscle is found only in the heart. The fibers of cardiac muscles are branched, interconnecting with more than one cell at a time. Nuclei are centrally located in the cells. Also observe the intercalated discs, dark-staining gaps that carry electrical impulses between cells.

Questions

1. **Why is it important for cardiac muscle cells to be branched and interconnected with other cardiac muscle cells?**

2. **Sketch and label the components of the three types of muscle tissue.**

   Smooth:

   Skeletal:

Cardiac:

Exercise 22.4   THE SKELETAL SYSTEM

The skeletal system consists of bone and cartilage, and provides support for the body and a basis for muscle attachment. Using Figures 22.8 and 22.9 as your guide, learn the major bones of the skeleton [Krogh section 25.8].

1. Locate the **axial skeleton**, which includes the skull, thoracic cavity (sternum and ribs), and vertebral column. The vertebral column (Figure 22.9) consists of 7 cervical, 12 thoracic, and 5 lumbar vertebrae; it terminates in the sacrum and coccyx. The coccyx is the rudiment of a post-anal tail, one of the four hallmark chordate characteristics. Discs of cartilage between the vertebrae allow flexibility of the spine and protect vertebrae from stress and compression. Such cartilage joints are also found connecting the ribs with the sternum and at the symphysis pubis, the joint between the two coxae forming the pelvic girdle.

2. Locate the **appendicular skeleton**. As the name implies, it includes the paired appendages (arms and legs), the pelvic girdle, and the pectoral girdle. In living mammals, cartilage in the joints cushions the bone and keeps the joints moving freely. Destruction of this cartilage may cause arthritis.

3. Next, examine the anatomy of the long bone, using Figure 22.4 as your guide. The bone consists of a hollow shaft called a **diaphysis** with "knobs" at each end, called **epiphyses**. The diaphysis is composed of dense bone, and the epiphyses are composed of spongy bone. The hollow cavity of the diaphysis contains red bone marrow that functions in the production of blood cells. The spongy bone of the epiphyses is made up of bars and plates, forming an irregular grid-like pattern. Yellow bone marrow, which serves as fat reserves, fills the cavities in the epiphyses.

   The bone is lined by a tough, thin layer of connective tissue called the **periosteum**. The periosteum contains nerves and blood vessels that enter the bone.

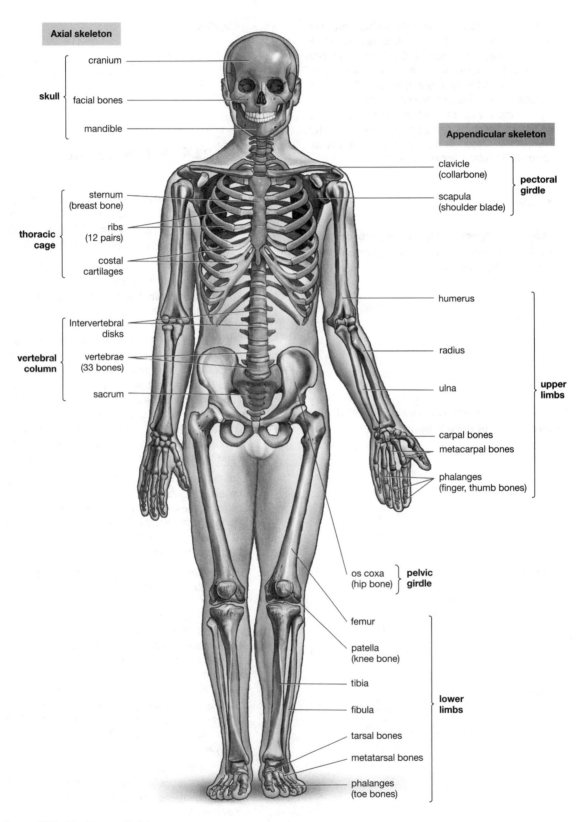

Axial skeleton

skull
- cranium
- facial bones
- mandible

thoracic cage
- sternum (breast bone)
- ribs (12 pairs)
- costal cartilages

vertebral column
- Intervertebral disks
- vertebrae (33 bones)
- sacrum

Appendicular skeleton

pectoral girdle
- clavicle (collarbone)
- scapula (shoulder blade)

upper limbs
- humerus
- radius
- ulna
- carpal bones
- metacarpal bones
- phalanges (finger, thumb bones)

pelvic girdle
- os coxa (hip bone)

lower limbs
- femur
- patella (knee bone)
- tibia
- fibula
- tarsal bones
- metatarsal bones
- phalanges (toe bones)

FIGURE 22.8   The Human Skeleton.

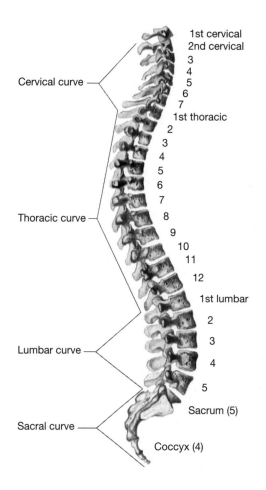

Cervical curve

1st cervical
2nd cervical
3
4
5
6
7
1st thoracic
2
3
4
5
6
7
8
9
10
11
12

Thoracic curve

Lumbar curve

1st lumbar
2
3
4
5
Sacrum (5)

Sacral curve

Coccyx (4)

FIGURE 22.9   Lateral View of the Human Vertebral Column.

Questions

1. **What is the function of the cartilage joints between the ribs and sternum and the symphysis pubis?**

2. **What is the function of the cartilage that lines shoulder, hip, and knee joints?**

23

The Nervous System and Sensory Perception

The brain is the only known entity in the universe capable of contemplating itself.
T.M. Graham, 1998

In order to survive, an organism must be able to perceive and respond to signals called **stimuli** from its environment. Animals accomplish this through the action of the nervous system. An organism's ability to respond to stimuli is called irritability, and it is one of the defining features of life. The nervous system also facilitates communication between cells and organs within an organism, enabling the billions of cells comprising the body to act as a single, highly integrated unit.

There are two major divisions of the nervous system—the **central nervous system** (CNS), consisting of the brain and spinal cord (Figure 23.1), and the **peripheral nervous system** (PNS). The PNS is subdivided into the **afferent division** and the **efferent division**. The afferent division contains **sensory neurons** that receive stimuli from the environment and carry these stimuli, in the form of nerve impulses, to the CNS where they are "processed." Nerve impulses generated in the CNS in response to the stimuli are carried by **motor neurons** in the efferent division of the PNS to **effectors** such as glands or muscles that carry out the proper response to the stimuli. **Interneurons** connect neurons in the CNS **[Krogh sections 26.1 and 26.2]**.

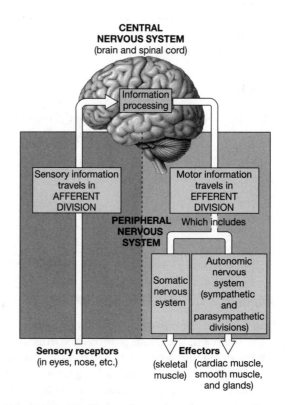

FIGURE 23.1   The Nervous System.

267

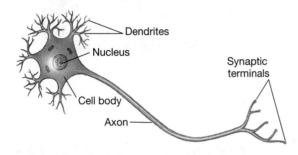

FIGURE 23.2   A Neuron.

Exercise 23.1   THE NEURON

1. The nerve cell, or **neuron** (Figure 23.2), is the functional unit of the nervous system. Observe a prepared slide of neurons on low and high power with a microscope. A nerve cell consists of branched **dendrites** that receive nerve impulses and carry the impulses toward the cell body. The long **axon** carries impulses away from the cell body. The cell body contains the nucleus and other organelles. When the impulse reaches the end of the axon at the **synaptic terminals**, it is passed on to another neuron or to an effector organ.

Sketches

1. **Sketch and label the components of the neuron you see with your microscope.**

Exercise 23.2   REFLEX RESPONSES

Many nervous responses are very complex, involving the brain and numerous neurons and effectors. A reflex arc is quite simple in comparison, involving one sensory neuron and one motor neuron. Have you ever heard the phrase, "knee-jerk reaction"? It implies that an action was made without thinking, and our bodily reflexes occur the same way. Reflexes are carried by sensory neurons to the spinal cord; then they are carried right back to muscles by a motor neuron—the impulse in most reflexes is not sent to the brain. Reflex responses can then initiate a cascade of other responses: You may not have to think about pulling your hand away from a hot stove, but your racing heart and uttered expletives are controlled by more complex reactions.

A literal knee-jerk reaction, called the patellar reflex, is shown in Figure 23.3. Physicians use the patellar reflex to help determine whether or not the nervous system is functioning properly. Working in pairs, you will observe the patellar reflex.

1. Sit on the lab bench, with your legs hanging comfortably over the side.

2. Using a reflex hammer, another student will gently strike the patellar tendon that is located just below the kneecap.

3. Exchange positions and repeat Steps 1 and 2.

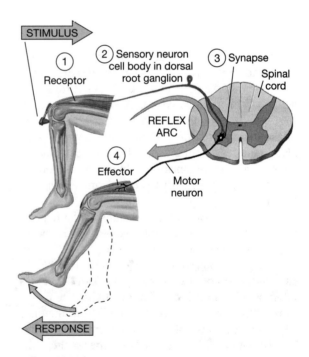

FIGURE 23.3   The Patellar Reflex. 1. Stimulus (tapping) arrives and a receptor is activated. 2. The signal from the receptor reaches a sensory neuron cell body, in the dorsal root ganglion near the spinal cord; the signal moves to a sensory neuron/motor neuron synapse in the spinal cord. 3. Information processing of the event takes place, prompting a signal to be sent through the motor neuron. 4. The motor neuron signal stimulates the effector (the quadriceps muscle) to contract. Note that CNS processing for this reaction was handled entirely in the spinal cord; the brain was not involved.

Questions

1. Why is the reflex action so quick?

Exercise 23.3 VOLUNTARY RESPONSES

(Adapted from Gunstream, 2001)

The reflex reaction facilitates a very quick response to a stimulus. In contrast, voluntary responses exhibit a delay between the stimulus and the response because of the involvement of synapses between one or more interneurons. In this exercise you will observe a more complex nervous reaction, in which a subject catches a ruler that is dropped, and this response time is measured in milliseconds (ms) on the ruler.

Caffeine has been reported to increase alertness, but what effect does it have on voluntary muscle activity? For this exercise, your class will be divided into two groups. People in the experimental group will drink caffeinated beverages, and people in the control group will drink caffeine-free beverages. (People who have consumed caffeine prior to lab will be disqualified.) You will not know which group you have been placed in until all measurements for the experiment have been completed. Measurements from the control group will be compared to the measurements from the experimental group to determine whether or not caffeine has an effect on voluntary responses. The control group measurements, on their own, will also reveal whether or not *conditioning* (practice) improves response time.

1. Because of time limits in the lab, you and your classmates have to get a jump-start on today's procedure. As soon as the lab period begins, drink two servings of the beverage your instructor gives you. Record the start time in Table 23.1.

2. Pair up with a person who is in the opposite group. Your partner will stand and hold a reaction time ruler by its top end. You will sit in a chair while holding your thumb and forefinger 2 cm apart, ready to catch the ruler. The bottom edge of the ruler should be even with your thumb and forefinger, as shown in Figure 23.4a. Tell your partner that you are ready, and watch the ruler.

TABLE 23.1 VOLUNTARY RESPONSE MEASUREMENTS		
Start time:		
Time elapsed (min)	**Your response time (ms)**	**Partner's response time (ms)**

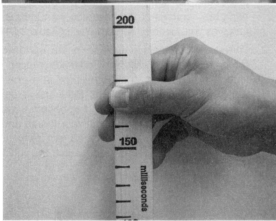

FIGURE 23.4   Measuring Response Time. Top: Subject prepares to catch the ruler. The thumb and forefinger are held about 1 inch apart just below the "thumb line" mark at the bottom of the ruler. Bottom: After the subject catches the ruler, the distance is measured in milliseconds (ms) at the upper edge of the thumb. In this case, the response time is about 180 ms.

3. When your partner drops the ruler, catch it between your thumb and forefinger as quickly as you can. The distance the ruler falls is a direct measure of your response time in milliseconds (ms). Estimate this measurement as carefully as possible, and be sure to measure it consistently each time, measuring to the nearest 5 ms. For example, Figure 23.4b shows a reaction time of 180 ms. Repeat this procedure a few more times.

4. Exchange places with your partner and repeat Steps 2 and 3.

5. Now that you know how to conduct the test, design an experiment to test the effects of caffeine on voluntary muscle activity. (Be sure to measure response time in both the experimental and control groups.) Answer Question 1.

6. Conduct the test according to your experimental design, and record your data in Table 23.1.

7. Graph your data in Figure 23.5. Use the symbols provided in the key.

Questions

1. Write the components of your experimental design:

Hypothesis:

Prediction:

Experimental group:

Control group:

Independent variable:

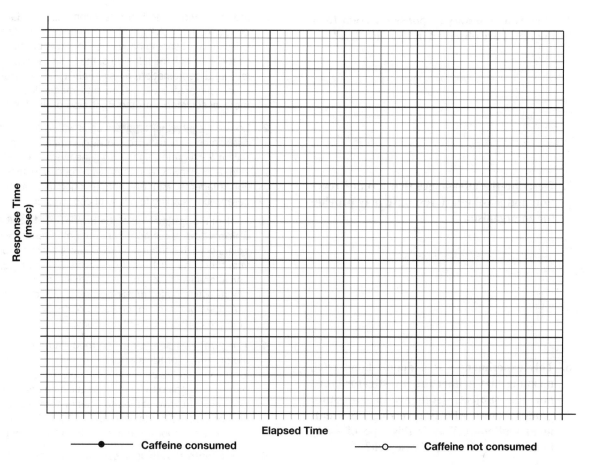

Response Time (msec) (vertical axis)

Elapsed Time (horizontal axis)

———●——— Caffeine consumed   ———○——— Caffeine not consumed

FIGURE 23.5   Graph of Response Time.

Dependent variable:

Caffeine dose for test subjects:

Time elapsed between measured responses:

2. **Were your predictions correct? What is the effect of caffeine on response time? Explain your results.**

3. **What is the effect of conditioning (practice) on response time?**

4. **How is a voluntary response different than the reflex arc observed in Exercise 23.2?**

5. **Describe the advantages that a reflex reaction and a voluntary response may provide for an organism.**

6. **This experiment was a double-blind trial, meaning that neither the investigator (you) nor the subjects (students in the other group) knows who is receiving the treatment (caffeine). Why is this type of study important in biomedical research?**

Exercise 23.4   THE BRAIN

Wear gloves and goggles.

The **brain** serves as the body's command center, with its estimated 100 billion neurons receiving and interpreting untold numbers of signals per second. Using Figure 23.6 as your guide, examine the major parts of whole and half sheep brains **[Krogh section 26.6]**.

1. First examine the whole brain. The **cerebrum** is the largest part of the brain, divided by **the longitudinal cerebral fissure**, the line separating the left and right halves. The cerebrum contains the centers for memory and intelligence. You may have heard that someone is "right brained" or "left

brained." Right-hand coordination, spoken language, number skills, written language, scientific skills, and reasoning have been found to be more associated with the left side of the brain. However, left-hand coordination, musical ability, art awareness, spatial abilities, insight, and imagination are more associated with the right side of the brain.

2. Locate the **cerebellum** at the back of the brain. It is associated with balance, equilibrium, and coordination. It integrates signals from the eyes, muscles, skin, and elsewhere. Excessive alcohol consumption interferes with the normal operation of the cerebellum, leading to a loss of coordination.

3. Look at the ventral side of the brain. The **medulla oblongata** is the elongate stem that leads to the spinal cord. It contains reflex centers that control vital functions such as the heartbeat, breathing, and digestive function. The medulla is also the site at which major motor nerves cross to opposite sides of the brain; one side of the brain controls the opposite side of the body.

4. The **pons** lies anterior to the medulla and is the major center for transfer of information back and forth between the cerebrum and cerebellum. If an organism is going to function as a single unit, it is necessary to have elaborate lines of communication between different aspects of the CNS. The pons is also involved in the control of breathing.

5. The **olfactory bulb** is associated with the sense of smell, and receives stimuli from the olfactory nerves. It is bilobed and located near the front of the sheep brain. Interestingly, it has been suggested that of all of the senses, smell has the greatest ability to evoke memories.

6. The **optic nerves** carry visual stimuli from the eyes to the cerebrum, where they are interpreted. They are also located near the front of the brain, and lead back to the **optic chiasma**, where they cross. Stimuli from your left eye are interpreted in the right cerebral hemisphere, and vice versa. Your specimen may not have the optic nerves, but you should still be able to identify at least part of the optic chiasma.

7. Examine the half brain so that you can see its internal structures. You will see the **cerebral cortex**, cerebellum, pons, and medulla in saggital section. You will also observe the **corpus callosum**, which contains many axons carrying impulses between the left and right cerebral hemispheres.

8. Locate the **thalamus**, which serves as a "switching station" by intercepting much of the information coming into the brain and directing it to the proper location.

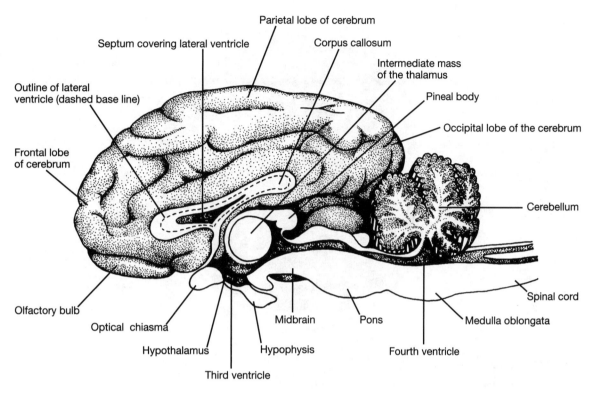

Parietal lobe of cerebrum

Septum covering lateral ventricle

Corpus callosum

Intermediate mass of the thalamus

Pineal body

Outline of lateral ventricle (dashed base line)

Occipital lobe of the cerebrum

Frontal lobe of cerebrum

Cerebellum

Olfactory bulb

Spinal cord

Optical chiasma

Midbrain

Pons

Medulla oblongata

Hypothalamus

Hypophysis

Fourth ventricle

Third ventricle

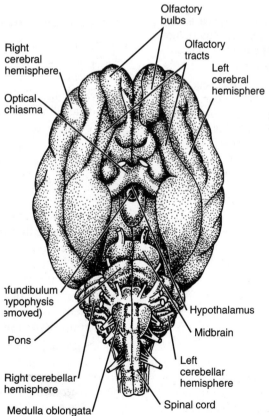

Olfactory bulbs

Right cerebral hemisphere

Olfactory tracts

Left cerebral hemisphere

Optical chiasma

Infundibulum (hypophysis removed)

Hypothalamus

Midbrain

Pons

Left cerebellar hemisphere

Right cerebellar hemisphere

Medulla oblongata

Spinal cord

FIGURE 23.6   a. Lateral View of Half of Sheep Brain b. Ventral View of the Sheep Brain.

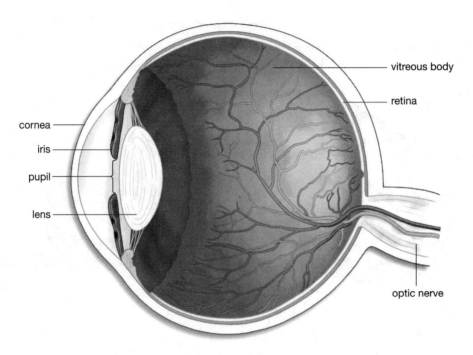

cornea

iris

pupil

lens

vitreous body

retina

optic nerve

FIGURE 23.7   Anatomy of the Human Eye.

9. The **hypothalamus**, as the name implies, lies directly under the thalamus. It monitors such things as hunger, thirst, body temperature, water content, blood sugar levels, and sex urge.

10. The **pituitary gland** functions in conjunction with the hypothalamus in controlling the secretion of various hormones. Because it lies below the rest of the brain, it often is broken off when the brain is removed, and you may not be able to see it on your specimen. Interestingly, sheep have a relatively large pituitary gland. The pituitary gland of humans is about the size of a pea.

Sketches

1. **Sketch the half brain, labeling the function of each part.**

Exercise 23.5   THE EYE

(Adapted from Gunstream, 2001)

Wear gloves and goggles.

The eye (Figure 23.7) is a fascinating and complex organ that allows us to have acute three-dimensional color vision.

The eye has fluid-filled chambers. Excess buildup of fluid in the anterior chamber causes *glaucoma*, which may damage the eye.

The eye works much like a camera. The amount of light that passes through the front of the eye is regulated, just as a shutter regulates the amount of light entering a camera. The lens bends incoming light rays so that they are all in focus at the same point on the back of the eye, similar to a camera lens. An image forms in the eye and is sent to the brain **[Krogh section 26.12]**.

1. Obtain a sheep eye and notice the abundant fat tissue around it, which provides protection. Locate the **cornea**, the clear, bulging region. Locate the **optic nerve** on the rear of the eyeball. A blow to the back of the head sometimes causes a person to "see stars." This is a result of excessive stimulation to the neurons associated with vision.

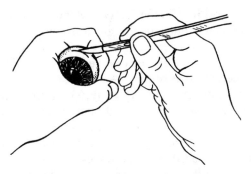

(a) Holding the eye as shown, make a small cut through the sclera with a sharp scalpel about 0.5 cm from the edge of the cornea.

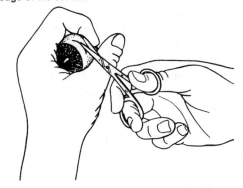

(b) Insert the point of a scissors into the cut and continue to cut through the wall around the eye.

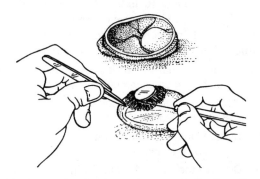

(c) After pouring the vitreous humor onto the dissecting pan, carefully separate the lens from the ciliary body with a dissecting needle.

FIGURE 23.8   Dissection of the Sheep Eye.

2. As shown in Figure 23.8, make a small vertical incision, about 0.5 cm long, through the **sclera**. Using your scissors, cut around the circumference of the eyeball, separating the anterior and posterior halves. Notice the consistency of the fluid (vitreous humor) in the **vitreous body**.

3. Examine the anterior portion of the eye. Locate the **iris** and the **ciliary body**, which looks like a thick black ring. Using a dissecting needle, separate the **lens** from the ciliary body. Hold the lens up near your eye, and look through it. Observe the dark, circular iris, which is the pigmented ring that surrounds the pupil.

4. Next locate the **retina** on the inner lining of the back of the eye. The retina is easily removed from the dark, underlying membrane called the **choroid**. The choroid contains blood vessels that supply the eye and absorb excess light that passes through the retina. In many animals (other than humans) there is an iridescent blue-green layer associated with the choroid, called a **tapetum**. Note the tapetum on the sheep eye. The tapetum reflects excess light back thorough the retina. This increases night vision and causes the characteristic green glow of animals' eyes when a light reflects off them. Likewise, the "red eye" of humans in flash photographs is a result of light reflected off the retina. The reflected light is red because of the many vessels supplying blood to the retina.

5. The blind spot: Remove your gloves and goggles. Draw a 2-cm circle in the center of a white piece of paper. Cover one of your eyes and hold the paper at arm's length in front of your uncovered eye. Look at a fixed point (such as a wall) in front of you, and slowly move the paper up and down. Do not follow the circle with your eye, but keep staring at the fixed point in front of you. After a few attempts, you will notice that the circle disappears when it is at a certain height.

Questions

1. **Describe the image that you observe through the sheep lens.**

2. **What causes the blind spot in your vision?**

3. Why is there no blind spot when you look with both eyes?

4. Notice the smooth texture and convex shape of the cornea and lens. What would happen to images if the cornea or lens had an irregular shape (a condition called astigmatism)?

Exercise 23.6　THE EAR

(Adapted from Gunstream, 2001)

The ear has three regions (Figure 23.9). The **outer ear** is composed of the **pinna, ear canal**, and **tympanic membrane** (eardrum). The pinna is the external structure that is composed of skin overlaying cartilage. Its function is to direct sound waves into the ear canal; the shape of an animal's ear affects how sounds are heard **[Krogh section 26.11]**.

Sound waves entering the ear cause many reactions that proceed like falling dominoes. The sound waves strike the eardrum, causing vibration of the eardrum at the same frequency as the sound waves. The vibration of the eardrum passes along to the three bones in the **middle ear: the hammer, anvil**, and **stirrup**.

The eustachian tube connects the middle ear with the throat so that air can enter or leave the middle ear. This allows pressure in the middle ear to be equalized when a person swallows or yawns. Unfortunately, in addition to air, bacteria and virus particles can also pass through the eustachian tubes, which can cause *otitis media,* or middle ear infection.

The vibration is then transferred to the **cochlea** of the **inner ear**. The vestibule has an **oval window** and a **round window**. The vibration of the stirrup in the oval window is transferred to the membrane of the round window, which leads to movement of fluid in the cochlea, which stimulates **hair cells** embedded in the cochlear membrane. These hair cells generate signals that are carried to the brain along the **auditory nerve**, where the stimulus is perceived as sound.

Vibrations occurring at different frequencies produce the various sounds that are perceived by the brain. Mammal species have different hearing capabilities because they can detect different frequencies caused by sound waves.

Within the inner ear are **semicircular canals**, which also have hair cells. When the head is tilted forward or backward, fluid moves the hair cells within the semicircular canals, causing signals to be sent to the brain. When these messages are "translated" they tell the brain the position of the head. When a person twirls around, the fluid in the semicircular canals spins also, moving the hair cells. A sudden stop will cause the hair cells to become even more displaced, causing dizziness and the sensation of unbalance.

There are two basic types of hearing loss. In **conduction deafness,** hearing is impaired because sound impulses are not sent from the middle ear to the inner ear. It usually involves hearing loss at all sound frequencies [See Krogh's *Too Loud* sidebar, page xx]. In **nerve deafness**, hearing is impaired because of damage to hair cells or the auditory nerve in the inner ear. Nerve deafness may involve hearing loss at only certain frequencies, such as the deafness that can occur with age *(Fox, 1993)*.

1. On the mammal skull on display, notice the opening of the ear canal through the skull bone.

2. Observe the middle ear bones on display. Sketch them (see Question 2).

3. Using Figure 23.9 trace the path of a vibration caused by a sound wave.

4. Conduct the **Rinne test** to determine if your lab partner has hearing loss. He will sit in a chair, one ear plugged with cotton, while you conduct the test. Strike a tuning fork on the heel of your hand. As it rings, hold the edge of the fork 8 inches from his unplugged ear. He will raise his hand when he hears the ringing sound, and will lower his hand when the ringing stops. Using a stopwatch, note how long he hears the sound.

5. When the pitch is no longer heard, gently touch the stem of the fork to the temporal bone behind his ear. Record whether he hears the sound again. Refer to Table 23.1 to determine what type of hearing loss he has, if any. When your own hearing is tested, mark an "x" in the appropriate row in the table.

(a) Anatomy of the ear

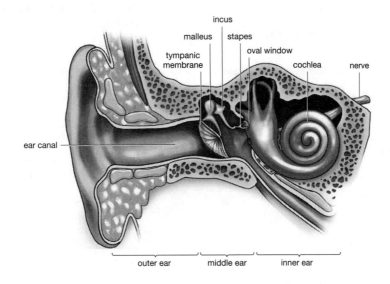

(b) From air vibration to nerve signal

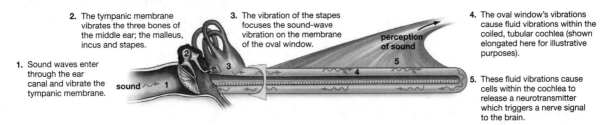

1. Sound waves enter through the ear canal and vibrate the tympanic membrane.

2. The tympanic membrane vibrates the three bones of the middle ear; the malleus, incus and stapes.

3. The vibration of the stapes focuses the sound-wave vibration on the membrane of the oval window.

4. The oval window's vibrations cause fluid vibrations within the coiled, tubular cochlea (shown elongated here for illustrative purposes).

5. These fluid vibrations cause cells within the cochlea to release a neurotransmitter which triggers a nerve signal to the brain.

FIGURE 23.9   Anatomy of the Human Ear.

TABLE 23.2 RINNE TEST			
Relative amount of time sound is heard 8 inches from ear	**Sound reappears through temporal bone**	**Type of hearing loss**	**Your hearing**
Long	No	None	
Moderate	Yes	Slight conduction deafness	
Short	Yes	Moderate conduction deafness	
None	No	Nerve deafness	

Questions

1. What parts of the ear lie inside the skull?

2. Sketch and label the middle ear bones.

3. What is the path of sound waves through the ear?

4. What anomalies of the ear might cause conduction deafness?

5. What methods might be used to restore hearing?

6. How might athletes (skaters, gymnasts) avoid losing their balance?

7. Compare and contrast the structure of a cat's pinna with that of a human. Why might cats have a greater hearing advantage?

The Cardiovascular System

The cardiovascular system consists of the heart, blood, and blood vessels, which transport nutrients, dissolved gases, hormones, and wastes to and from tissues of the body **[Krogh sections 28.1–28.5]**. Figure 24.1 illustrates the flow of blood through the body. The **heart** is the pumping organ of the circulatory system. The **pulmonary circulatory system** transports deoxygenated blood to the lungs to "pick up" oxygen, and then transports it back to the heart. Then the freshly oxygenated blood enters the **systemic circulatory system**. In systemic circulation, freshly oxygenated blood is sent to the rest of the body. The blood delivers oxygen, nutrients, and other essential materials to tissues through **arteries**, which branch into **arterioles** and then into tiny **capillaries**. Carbon dioxide and other wastes are "picked up" and removed from the tissues, and carried by blood back toward the heart through **venules** that enlarge into **veins**. Figure 24.2 shows blood vessels in the human body.

Exercise 24.1   THE HEART

Wear gloves and goggles.

1. Examine a whole sheep heart, and identify the four chambers (Figure 24.3). The apex or "lower

The pulmonary and systemic circulation networks

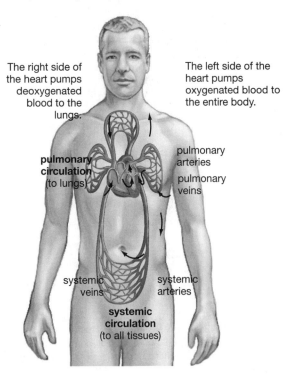

The right side of the heart pumps deoxygenated blood to the lungs.

The left side of the heart pumps oxygenated blood to the entire body.

pulmonary circulation (to lungs)

pulmonary arteries

pulmonary veins

systemic veins

systemic arteries

systemic circulation (to all tissues)

FIGURE 24.1   The Flow of Blood through the Body.

279

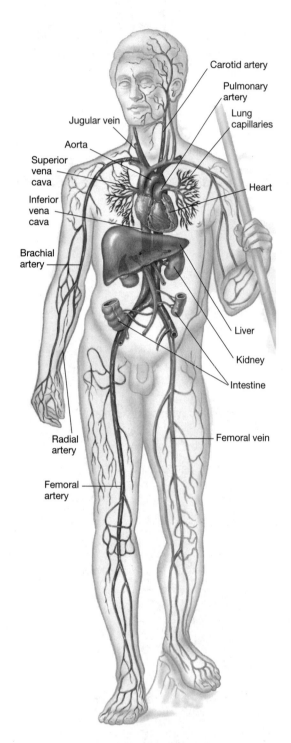

Carotid artery

Pulmonary artery

Lung capillaries

Jugular vein

Aorta

Superior vena cava

Inferior vena cava

Heart

Brachial artery

Liver

Kidney

Intestine

Radial artery

Femoral vein

Femoral artery

FIGURE 24.2   Circulatory System of the Human Body. For simplification, veins are shown only on one side, and arteries are shown only on the other. Lung capillaries are shown greatly enlarged.

tip" of the heart is part of the left ventricle. Note the coronary arteries on the heart. These arteries supply the heart muscle with oxygen-rich blood **[Krogh section 28.2]**.

The right side of the heart is also on the animal's right side; if you place the heart so that its ventral side is facing you, the left side of the heart will be on your right.

2. Place the whole sheep heart on your lab bench so that its ventral side is facing you, with the ventricles closest to you. Now do the same with a bisected sheep heart, but open the halves of the heart so that you can identify its internal anatomy (use Figure 24.4).

3. Note that the ventricles are large, thick-walled muscular chambers. The left ventricle has the thickest walls. Just above the right and left ventricle are the right and left atria (singular, *atrium*) respectively.

4. Begin tracing the sequence of the flow of blood through the heart. Blood is carried by the **inferior vena cava** and **superior vena cava** to the **right atrium**. You should identify these vessels by placing a blunt probe up through the right atrium. You may see part of the vessel, but in many specimens the vessels were cut very close to the atria. The inferior vena cava brings in blood from the systemic circulation of the lower part of the body, and the superior vena cava brings in blood from the systemic circulation of the upper part of the body . When the atria contract, blood passes from the right atrium through the **right atrioventricular (AV)** valve into the **right ventricle**.

5. When the right ventricle contracts, blood pushes back up against the right AV valve, causing it to close. With a blunt probe, lift the flaps of the AV valves, and notice how they can fit together. When the valve works properly, it is held in place by strands of connective tissue called **chordae tendinae**. The closing of the right AV valve prohibits blood from backing up into the right atrium.

6. Blood is forced out of the right ventricle into the **pulmonary artery**. Gently place the blunt probe up through the right ventricle, so that you can identify which vessel is the pulmonary artery. Blood enters the pulmonary circulatory system here.

7. Upon entering the pulmonary artery, the blood passes through another one-way valve called a **semilunar (SL) valve** (not shown in Figure 24.4). The valve gets its name from its crescent-moon shape. With the blunt probe, open the flaps of the SL valve, and again note how they "catch"

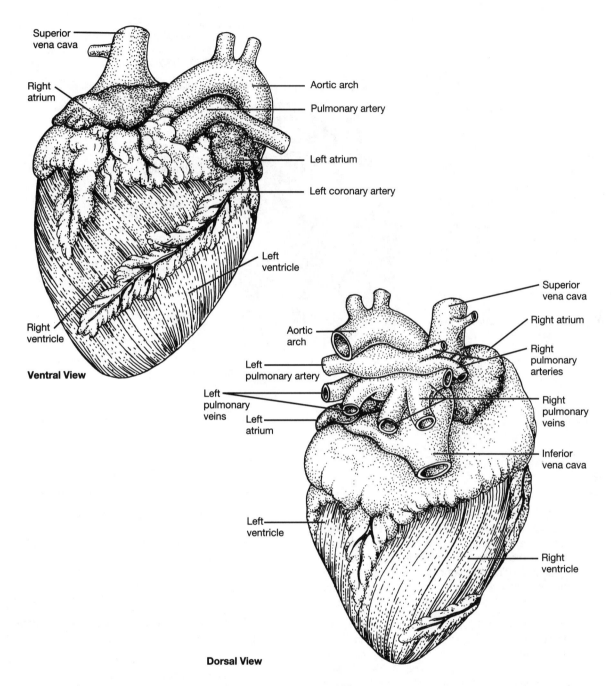

FIGURE 24.3   External Anatomy of the Sheep Heart, Dorsal and Ventral View.

blood. When the SL valve closes, blood can't flow down into the right ventricle. The pulmonary artery carries the deoxygenated blood from the heart to the **lungs**, where carbon dioxide is released from the plasma and oxygen is picked up by hemoglobin in red blood cells.

8. The freshly oxygenated blood is returned to the heart though the **pulmonary veins**. The pulmonary veins empty their contents into the **left atrium**. Place the blunt probe up through the left atrium, so that you can identify the pulmonary veins.

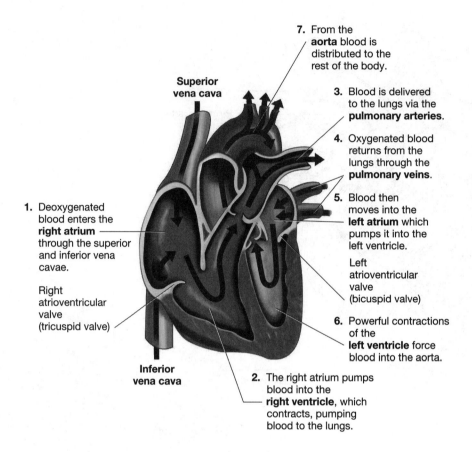

7. From the **aorta** blood is distributed to the rest of the body.

3. Blood is delivered to the lungs via the **pulmonary arteries**.

4. Oxygenated blood returns from the lungs through the **pulmonary veins**.

5. Blood then moves into the **left atrium** which pumps it into the left ventricle.

Left atrioventricular valve (bicuspid valve)

6. Powerful contractions of the **left ventricle** force blood into the aorta.

Superior vena cava

1. Deoxygenated blood enters the **right atrium** through the superior and inferior vena cavae.

Right atrioventricular valve (tricuspid valve)

Inferior vena cava

2. The right atrium pumps blood into the **right ventricle**, which contracts, pumping blood to the lungs.

FIGURE 24.4   Internal Anatomy of the Human Heart.

9. When the left atrium contracts, the blood passes through the **left AV valve** into the **left ventricle**. When the left ventricle contracts, blood is forced through the **aorta** and travels through the systemic circulation. Place the blunt probe up through the left ventricle to identify the aorta. You will see another **SL valve** at the beginning of the aorta.

10. Listen to the "lub-dub" sounds of your own (or someone else's) heartbeat with a stethoscope by placing the stethoscope diaphragm over the heart. The simultaneous closing of the right and left AV valves causes the "lub" sound (when the ventricles contract). The simultaneous closing of the SL valves causes the "dub" sound (when the ventricles relax). When a physician uses a stethoscope, a heart valve defect can be detected by listening for any abnormal heartbeat sounds **[Krogh section 28.3]**.

   It is important to note that *both atria contract simultaneously, and both ventricles contract simultaneously*. After the ventricles relax, the AV valves

"fall open," and the blood in the atria dumps into the ventricles. In fact, about 80% of the blood from the atria passes into the ventricles even before the atria contract.

Questions

1. **List the pathway by which blood travels through the heart and body.**

2. Mammals have a four-chambered heart, but amphibians and reptiles have a three-chambered heart, with one ventricle that receives blood from the body and lungs at the same time. What might be the advantage of having a four-chambered heart?

Exercise 24.2   PULSE RATE AND BLOOD PRESSURE

(Adapted from Gunstream, 2001)

Measuring how hard the heart works is an important indicator of cardiac health. When the heart ventricles contract, blood is pushed out of the heart with high pressure, creating a pulsation of blood. The number of pulses per minute is called the **pulse rate**.

A high pulse rate indicates that the heart is working hard to pump blood through the body. Higher pulse rates are common in animals that are exercising or that are under some other stress. In resting human adults, the normal pulse rate is 65–80 beats per minute, and in children it is higher. During exercise, the pulse rate increases as the heart works to pump blood to skeletal muscles. During rest, a well-conditioned athlete's pulse rate may be as low as 40 beats per minute.

The arteries carrying this blood must be resilient enough to withstand the pressure that pulsating blood exerts, called the **blood pressure**. Blood pressure can be measured by temporarily stopping blood flow through the brachial artery (Figure 24.2), then by noting the pressure at which certain arterial sounds are made when the blood begins to flow again. These arterial sounds are like the sounds you hear when you use a garden hose and nozzle; when the nozzle stops the flow of water, no sounds are heard. But as you begin to release the nozzle lever, the turbulent flow of water through the hose creates an audible sound, until finally the pressure is low enough that the water stops making sounds within the hose.

Chronic high blood pressure, called **hypertension**, is dangerous because it may cause damage to the heart and arteries. People with hypertension have an increased risk of heart damage, heart disease (buildup of plaque on the arteries), and stroke (rupturing or blockage of blood vessels in the brain).

1. With your index and middle fingers, find the pulse in your carotid or radial artery (see Figure 24.2). Record your pulse rate in Table 24.1. Then calculate average pulse rate for people in your class who exercise regularly, and record these in Table 24.1. Do the same for those who do not exercise regularly.

2. Watch your instructor demonstrate how to take a blood pressure reading before completing the following steps (see Figure 24.5).

3. The subject's arm should be resting on top of the lab bench. With your index and middle fingers, find the pulse in the subject's brachial artery.

4. Apply a blood pressure cuff to the arm, and then tighten the cuff by squeezing the cuff bulb until the meter has a reading greater than 150 mm. Tightening the cuff constricts the brachial artery.

Never let the student's arm remain compressed longer than 30 seconds!

While listening for the blood flow in the student's brachial artery with a stethoscope, slowly release the air by slightly opening the cuff valve.

5. Record the **systolic pressure** in Table 24.1 when you hear the blood start to pulse through the artery. Record the **diastolic pressure** when you no longer hear the pulsating sound. The systolic pressure is heard as the ventricles in the heart contract, increasing blood pressure in the arteries and forcing the blood through the small

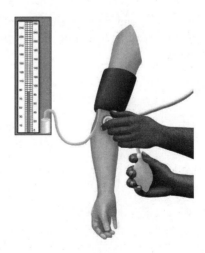

FIGURE 24.5   Measuring Blood Pressure.

TABLE 24.1 PULSE RATE AND BLOOD PRESSURE READINGS		
	Pulse rate (beats/min)	**Blood pressure (systolic/diastolic) [mm Hg]**
Your readings		
Average readings for regular exercisers		
Average readings for non-exercisers		

opening of the brachial artery. As the brachial artery continues to open, the pressure decreases so that blood no longer moves forcibly through the opening, and no sounds are heard. A normal systolic pressure is 120 mm (±10 mm), and a normal diastolic pressure is 80 mm (±10 mm).

6. Record your blood pressure reading in Table 24.1. Then calculate the average blood pressure readings for exercisers and non-exercisers in your class.

Questions

1. **What might cause an unusually high pulse reading?**

2. **What might cause an unusually high blood pressure reading? A low blood pressure reading?**

3. **What is the difference in the average pulse rate readings between exercisers and non-exercisers? Explain why this difference occurs.**

4. **What is the difference in the average blood pressure readings between exercisers and non-exercisers? Explain why this difference occurs.**

25

Rat Dissection

In this chapter you will dissect a white laboratory rat (*Rattus norvegicus*) to learn about basic mammalian anatomy. The white rat has played extremely important roles in biological, psychological, and medical research. Research conducted with laboratory rats has saved countless millions of lives, and has improved the quality of life for millions more. The laboratory rat was derived from wild strains of the Norway rat from the Old World. Other names for this rat are the common rat, brown rat, water rat, and sewer rat. These rats originally came from Asia and have been intimately associated with humans for a very long time. They spread across Europe centuries ago and arrived in the New World on trading ships during colonization.

The Norway rat is omnivorous and opportunistic in its feeding habits. It eats meat, insects, wild plants, and seeds, but shows a preference for stored grain (Figure 25.1). Every year, rats destroy millions of dollars' worth of grains and other goods. Rats have also been the source of much human suffering, serving as reservoirs for many serious human diseases, such as plague and typhus. It has been estimated that rat-borne diseases have cost more human lives in the past thousand years than all wars combined (*Sealander, 1979*).

FIGURE 25.1   The Norway Rat, *Rattus norvegicus*. It wreaks havoc in grain stores.

Rats used in laboratory exercises are purchased from biological supply companies; they are well preserved, and do not carry any pathogens.

The rat is useful in the general biology laboratory because it is an excellent representative of mammalian anatomy. The internal anatomy of the rat is, at least superficially, virtually identical to that of humans, with two exceptions. Rats lack a gallbladder and have a pronounced cecum. Humans have a gallbladder and a reduced cecum, called the appendix.

Exercise 25.1 EXTERNAL ANATOMY

Wear gloves and goggles.

You must wear eye protection at all times when working with the rat, because the preservatives can cause eye irritation.

Do not pour the preservative down the sink! It must be disposed of properly in a container provided by your instructor.

1. Identify the dorsal and ventral surfaces, and the anterior and posterior ends. Note the prominent post-anal tail of the rat, which helps the rat maintain its balance. On the ventral side, locate the mammary papillae, or teats.

2. Using your scissors, make a lateral incision on each side of the mouth as far back as you can cut. Fully open the mouth of the rat (Figure 25.2). You will observe the **tongue** with its associated **taste buds**. Under the taste buds are nerve cells that send impulses to the brain. Notice the **teeth**, which are part of the digestive system. The digestive process begins with breaking food materials by grinding or chewing, a process called *mastication*.

3. Observe the roof of the mouth. The anterior portion is the **hard palate**, and the posterior portion is the **soft palate**. The nasal cavities lie just above the hard and soft palates. The **nasopharynx** is in the soft palate near the rear of the mouth, and it connects the mouth with the nasal cavity. Cut away a part of the soft palate at the nasopharynx to expose a portion of the nasal cavity. Place a blunt probe into a nasal cavity so that you can trace its path to a nostril.

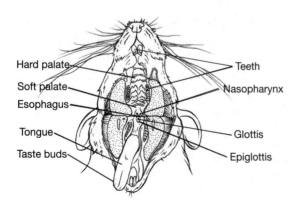

FIGURE 25.2 Rat Mouthparts.

4. In the back of the mouth, find the opening to the **esophagus**, which carries food to the stomach. Just ventral to the opening of the esophagus, notice a slit-like opening called the **glottis**. The glottis is the opening to the **trachea**, which carries air to the lungs. The **epiglottis** is the flap of tissue next to the glottis that prevents food from entering the trachea when the animal swallows. As food is swallowed, the epiglottis covers the glottis, so that the food passes into the esophagus, instead of into the trachea. Now place the jaws back together, as they would be situated when the mouth is closed. Notice how well the epiglottis fits into the nasopharynx.

Exercise 25.2 INTERNAL ANATOMY

Continuous with Exercise 25.1

BODY CAVITIES

1. The rat's body cavity is divided into two regions, an anterior **thoracic cavity** and a posterior **abdominal cavity**. The thoracic cavity contains the heart and lungs and is protected by the rib cage. The abdominal cavity contains most of the digestive organs and all of the organs of the urogenital system. Use your thumb and forefinger to pinch the abdominal wall of your rat (on the ventral side). Take care not to cut through the organs below the skin! Make cuts with your scissors as shown in Figure 25.3. Start by making a 1-cm horizontal incision at (1), then make incisions at (2), (3), and (4). Pour the excess preservative from your rat into the container provided. Notice that the two cavities are separated by a sheet of muscle called the **diaphragm**, which controls the expansion of the thoracic cavity during breathing. The esophagus, inferior vena cava, and aorta pass through the diaphragm.

2. Next, make incisions at (5), (6), and (7) as shown in Figure 25.3. Pull back the walls of the thoracic cavity and pin them to your dissecting tray so that your rat appears as shown in Figure 25.4. You are now ready to conduct a thorough examination of the body cavities.

VISCERA OF THE NECK

3. The **lymph nodes** are associated with the immune system and are involved in the production of white blood cells, which "fight" infections in various ways. The lymph nodes often become swollen when you are ill, especially with a

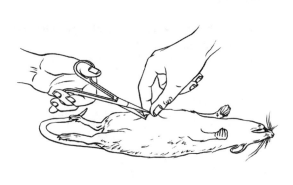

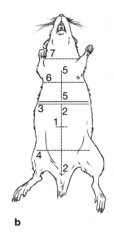

FIGURE 25.3   Incisions for Rat Dissection. Lift the skin as shown in (a) and make the incisions as shown in (b).

a

b

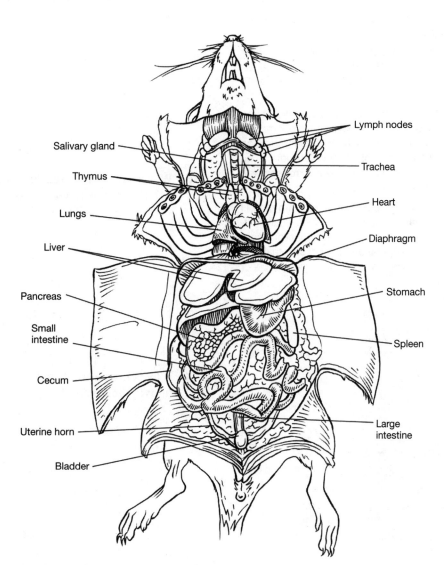

FIGURE 25.4   Internal Anatomy of the Rat.

bacterial infection. The **thymus gland** is where T cells (immune system cells) mature, especially in younger animals, and secretes hormones involved in the immune response. The thymus is bilobed and located on both sides of the neck (Figure 25.5).

4. Behind the glands and muscles you will find the **trachea**, or windpipe. The trachea is lined with rings of cartilage. The rings of cartilage keep the trachea from collapsing as a food bolus passes down the **esophagus** that lies dorsal to the trachea.

VISCERA OF THE THORAX

5. Follow the trachea to where it branches into the two **bronchi**, one leading to each **lung**. Look at the position of the lungs in the body cavity. Notice that the **heart** is surrounded by a thin layer of tissue called the **pericardium**. Remove the pericardium, and identify the atria and ventricles.

6. If your instructor wants to continue the dissection next week, close the body cavities of your rat and place the animal in a plastic bag. Seal the bag and label it with your initials and lab period, so that you can easily find it next time. (Proceed with Step 7 before you leave class.)

7. When your gloves are off and your lab bench is clean, use a stethoscope to listen to your partner's breathing sounds. As the diaphragm contracts, the lungs expand and air moves into them. Listen to each lobe of your partner's lungs.

Questions

1. **Which hallmark chordate characteristic(s) do you see in the rat?**

2. **Which of these are not found in adult humans?**

3. **Where are the rat's sensory structures located? What fundamental evolutionary innovation does this illustrate?**

4. **Why is it an advantage for the epiglottis to fit over the trachea?**

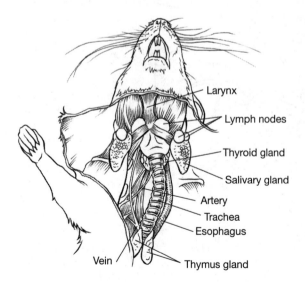

Larynx

Lymph nodes

Thyroid gland

Salivary gland

Artery

Trachea

Esophagus

Thymus gland

Vein

FIGURE 25.5   Neck Viscera.

5. When a physician listens to your lungs, where is the stethoscope placed? Why?

Exercise 25.3   THE DIGESTIVE SYSTEM [Krogh sections 28.8–28.10]

Continuous with Exercise 25.1

1. Using Figure 25.4, find the **stomach** in the anterior portion of the abdominal cavity. Find the point of entry of the esophagus into the stomach. Find the point at which the stomach joins the **small intestine**. Lift the stomach, and notice that it is bound to the other intestinal organs by a thin, translucent membrane called the **mesentery**. The mesentery contains the many blood vessels supplying the intestine. The intestine is highly vascularized because the nutrients resulting from the digestion of our meals must be absorbed into the circulatory system for transport to all tissues of the body.

2. Find the **pancreas**, a diffuse organ located in the mesentery between the stomach and the first loop of the small intestine. The pancreas produces digestive enzymes that are secreted into the small intestine.

3. The largest organ in the abdominal cavity is the dark, lobed **liver**. A primary function of the liver is the production of bile that aids in the breakdown of fats. In most mammals, the bile is stored in the gallbladder. Because rats have no gall-bladder, the bile is secreted directly from the liver into the small intestine.

4. Another prominent organ is the tongue-shaped **spleen**, which also has a dark pigment in most specimens. Locate the spleen that lies dorsal to the stomach on the rat's left side. The spleen produces certain types of leukocytes, filters the blood, and destroys and removes old red blood cells. The spleen is not a vital organ in adults, but is essential in infants because it produces red blood cells early in life.

5. Follow the small intestine until you find the **cecum**, a large pouch situated at the junction of the small intestine and the **large intestine** (colon). In many animal species, the cecum contains microbes that secrete enzymes that help to digest plant matter. In humans, the cecum is greatly reduced, with the long, finger-like appendix extending from it. Our appendix is a *vestigial organ*, a remnant of a more prominent cecum presumably possessed by our ancestors.

INTESTINE CROSS SECTION

6. Obtain a slide of a cross section of a human small intestine, and observe it with a compound microscope. The intestine is made of several different types of cells, which comprise several different types of tissues of the four intestinal layers. The hollow part of the intestine is called the **lumen**, where digestion occurs (Figure 25.6). Also see Figure 22.2.

7. The **mucosa** is the innermost layer of the intestine. Observe this on high power, and note that it is a single layer of simple columnar epithelial cells. This thin layer of epithelial tissue serves as the site of nutrient absorption. It is also the first line of defense against the entry of pathogens from the lumen. The mucosa consists of many finger-like extensions called **villi** (singular, *villus*) that increase the surface area available for digestion and absorption. Interspersed within the mucosal lining are specialized cells called **goblet cells**, which appear as clear, goblet-shaped bubbles. They secrete mucus, which protects the mucosal layer from its own digestive enzymes.

8. The next layer is the **submucosa**, which consists of loose connective tissue interspersed with blood vessels and nerves. Nutrients are absorbed across the mucosa and enter the capillary network in the submucosa. The submucosa is an excellent place to observe the structure of blood vessels.

9. The next layer is the **muscularis**, which consists of an inner layer of circular smooth muscle and a layer of longitudinal smooth muscle. Contraction of the circular smooth muscle is responsible for "segmentation movements" that churn the food materials around within a limited area of the intestine, much like a washing machine. Contraction of the longitudinal smooth muscle is responsible for **peristalsis**, waves of contraction that move the food along the digestive tract.

10. The outer layer is the **serosa** and consists of loose connective tissue, including adipose tissue, and blood vessels. The serosa is bound to the mesentery, which holds the intestine in place.

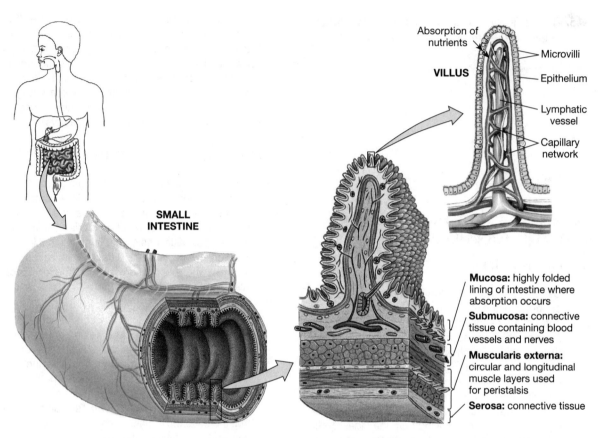

Absorption of nutrients

VILLUS

Microvilli

Epithelium

Lymphatic vessel

Capillary network

SMALL INTESTINE

Mucosa: highly folded lining of intestine where absorption occurs

Submucosa: connective tissue containing blood vessels and nerves

Muscularis externa: circular and longitudinal muscle layers used for peristalsis

Serosa: connective tissue

FIGURE 25.6   Structure of the Small Intestine.

Questions

1. **Why is increased surface area in the intestine an advantage for organisms?**

2. **Why are there so many blood vessels adjacent to the serosa? What is their function?**

3. **Sketch the layers of the intestine cross section.**

4. **On the basis of the presence of the appendix, what might you infer about dietary habits of our evolutionary antecedents? Is this information consistent with what you learned about human evolution in Exercises 12.3 and 12.4?**

Exercise 25.4   THE EXCRETORY SYSTEM

Continuous with Exercise 25.1

The excretory and reproductive systems make up the urogenital system. We will begin by looking at the morphology of the excretory system (Figure 25.7), whose function is to rid the body of metabolic wastes (urine). The **kidney** is the principal excretory organ in the body. The mammalian kidney is *retroperitoneal*, meaning it lies outside the body cavity, dorsal to the peritoneum **[Krogh sections 28.11–28.13]**.

Remember that chordates are eucoelomates, possessing a true body cavity, with a peritoneum of mesodermal origin that lines the body cavity.

1. Lay the intestine of the rat aside so that the dorsal wall of the body cavity is exposed. The thin, translucent membrane lining the body cavity of the rat is the peritoneum. You will have to cut the peritoneum in order to fully expose the kidney. Locate the **renal vessels** that carry blood to and from the kidney. Find the thin **ureter**, which carries urine from the kidney to the **bladder** where it is stored until it is voided from the body. Trace the path of the ureter from the kidney to the urinary bladder.

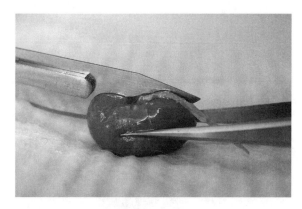

FIGURE 25.8   Kidney Dissection. Caution! Use forceps and a scalpel to make the cross-section shown.

2. Carefully remove one of the kidneys with your scissors. Cut the kidney as shown in Figure 25.8. Examine the two halves of the kidney under a stereoscopic microscope. The functional unit of the kidney, the **nephron**, is located in the **renal cortex**, with its loop of Henle and collecting duct extending into the **renal medulla**. The collecting ducts carry urine into the **renal pelvis**, where it is carried out of the kidney through the ureter (Figure 25.9).

Sketches

1. **Sketch and label the cortex, medulla, renal pelvis, and ureter of the kidney.**

2. **Sketch and label the positions of the kidney, peritoneum, ureters, and bladder of your rat.**

3. **Where do the ureters enter the body cavity?**

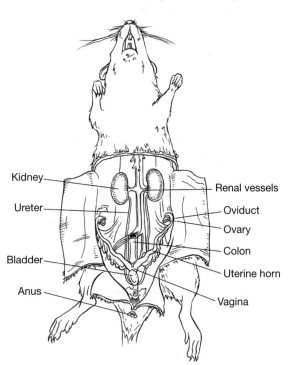

FIGURE 25.7   Female Urogenital System.

(a)

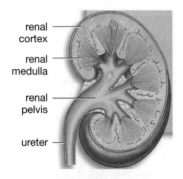

renal cortex
renal medulla
renal pelvis
ureter

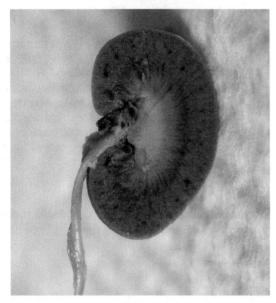

FIGURE 25.9 Top: Kidney Cross Section. Bottom: Cross Section of a Rat Kidney. Compare this photo to the features labelled in the top figure.

Exercise 25.5 THE REPRODUCTIVE SYSTEMS [Krogh sections 30.1–30.3]

Continuous with Exercise 25.1

MALE REPRODUCTIVE SYSTEM

1. If you have a male rat, carefully make a longitudinal incision down the length of the **scrotum**, exposing one **testis** (Figure 25.10). Carefully free the testis from the scrotum. Locate the **epididymis**, where sperms mature and are stored. The epididymis is often protected by fat tissue that is yellow in color. During ejaculation, sperms pass through the vas deferens and are joined by secretions from the seminal vesicles, prostate gland, and bulbourethral gland. The **seminal vesicles**

are often darker than the surrounding organs, and they are granular in appearance. They secrete energy-rich fructose that nourishes the sperm on their journey toward the egg. The **prostate** gland is white and has small lobes lying around the urinary bladder. The prostate secretes a basic solution that helps to neutralize the acidity of the vagina, enabling sperm to survive longer in the female reproductive tract.

2. Next locate the urogenital opening, and cut upward through the opening to expose the **penis**. Like the rat's, the penis of many mammals is supported by cartilage or bone. Collectively, the sperm along with its secretions make up the *seminal fluid*. Like urine, seminal fluid leaves the penis through the urogenital opening.

3. Exchange your male rat with a dissected female rat from another lab group, so that you can identify the female reproductive organs.

THE FEMALE REPRODUCTIVE SYSTEM

4. If you have a female rat, identify the various parts of the female reproductive system (Figure 25.7). Lay the small intestine aside and locate the **ovaries**, which are embedded in fat in the dorsal wall of the body cavity. At ovulation the ovaries release eggs that are swept up by the **oviducts** (fallopian tubes). The oviducts have finger-like projections called *fimbrae* that envelop the ovary and guide the egg into the tube.

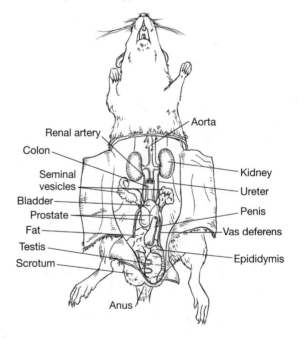

Renal artery
Colon
Seminal vesicles
Bladder
Prostate
Fat
Testis
Scrotum
Anus
Aorta
Kidney
Ureter
Penis
Vas deferens
Epididymis

FIGURE 25.10 Male Urogenital System.

5. After fertilization, the embryos implant in the **uterus**, which has two halves, or horns, leading up to each ovary. Animals with uterine horns can bear many young during a single pregnancy. If the uterine horns have swellings in them, the rat may be pregnant. If so, cut a uterine horn to reveal the fetuses. Rats have a gestation period of approximately 22 days, and have an average of six young per litter. Powerful uterine contractions expel the young rats through the **vagina**. After a baby's head is expelled, the female rat will assist by pulling the baby out with her mouth. Then she allows the newborns to nurse.

6. Exchange your rat with a dissected male rat from another group, so that you can identify the male reproductive organs.

Questions

1. **For each mammalian organ system, list the organs and their function.**

   Respiratory:

   Digestive:

Excretory:

Reproductive:

Circulatory:

Immune:

UNIT 7
Ecology

26

Population Ecology

*... humanity threatens to do what living systems
never do, namely to suffocate in itself.*
Konrad Lorenz, 1973

Ecology is the study of interactions among living things and between living things and their nonliving environment. Ecology is a holistic discipline; life is highly organized in a hierarchical fashion (Figure 26.1), and an ecologist must understand something of every level of this organization. At the conclusion of these exercises, you should appreciate how living things are interwoven; and hopefully, you will begin to realize the place that humans occupy in the biosphere.

A **population** consists of all members of a species living together within a specific geographic region. To a great extent, population ecology deals with the study of population dynamics—how a population is structured and how its size fluctuates over time. In 1798, Thomas Malthus pointed out that animals tend to produce many more offspring than will survive to adulthood. He observed that populations tend to increase at an exponential rate, while resources increase at an arithmetic rate, at best (Figure 26.2).

Thomas Malthus was an English clergyman and philosopher who wrote *An Essay on the Principle of Population*. It not only provided much of the framework for population ecology but also provided Darwin and Wallace with a foundation for their theory of organic evolution by natural selection.

Populations can show an exponential growth rate (Figure 26.3a) for a short period of time under ideal conditions. Such conditions include a habitat free of predators and disease where there is no shortage of food, water, or space. Under such conditions the birth rate is high, the death rate is low, and population growth is limited only by the reproductive physiology of the organisms. Obviously, no population can grow this way forever. Something will eventually check the growth—predators, disease, or a shortage of food, water, or shelter. Such factors that lead to high mortality and/or a low birth rate are called **density-dependent factors**, or **environmental resistance**. A population may start exhibiting exponential growth, but eventually a density-dependent factor will "kick in." Therefore, the dynamics of most populations may be described by a logistic growth curve (Figure 26.3b). The upper limit is called the **carrying capacity** of an environment, or *K*. In reality, the population size does not stay exactly at the carrying capacity, but fluctuates near the carrying capacity over time (Figure 26.3c) (*Robinson and Bolen, 1984*) **[Krogh sections 31.2 and 31.3]**.

The scenario that we have talked about so far applies to **K-selected equilibrium species** that are limited by a carrying capacity defined by density-dependent factors. K-selected equilibrium species tend to produce relatively few offspring, and they exhibit high degrees of parental care. There are also **r-selected opportunistic species** whose population growth is density-independent. These are limited only by their reproductive rate, r. These species provide no parental care, and they produce as many offspring as physiologically possible. In this case, the population will exhibit exponential growth until food runs out or environmental conditions otherwise turn bad, and then it will crash. This reproductive strategy is sometimes called "boom or bust." Figure 26.4 summarizes the differences in K-selected and r-selected species.

Exercise 26.1   DYNAMICS OF AN ANIMAL POPULATION

Extended 5- to 11-week exercise

In this exercise you will monitor the population dynamics of the common fruit fly, *Drosophila melanogaster*. This is the same species you used for a monohybrid cross in the genetics lab (Exercise 10.1). Two days after emerging from their pupae, female fruit flies begin laying eggs, and a

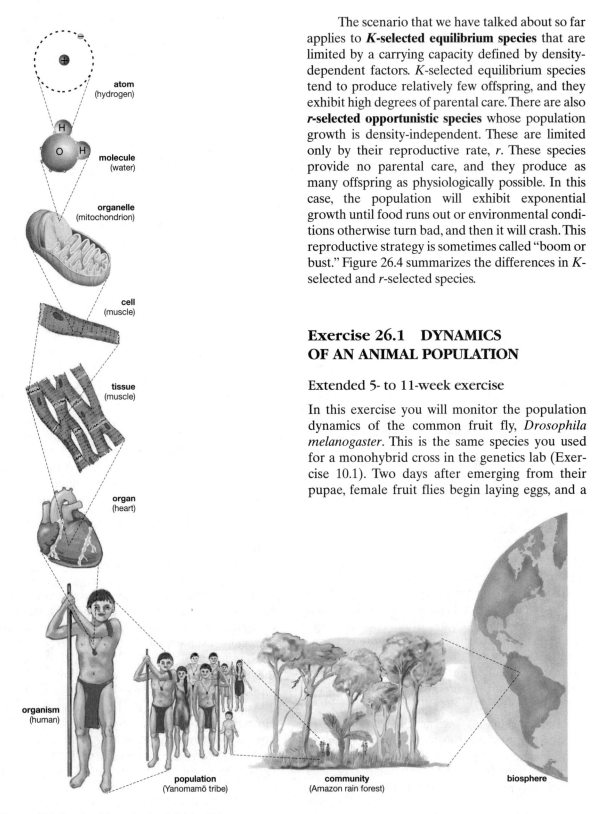

atom
(hydrogen)

molecule
(water)

organelle
(mitochondrion)

cell
(muscle)

tissue
(muscle)

organ
(heart)

organism
(human)

population
(Yanomamö tribe)

community
(Amazon rain forest)

biosphere

FIGURE 26.1 Levels of Organization in Living Things.

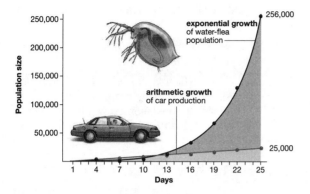

FIGURE 26.2 Arithmetic Growth v. Exponential Growth. In arithmetic growth, the same number of units is added to the population in each time interval, so that the growth rate is constant. In exponential growth, the number of units added to the population depends on the population size, so that the rate of growth increases with each interval.

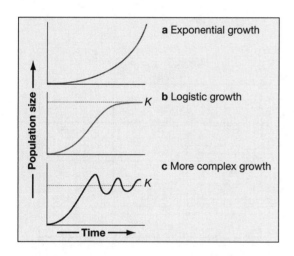

FIGURE 26.3 Logistic growth (b) is a pattern seen in natural populations, which starts like exponential growth (a), but slows because of density-dependent factors. In reality, K-selected populations actually fluctuate around the carrying capacity, as shown in (c).

new generation appears in only two weeks. Fruit flies live for several weeks and can produce many offspring, and you'll be counting them.

Week 1

1. Prepare 10 vials for your fly cultures: using the small plastic cup provided, pour equal parts of *Drosophila* medium and tap water into each vial. Drop 5–10 grains of yeast onto the medium in each vial. On new labels, write your initials and the data, and place the labels on the vials.

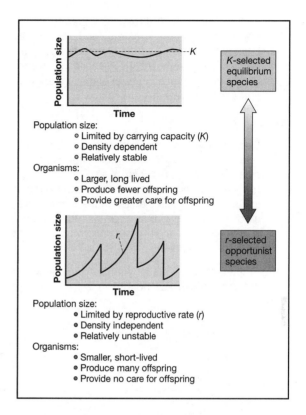

FIGURE 26.4 K-selected and r-selected Species.

2. Anesthetize the flies from the stock culture: dip the absorbent end of a wand into the Flynap™ bottle. Gently tap the vial containing the flies on your lab bench, push the stopper aside with your index finger, and quickly place the absorbent end of the wand below the stopper. Try to avoid getting Flynap™ on the foam stopper. Lay the vial on its side on your lab bench to prevent the sleeping flies from getting stuck in the medium. Watch the flies closely, and remove the foam stopper immediately after all the flies have stopped moving. If you are anesthetizing the flies in the stock culture vial, do not leave the wand in it longer than 4 minutes. If you are using an empty vial (without culture medium) do not leave the wand in it longer than 2 minutes.

DO NOT OVER-ANESTHETIZE THE FLIES!

3. Pour the sleeping flies onto a small, white piece of paper. Use a dissecting microscope

and soft-bristle paintbrush to separate males from females. Place 1 male fly and 1 female fly into each of your new vials. (You will have extra vials in case some of the flies do not reproduce.) Put a foam stopper in place in each vial, and store the vials in a safe place.

Weeks 3–11

4. Every two weeks, heavily anesthetize the flies in **one** of your vials. Be sure to lay the vial on its side when using the Flynap™. Pour the flies onto a white card and count them. Record this number in Table 26.1. Place the flies in the morgue, discard your empty vial as your instructor directs, and continue storing the remaining vials until you finish the experiment.

5. When gathering data for your last sample, weigh 100 flies and divide by 100 to get an average weight per fly (to the nearest mg). Record this weight in Table 26.1.

6. At the end of the experiment, graph the class data in Figure 26.5.

7. Calculate r, which is the rate of growth for the population. It is actually a measure of the relative change of population growth between time intervals in Table 26.1, according to the following formula:

$$r = \frac{\text{flies at time}_{(n+2 \text{ weeks})} - \text{flies at time}_n}{\text{flies at time}_n}$$

where time$_{(n+2 \text{ weeks})}$ refers to the number of flies at the present time interval, and time$_n$ refers to the number of flies at the previous time interval.

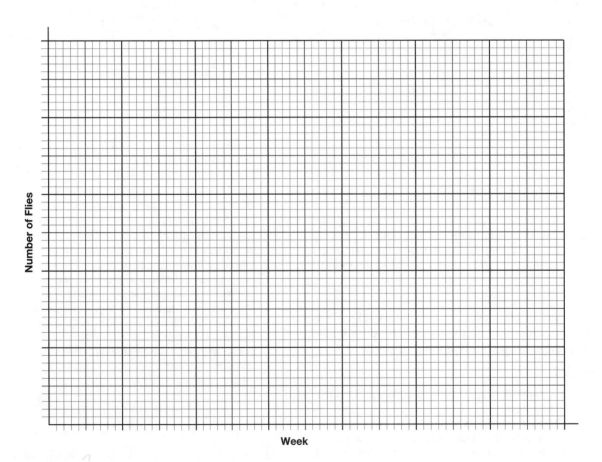

Number of Flies

Week

FIGURE 26.5 Graph of *Drosophila* Population Dynamics.

TABLE 26.1 *DROSOPHILA* POPULATION DYNAMICS		
Start date:		
Elapsed time (weeks)	**Number of flies**	*r*
2		
4		
6		
8		
10		
Average fly weight:		

Questions

1. Do the dynamics of your fly population best conform to exponential growth (Figure 26.3a), logistic growth (Figure 26.3b) or more complex growth (Figure 26.3c)?

2. Are these flies best described as an *r*-selected or *K*-selected species?

3. What factor(s) do you think limited the growth of your fly population?

4. What factor(s) might limit fly populations in nature?

5. Using the maximum *r* value, estimate the number of flies that a pair of flies could give rise to after 6 months, 1 year, and 5 years, assuming that they were allowed to reproduce at their full potential. You may predict the population size for any point in time with the formula

$$y = e^{rx}$$

where

y = the numbe r of individuals,
e = 2.71828 (base of natural logarithms),
r = the observed intrinsic growth rate of the population, and x = time.

6. How much would all the flies weigh at 6 months, 1 year, and 5 years?

Exercise 26.2   DYNAMICS OF A PLANT POPULATION

(Adapted from DeBuhr, 1994)

Extended 5- to 15-week Exercise

Duckweed (*Lemna* spp.) is a common, aquatic flowering plant found on the surface of ponds, marshes, and other stagnant bodies of water throughout North America. Duckweed is the smallest flowering plant known, and although it reproduces sexually, its rapid reproduction and high reproductive potential is because of its ability to reproduce asexually. Because they are so prolific, duckweed may serve as an important source of primary production in an aquatic ecosystem. In this exercise you will monitor the population dynamics of duckweed in an aquarium.

Week 1

1. Determine the surface area of your aquarium in cm^2. Place the aquarium near a window or under a grow light.

2. Fill the aquarium half full with tap water and gently place 10 duckweed plants in it, using forceps. Insert an airstone into the water, attach it to an air pump, and turn the pump on.

Week 2–15

3. Each week, count the number of plants in your aquarium. Record your results in Table 26.2. At some point, the plants are going to become too numerous to count. When this happens, discuss with your classmates about how to design a method of accurately estimating the number of plants in your aquarium.

4. Upon completion of the experiment, graph your data in Figure 26.6.

5. Calculate *r*, which is the rate of growth for the population. It is actually a measure of the relative change of population growth between time intervals in Table 26.2, according to the following formula:

$$r = \frac{\text{plants at time}_{(n+1 \text{ week})} - \text{plants at time}_n}{\text{plants at time}_n}$$

where $\text{time}_{(n+1 \text{ week})}$ refers to the number of plants at the present time interval, and time_n refers to the number of plants at the previous time interval.

TABLE 26.2 *Lemna* Population Dynamics		
Start date:		
Elapsed time (weeks)	**Number of plants**	***r***
1		
2		
3		
4		
5		
6		
7		
8		
9		
10		
11		
12		
13		
14		
15		

Questions

1. **Describe your method of estimating the number of duckweed plants.**

2. **Do the dynamics of your duckweed population best conform to exponential growth**

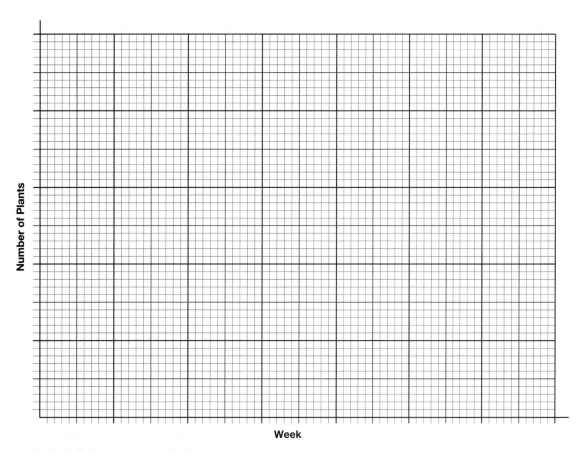

FIGURE 26.6 Graph of *Lemna* Population Dynamics.

(Figure 26.3a), logistic growth (Figure 26.3b), or more complex growth (Figure 26.3c)?

3. **Are these plants best described as an *r*-selected or *K*-selected species?**

4. **What factor(s) do you think limited the growth of your duckweed population?**

5. **Compare and contrast the limiting factors listed in Question 4 above with those that you identified in Question 3 for your fly population. List the differences and give the reason(s) why they occur.**

6. **What factor(s) might limit duckweed populations in nature?**

7. **What is the carrying capacity of duckweed in your "community"?**

8. **What factors might influence the initial growth rate of your duckweed population?**

9. **How might the growth rate be increased?**

10. **How might the growth rate be decreased?**

11. **Name at least 2 ways in which the carrying capacity might be manipulated.**

12. **State a hypothesis concerning one of the factors that might influence the carrying capacity.**

13. **At the peak growth rate exhibited by your duckweed population, how long would it take the duckweed to completely cover an area the size of a football field, starting with your 10 plants? A football field is 44,593,459 cm^2. Hint: You may calculate this empirically using the formula in Question 5 in Exercise 26.1. You may either solve for x or you can just plug several x values and solve for y to get the number of plants expected at various time intervals, and then approximate the time required for the plants to fill an area the size of a football field.**

Exercise 26.3   LIFE TABLE FOR A HUMAN POPULATION

(Adapted from an exercise by R. E. Clopton)

Extended 2-week exercise OR only 1 week if data is already available

A **life table** tells you how likely it is that an individual of a certain age in a population will be alive at a particular time. An example of population structure for 17th century London is given in

Table 26.3. As you can see, the percent of people surviving decreased as the sample population aged. But this data can also be used to construct a life table, which will show life expectancy for people in this sample. Life tables are used by life insurance companies, historians, and population ecologists, just to name a few. In this exercise, you will construct a life table for a human population using data from cemetery headstones or records.

What are the dynamics of a population that affect survivorship? There are a variety of questions that you can investigate with life table data. Aside from calculating overall life expectancy for the entire sample, you could investigate whether there are differences in survivorship of males and females, or differences in survivorship between different age groups in your historical period. Another interesting phenomenon to investigate is infant mortality. Don't overlook the potential role of natural disaster and war in the life expectancy for your sample [Krogh section 31.4].

1. Obtain records from local cemeteries, or go out to a cemetery and gather the data with your lab partners. Work with your lab partners to gather

TABLE 26.3 POPULATION STRUCTURE (SURVIVORSHIP) FOR PEOPLE BORN IN LONDON, 1662. (DATA FROM WILCOX, 1937)

London, 1662

Age	Percent surviving
0	100
6	64
16	40
26	25
36	16
46	10
56	6
66	3
76	1
86	0

data for people born during your chosen period (your instructor may assign a certain period to you). Periods include individuals who were born from 1650–1750, 1751–1850, 1851–1900, and 1901–1920.

SHOW RESPECT FOR THE DEAD! DO NOT WALK ON OR ACROSS GRAVES

For each individual, record the date of birth, date of death, and gender in Table 26.6. For convenience, record data from one gender at the top, and data from the other gender at the bottom. If you are unsure of the gender (for names such as Marion or Robin) exclude the individual from the sample. If the cemetery headstones do not have first names recorded, then you will not be able to investigate the effects of gender on life expectancy. Instead, you will have to use Table 26.9 for combined genders; after reading the rest of this page, skip to Step 13.

A hypothetical data set based on the actual London, 1662 population is in Table 26.3. A life table was prepared from this data (Table 26.4) to provide you with an example to follow when you construct life tables for your own data. Dates of death were made up so that the "Age at Death" column could be determined. The age at death shows how many people died before they reached a certain age. Table 26.4 shows that four people died before age one. This number is recorded in column 4 of Table 26.5.

Here are the columns of data you'll be using in the life tables:

Column 1: Age group (recorded in years)

Column 2: Number of people surviving at the end of the age group. In the 1–4 age group, this is actually the number of people surviving until age 5.

Column 3: Percent of people surviving at the end of the age group (survivorship).

Column 4: Number of people that didn't survive until the end of the age group (i.e., people that died within the age group specified).

Column 5: Median age of age group. For convenience, you may use the median age of the age group as the age at death for **everyone** that died in that age group.

Column 6: Total years lived by those that died. Confusing? Hang in there; you'll need this number later. For example, in Table 26.5 you will multiply the numbers in column 4 and column 5 to

TABLE 26.4	HYPOTHETICAL DATA FOR LONDON, 1662				
Time period: London, 1662					
#	Name	Date of birth	Date of death	Age at death	Gender
1		1662	1662	<1	
2		1662	1662	<1	
3		1662	1662	<1	
4		1662	1662	<1	
5		1662	1663	1	
6		1662	1664	2	
7		1662	1666	4	
8		1662	1667	5	
9		1662	1669	7	
10		1662	1671	9	
11		1662	1673	11	
12		1662	1675	13	
13		1662	1678	16	
14		1662	1680	18	
15		1662	1684	22	
16		1662	1695	33	
17		1662	1699	37	
18		1662	1705	43	
19		1662	1724	62	
20		1662	1744	82	

get the number for column 6 (number of people that died × the median age of the age group), then you'll use this in column 7.

Column 7: Show your calculation here. Subtract the number you contrived in column 6 from the total number of years lived by all those in the sample (this is from the "Total" line at the bottom of the table). When you calculate this formula for the next age group, be sure to use the answer you calculated from the row just above it. See Table 26.5 for an example.

Column 8: Life expectancy: Divide the number from column 7 by the number surviving (column 2).

TABLE 26.5   HYPOTHETICAL AVERAGE LIFE EXPECTANCY FOR LONDON, 1662 (COMBINED GENDERS)

Location and time period: London, 1662

Sample size: 20
Infant mortality rate: 4 ÷ 20 = 20%

1	2	3	4	5	6	7	8
Age group (years)	Number surviving at end	Percent surviving	Number that died	Median age of age group	Total years lived by those that died	Calculation	Life expectancy
<1	16	80	4	0.5	2	372 − 2 = 370	370 ÷ 16 = 23
1–4	13	65	3	2.5	7.5	370 − 7.5 = 362.5	362.5 ÷ 13 = 28
5–9	10	50	3	7.5	22.5	362.5 − 22.5 = 340	340 ÷ 10 = 34
10–14	8	40	2	12.5	25	340 − 25 = 315	315 ÷ 8 = 39
15–19	6	30	2	17.5	35	315 − 35 = 280	280 ÷ 6 = 47
20–24	5	25	1	22.5	22.5	280 − 22.5 = 257.5	257.5 ÷ 5 = 52
25–29	5	25		27.5			
30–34	4	20	1	32.5	32.5	257.5 − 32.5 = 225	225 ÷ 4 = 56
35–39	3	15	1	37.5	37.5	225 − 37.5 = 187.5	187.5 ÷ 3 = 63
40–44	2	10	1	42.5	42.5	187.5 − 42.5 = 145	145 ÷ 2 = 73
45–49	2	10		47.5			
50–54	2	10		52.5			
55–59	2	10		57.5			
60–64	1	5	1	62.5	62.5	145 − 62.5 = 82.5	82.5 ÷ 1 = 83
65–69	1	5		67.5			
70–74	1	5		72.5			
75–79	1	5		77.5			
80–84	0	0	1	82.5	82.5		
Total					372		

Average life expectancy: 372 ÷ 20 = 19

TABLE 26.6	RAW DATA FOR LIFE TABLE CONSTRUCTION				
Time period:					
#	Name	Date of birth	Date of death	Age at death	Gender
1					
2					
3					
4					
5					
6					
7					
8					
9					
10					
11					
12					
13					
14					
15					
16					
17					
18					
19					
20					
21					
22					
23					
24					
25					
26					
27					
28					
29					
30					

TO ANALYZE DATA FOR GENDER DIFFERENCES IN ONE TIME PERIOD

2. Show all your calculations in the tables. Data will be analyzed for infants (<1 year) and then in groups of 5-year intervals. Record the total number of males from your raw data in Table 26.6 as the sample size in Table 26.7. This number represents every male born in that time period. The difference between the sample size and the number surviving at the end of age group <1 represents male infant mortality for the time period. For example, if 5 males out of 100 died, the infant mortality rate would be 5%.

3. Using the "Age at Death" column from your raw data in Table 26.6, count the number of males who died before their first birthday and put this number in column 4 of Table 26.7. Fill in the entire column in this manner, continuing with the number of males who died in the next age group, before their fifth birthday, and so on. For example, Table 26.4 lists 4 people who died before age 1, 3 who died between ages 1–4, 3 who died between ages 5–9, 2 who died between ages 10–14, and so on. Check your work by adding up the numbers in column 4. This number should equal your sample size.

4. Subtract the number in column 4 from the sample size and record this number in column 2 (number surviving) for age group <1; this is the number of people surviving until age 1.

5. For age group 1–4, subtract the number in column 4 from the number in column 2 **from the previous age group (<1)**, and record this in column 2 for age group 1–4. This is the number of survivors at the end of age group 1–4. Continue this for each age group (5–9, 10–14, etc.) until all of column 2 is filled.

6. Next convert the number of surviving individuals for each age group to percentage values by dividing the number in column 2 by the sample size. Record these percentage values in column 3. This gives you the **population structure** (survivorship) for your sample population.

7. For each time period, multiply the number of people who died (column 4) by the median age for the time period (column 5). Record the product in column 6.

8. The **average life expectancy** for all males in your historical period is the average age at death of the males in your sample. Add the numbers in column 6 of Table 26.7 and record the total in the space

provided at the bottom of the table. Then divide this total by the sample size. Record this number as overall life expectancy in Table 26.7. For example, our hypothetical population in Table 26.5 has an overall life expectancy of 372 ÷ 20 = 19.

9. Calculate the average life expectancy for each age group. (Follow the example given in Table 26.5.) For age group <1, subtract the number in column 6 from the total at the bottom of the Table 26.7; record this in column 7.

10. Divide the answer from column 7 by the number given in column 1; record this in column 8. This is the life expectancy for individuals at the end of the age group; in other words, for those surviving to age one. In contrast, **Krogh Table 31.2** shows the average remaining lifetime in years, which is the life expectancy minus the age of the individual.

11. Repeat Steps 9 and 10 for all the age groups.

12. Repeat Steps 2–11 for females in your sample; record the numbers in Table 26.8. Answer Question 1.

TO ANALYZE DATA FOR COMBINED GENDERS

13. Repeat Steps 2–11 again, but this time combine data for both genders. Record the numbers in Table 26.9.

TO COMPARE LIFE EXPECTANCY WITHIN A GENDER IN THE SAME TIME PERIOD

14. Example: Suppose you are testing the prediction that complications associated with childbirth increase the mortality of females. If this were the case, you would expect the average life expectancy for females beyond their childbearing years to be greater than that of women in their childbearing years. Split the data for the women into two age groups: 15–19, and 50–54. Then compare these 2 female age groups to the same age groups for males. This provides a control.

15. Compare these two groups to see if life expectancy is lower for the childbearing age group. Answer Questions 2 and 3.

TO COMPARE DATA FROM TWO DIFFERENT TIME PERIODS

16. Now, test your own hypothesis, using similar methods and comparing life expectancy between 2 or more groups in different time periods. Compare the life expectancy of an age group (of males, females, or both genders) by sharing data

TABLE 26.7 LIFE TABLE FOR MALES							
Location and time period:							
Sample size: **Infant mortality rate:**							
1	**2**	**3**	**4**	**5**	**6**	**7**	**8**
Age group (years)	Number surviving at end	Percent surviving	Number that died	Median age of age group	Total years lived by those that died	Calculation	Life expectancy
<1				0.5			
1–4				2.5			
5–9				7.5			
10–14				12.5			
15–19				17.5			
20–24				22.5			
25–29				27.5			
30–34				32.5			
35–39				37.5			
40–44				42.5			
45–49				47.5			
50–54				52.5			
55–59				57.5			
60–64				62.5			
65–69				67.5			
70–74				72.5			
75–79				77.5			
80–84				82.5			
85–89				87.5			
90–94				92.5			
95–99				97.5			
100–104				102.5			
Total							
Average life expectancy:							

TABLE 26.8	LIFE TABLE FOR FEMALES						
Location and time period:							
Sample size: **Infant mortality rate:**							
1	2	3	4	5	6	7	8
Age group (years)	Number surviving at end	Percent surviving	Number that died	Median age of age group	Total years lived by those that died	Calculation	Life expectancy
<1				0.5			
1–4				2.5			
5–9				7.5			
10–14				12.5			
15–19				17.5			
20–24				22.5			
25–29				27.5			
30–34				32.5			
35–39				37.5			
40–44				42.5			
45–49				47.5			
50–54				52.5			
55–59				57.5			
60–64				62.5			
65–69				67.5			
70–74				72.5			
75–79				77.5			
80–84				82.5			
85–89				87.5			
90–94				92.5			
95–99				97.5			
100–104				102.5			
Total							
Average life expectancy:							

312 Biology: A Laboratory Guide to the Natural World

TABLE 26.9 LIFE TABLE FOR COMBINED GENDERS							
Location and time period:							
Sample size: **Infant mortality rate:**							
1	2	3	4	5	6	7	8
Age group (years)	Number surviving at end	Percent surviving	Number that died	Median age of age group	Total years lived by those that died	Calculation	Life expectancy
<1				0.5			
1–4				2.5			
5–9				7.5			
10–14				12.5			
15–19				17.5			
20–24				22.5			
25–29				27.5			
30–34				32.5			
35–39				37.5			
40–44				42.5			
45–49				47.5			
50–54				52.5			
55–59				57.5			
60–64				62.5			
65–69				67.5			
70–74				72.5			
75–79				77.5			
80–84				82.5			
85–89				87.5			
90–94				92.5			
95–99				97.5			
100–104				102.5			
Total							
Average life expectancy:							

with other student groups. But first, generate a hypothesis and prediction(s) about the differences in life expectancy for different groups you choose to investigate. You may choose to compare single age groups, or to lump them into larger groups. Provide the rationale for your hypotheses in Question 4, then answer the remaining questions.

Questions

1. **Compare the overall life expectancy between males and females in the time period you investigated. If there is a difference, why do you think this is so?**

2. **Compare the life expectancy of the two female age groups you investigated.**

    Life expectancy of 15–19 age group:

    Females:

    Males:

    Life expectancy of 50–54 age group:

    Females:

    Males:

3. **Is childbearing an increased risk for women of the time period you investigated? If not, what other risk factors may have affected life expectancy? Why do we need to use males as the control group?**

4. **Write the experimental design of your comparison between the life expectancies of two different groups, based on ages, gender, or time period.**

    Hypothesis:

    Prediction:

Independent variable:

Dependent variable:

Experimental group(s):

Control group:

5. **What were the differences in life expectancy between your experimental groups?**

6. **Do your data support your hypotheses? Discuss your findings: Why do you think differences in mortality exist? What historical events may have led to these differences?**

7. **Look over your data (or compare it to another group's data) to see if there are any**

periods of unusually high mortality. How might you explain a spike of high mortality?

8. **What kind of cemetery is your data from? Is it affiliated with a particular religion or group? How might this affect the data?**

Exercise 26.4   IMPLICATIONS OF THE HUMAN POPULATION

Web exercise

On October 12, 1999, the human population passed a milestone: the 6 billion mark. The equivalent of a large city (200,000 people) continues to be added to the world daily.

Although humans are theoretically a *K*-selected equilibrium species, the global human population is currently exhibiting nearly exponential increase (Figure 26.7) **[Krogh section 31.4]**. When will we hit our carrying capacity? Once again we return to the frightening suppositions of Malthus:

> I may fairly make two postulata. First, That food is necessary to the existence of man. Secondly, That the passion between the sexes is necessary and will remain nearly in its present state. . . . Assuming then, my postulata as granted, I say, that the power of population is indefinitely greater than the power in the earth to produce subsistence for man.

Because humans are animals, we are governed by the same ecological principles as all other creatures. What is the fate of the human population? Are we headed for the problem Malthus predicted over 200 years ago?

In addition to natural resources, it takes a lot of monetary resources to support the human population (some natural resources are included in economic cost). The U.S. Department of Agriculture estimates the cost of raising a child in their annual *Expenditures on Children by Families* Report. For a child born in 2001, two-parent families (with an annual income between $39,100 and $65,800 per year) with two children will spend approximately $170,460 to raise each child, and that's only until age 17. Accounting for inflation, that expenditure increases to $231,470.

Using the Web sites listed below, as well as sites you find with a search engine, answer the following questions in a typed, double-spaced format. Be sure to cite your information properly (see Appendix 1).

http://www.census.gov

http://www.census.gov/main/www/popclock.html

http://www.census.gov/ipc/www/world.html

http://www.usda.gov

http://www.usda.gov/cnpp/Crc/crc2001.pdf

http://www.nationalgeographic.com/eye/overpopulation /overpopulation.html

http://www.earthday.net/footprint/index.asp.

http://rprogress.org

http://www.redefiningprogress.org/programs /sustainability/ef/efbrochure.pdf

http://www.earthtrends.wri.org

http://www.cdc.gov/nchs/releases/02news/womenbirths.htm

Questions

1. **The United States is one of few industrialized nations in the world that has a positive population growth rate. How much has the U.S. population grown from 1990 to 2000? Cite your reference.**

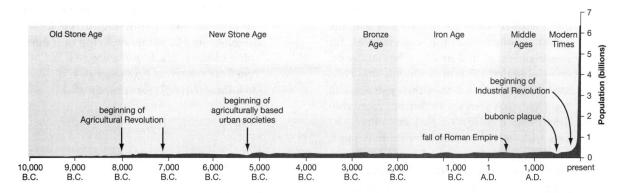

Figure 26.7   Human Population Increase through the Ages.

2. **List the two primary factors that caused this growth.**

3. **How many children were born in the United States in 2002? See** *http://www.cdc.gov/nchs /releases/02news/womenbirths.htm.* **Using this Web site, estimate the expenditures families make on housing and healthcare for these children. Assume that 25% live below poverty level, 35% are in the low-income bracket, 30% are in the middle-income bracket, and 10% are in the highest income bracket. In your answer, explain how you calculated your answers for children at each income level, and give your rationale for any estimates made. Again, cite all references used to answer this question.**

4. **What is the current world population? What is the world population growth rate? See** *http://www.census.gov/ipc/www/world.html.* **How can the population be increasing while the growth rate is simultaneously decreasing? (See Figure 26.4.) In your citation, be sure to give the date of retrieval from the World Wide Web.**

5. **Discuss the similarities and differences between the population dynamics observed in Exercises 26.1 and 26.2 and the dynamics of the human population. Properly cite the references used to answer this question.**

6. **Discuss what will happen to the U.S. human population if we continue to exhibit the current population trends. Again, cite your references.**

7. See *http://www.usda.gov/cnpp/Crc/crc2001 .pdf*. Scroll down to Table ES-1 and read the caption. Notice that the expenditures for households are calculated for a husband-wife, two-child family. Parents with only one child spend 24% more per child than parents with two children. Parents with three or more children spend 23% less on each child than parents with two children. Calculate how much money you would spend on 1 and 3 children, clearly explaining your method. Ignore cost of living increases, and assume you support the children until their 17th birthday. You must choose your income bracket.

8. How big is your "ecological footprint"? See *http://rprogress.org* and *http://www .earthday.net/footprint/index.asp*. Print out and attach a copy of your footprint calculation. What units are used with the ecological footprint?

9. Compare your ecological footprint to the footprint of someone living in the United Kingdom and to someone living in a developing country. See *http://www .redefiningprogress.org/programs /sustainability/ef/efbrochure.pdf*.

10. In your opinion, is the United States population going to be able to continue its current trend of consumption? In other words, are our habits of consumption sustainable? If so, how? If not, why not? Defend your answer with two detailed examples.

27

Community and Environmental Ecology

Civilization is a state of mutual and interdependent cooperation between human animals, other animals, plants, and soils, which may be disrupted at any moment by the failure of any of them.
Aldo Leopold, 1933

A community consists of all species living in a single region that may interact with one another. A community has such complex interactions that it seems to be more than the sum of its parts [**Krogh section 31.5**]. What are these parts? An ecologist is continually trying to find ways to quantify and describe communities, using five characteristics:

1. In describing **vegetation type**, ecologists simply ask what plants are there. Vegetation type is usually identified as one of a few well-known types; for example, deciduous hardwood forest, or short-glass prairie. The abundance and kinds of these primary producers will determine what animals live there.

2. **Trophic structure** refers to who is eating whom. The trophic structure of a community is often depicted as a food web, which illustrates how primary producers support primary consumers (herbivores), which in turn support secondary consumers (carnivores), and so on. Review the basic principles of trophic structure in Figure 7.1.

   The adjective *trophic* means "feeding." Trophic levels are feeding levels.

3. The many species existing together in a community exhibit different levels of **dominance**. Some could be characterized as **dominant species**; these species are highly successful at exploiting their environment, and their presence may determine the conditions under which other species in the community must exist. To a great extent, dominance is determined by a species' behavior. Note that dominant species are not always predominant. A **predominant species** refers to the species in a community with the greatest numbers.

   A predominant species is also called an *ecological dominant* in your Krogh textbook, **section 31.5**.

   A good example of a dominant species is a beaver. Through the activity of dam building, beavers can transform a stream community into a pond community; thus, beavers may be a dominant species in an environment, but they would never be predominant. Humans are the ultimate dominant species; we have an unprecedented physical impact upon our planet. We determine to an overwhelming extent the conditions under which

317

other species must exist—or all too often, cease to exist.

4. **Stability** refers to the constancy in the size of populations of a community. Stability decreases with environmental stress, such as drought and flooding. It is a dynamic attribute of living ecosystems, because even a small change in the stability of one population can have a dramatic effect on the stability of another.

5. **Biodiversity** refers specifically to differences characterizing the living world. There are three main types of biodiversity—species diversity, geographic diversity, and genetic diversity [**Krogh Figure 31.16**]. Biodiversity has become a very hot topic, and for good reason.

The term *biodiversity* was first introduced by the biologist Walter Rosen in 1986 and has become one of the most widely used ecology-related terms. A Web search using *biodiversity* as the keyword yielded over 3 million hits in 2003.

We are losing the miraculous biodiversity comprising our planet at an alarming rate. Tropical rain forests cover 6 to 7% of the Earth's surface but harbor an estimated half of the Earth's species! These rain forests are suffering an unrelenting assault mounted by the dominant species, *Homo sapiens*. Eighty-six acres of rain forest fall victim to fire, machete, and bulldozer every minute. Three species are lost to extinction each hour in the tropical rain forest alone *(Wilson, 1996)*.

Exercise 27.1   AQUATIC COMMUNITY TROPHIC STRUCTURE

Extended 2-week exercise

In this exercise, you will take an in-depth look at the composition and trophic structuring of a pond community or a stream community. You will collect invertebrates from an aquatic habitat, identify them, and then construct a food web that encompasses all of them. A food web is essentially a hierarchical arrangement revealing who eats whom, showing how energy flows through the trophic levels.

Recall that autotrophic organisms are at the base of the food web. These are the primary producers, the ones that capture solar energy and store it in the form of chemical bond energy through the process of photosynthesis. In any community, the primary producers are responsible for the capture of energy that will drive the whole system. **The source and abundance of primary production are the key factors that will determine the overall complexion of any given ecosystem**. Herbivores are the primary consumers, and carnivores are secondary, tertiary, and quaternary consumers. Many organisms are omnivores and therefore may occupy several different positions in the food web.

Ponds, marshes, and lakes tend to be rich ecosystems (Figure 27.1). These bodies of water serve as something of an "energy sink" with much energy coming in, but with less energy being carried out. There are a variety of sources of primary production in a pond ecosystem—algae, aquatic plants, and **detritus** (decaying organic matter that results mostly from leaves falling into the pond). When leaves fall into the pond in autumn, they carry much organic energy with them, as do dying aquatic plants. All of this vegetable matter goes to the bottom of the pond, where it forms a rich layer of detritus.

Primary consumers called detritivores eat this detritus; these consumers include crustaceans such as amphipods and isopods cladocerans, ostracods, and copepods, as well as some aquatic insects and annelids. Snails feed on algae and plants and are also important primary consumers in pond communities. Many carnivorous insects eat detritivores. These secondary consumers include backswimmers, water striders, water scorpions, dragonfly and damselfly larvae, some beetle larvae, and adult beetles

FIGURE 27.1   A Pond Ecosystem.

such as whirligig beetles and predaceous diving beetles.

Insect larvae are also called *nymphs* or *naiads*.

Vertebrates such as fish, birds, frogs, salamanders, snakes, and turtles often assume roles of higher-level consumers in a trophic web.

In the pond ecosystem, there is a continual cycling and recycling of energy. When organisms die, they end up on the bottom of the pond and mix with the detritus. The energy stored in their bodies is recycled through the trophic levels making up the pond community. Envision a continual, steady flow of energy into the pond ecosystem. Some of this energy leaves the ecosystem. Animals such as birds feed in the system for a time, and then fly on to greener pastures, and fishermen consume pond consumers for dinner. However, much of this energy stays there and is recycled until it dissipates.

Because cellular metabolism is less than 50% efficient, eventually the energy dissipates, so there must be a continual influx of energy to support the ecosystem.

At first glance, a stream community looks very similar to that of the pond; however, there are some very important differences (Figure 27.2). Algae and plants are found in streams, but they are restricted to sheltered areas where the water flow is reduced. Detritus does not accumulate the way that it does in a pond, because flowing water washes it away. Any organism living in the fast-flowing water of a stream will have to cope with the problem of "staying put." Many organisms in the stream

environment have unique adaptations to meet this challenge.

In many instances, the main source of energy in stream ecosystems is what falls in from above. This is especially true in the instance of streams whose banks are lined with trees. Any stream fisherman will tell you that a good place to cast a line is under the branches of a tree. Why do fish hang out under tree branches? It isn't for the shade.

Week 1

1. Before collecting invertebrate samples from the collecting site, you will separate into groups: vegetation, surface, and substrate/benthos. The vegetation group will use aquatic dip nets to sample in and near aquatic vegetation and algae. The surface group will collect invertebrates from the surface of the water using aquatic dip nets, and from the air over the water, using aerial nets. The substrate/benthos group will examine the bottom, picking invertebrates off stones and sticks. The substrate/benthos group may also use a dip net or scraper net, collecting invertebrates from the bottom mud or gravel. Each group will collect for a period of 10–20 minutes, as directed by your instructor. Following the instructions carefully is essential for having valid comparisons of data from one year to the next.

2. Proper use of the equipment will be demonstrated by your instructor. A shallow white pan should be filled half full with water. When a net of pond material is brought up, invert its contents into the pan. Separate the invertebrates into different types, placing them into water-filled collecting jars of appropriate sizes. Label the jars with the collecting site (name of pond or stream) and microhabitat (on the surface, among vegetation, attached to rocks, and so forth). When the collecting period is done, place lids on the collecting jars and clean the nets.

3. At the lab, count and identify the invertebrates using the picture key in Appendix 3, standard taxonomic keys, identification cards, or Web sites devoted to aquatic invertebrates. Counting the organisms is necessary if you wish to complete Exercise 27.2, 27.3, or 27.4. Total the number of individuals collected in the sample. Record your data in Table 27.1.

Web sites that might be helpful in identifying invertebrates:

http://www.wlu.ca/~wwwbiol/bio305/Database /Categories.htm
http://www4.ncsu.edu/~ajclevel/Ident.html
http://www.boquetriver.org/adoptaqinvt.html

FIGURE 27.2   A Stream Ecosystem.

TABLE 27.1 INVERTEBRATES OF AN AQUATIC COMMUNITY				
Invertebrate	**Number collected**	**Relative abundance**	**Life cycle stage**	**What it consumes**

Total number of individuals collected:

*http://www.riverwatch.ab.ca/how_to_monitor
/invert_identifying-ident.cfm*

Week 2

4. After all the data are compiled, record the relative abundance of each species. For example, if there were 20 amphipods collected and the total number of all the organisms listed in Table 27.1 is 200, then the relative abundance of amphipods would be **20** ÷ 200 = 0.10. Record this in Table 27.1.

5. Conduct some research about the life cycles of the invertebrates, and note how each is adapted for living in its environment. Pay particular attention to what each group feeds upon. (See Appendix 3.) Record your data in Table 27.1.

Questions

1. **What are the main sources of primary production in the community you sampled?**

2. **Compose a list of primary, secondary, and tertiary consumers from Table 27.1.**

3. **What organisms are dominant in the community?**

4. **What organisms are predominant in the community?**

5. **Using your list of organisms in Table 27.1, construct a food web for the community on a separate sheet of paper.**

Exercise 27.2 INVERTEBRATES AS WATER QUALITY INDICATORS

Continuous with Exercise 27.1

In this exercise, you will evaluate the sample collected in Exercise 27.1 in order to assess water quality, using the relative abundance of certain aquatic invertebrates. Aquatic invertebrates may be categorized as sensitive, moderately sensitive, or tolerant to pollution and degradation of water quality. Presence of excessive nutrients is known as *organic pollution* and may result from contamination with sewage, fertilizer, and other nutrient-rich substances.

Meaningful water quality data is produced by monitoring invertebrate populations over time. It is best to compare data from samples collected at roughly the same time each year. Increases or declines in the population of certain organisms may indicate deteriorating or improving water quality. Obviously it is good to see moderate numbers of sensitive species, with their abundance maintained year after year. If you observe a decline in the population of sensitive species over time, this is an indication of declining water quality. An increase in the population of sensitive species indicates an improvement in water quality. On the other hand, organic pollution causes excess

TABLE 27.2 RELATIVE ABUNDANCE OF POLLUTION INDICATOR SPECIES						
	Sensitive species					Tolerant species
	Caddis flies	Damselflies	Dragonflies	Mayflies	Stoneflies	Isopods
Number collected this year						
Relative abundance						
Number collected year ____						
Relative abundance						
Number collected year ____						
Relative abundance						
Number collected year ____						
Relative abundance						
Number collected year ____						
Relative abundance						
Number collected year ____						
Relative abundance						
Number collected year ____						
Relative abundance						

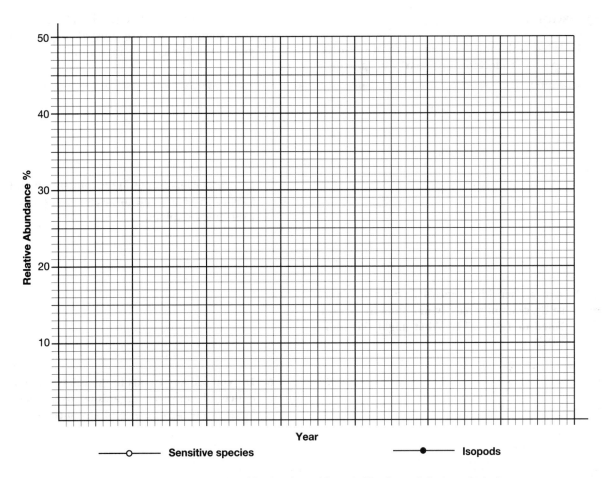

FIGURE 27.3   Graph of Relative Abundance of Sensitive Species and Isopods. Use the symbols shown in the key.

algal growth, and so isopod populations will thrive because they feed on this excess algae. Isopods are affected by chemical pollutants, however.

Sensitive species require very clean water that is free from pollution, excess amounts of nutrients, and other quality-degrading factors. Sensitive invertebrates include larvae of the following insects: caddis flies, dragonflies, damselflies, mayflies, and stoneflies. Moderately sensitive species can tolerate a small amount of pollution and excess nutrients, including amphipods and beetles such as whirligigs, predaceous diving beetles, and water pennies. Tolerant species are those that can withstand greater amounts of pollution and excess nutrients: crayfish, isopods, blackfly

larvae, planarians, leeches, midges, and most snails.

1. In Table 27.2, record the combined relative abundance values (that you calculated in Table 27.1) of the following species: caddis flies, damselflies, dragonflies, mayflies, and stoneflies.

2. In Table 27.2, record the relative abundance of isopods.

3. If you have data from previous years, incorporate it into Table 27.2.

4. In Figure 27.3, graph the relative abundance of sensitive species and isopods. Using the symbols shown in the figure, you will graph one line for the sensitive species, and one line for the isopods.

Questions

1. What is the relative abundance of sensitive species collected this year? In previous years?

2. Are all sensitive species (caddis fly, damselfly, dragonfly, mayfly, and stonefly larvae) listed in Table 27.1? What might this indicate?

Answer the following questions if you are comparing your data to data from previous years.

3. Are there any sensitive species present in your sample that were not present in earlier years? What might this indicate?

4. Are there any sensitive species absent from your sample that were present in previous years? What might this indicate?

5. Do you see any trends in the total number of species in your pond or stream community over time? Explain.

6. Do you see any trends in the relative abundance of sensitive organisms over time? What is the significance of any such trend?

7. Do you see any trends in the relative abundance of isopods over time? What is the significance of any such trend?

Exercise 27.3   EFFECTS OF ACID RAIN

Extended 4- to 26-week exercise

Continuous with Exercise 27.1

One damaging effect of human activity to biological communities is acid rain. Rain is normally slightly acidic, with a pH of 5.6; rain with a pH below 5.6 is characterized as acid rain. Atop Mt. Mitchell in the Smokey Mountains of North Carolina, fog with pH as low as 2.9 has been recorded *(Audesirk et al., 2002)*, which is lower than the pH of vinegar! Acid rain results from pollution of the air by emissions from coal- and oil- burning power plants and automobiles (Figure 27.4). Acid rain leads to destruction of vegetation, damage to statues and other monuments (those made of marble are particularly susceptible), and acidification of our lakes and streams.

Acidification of a body of water may ultimately lead to a decrease in numbers of aquatic organisms, and those most sensitive to pH are at the base of the food chain. Their destruction may have dire consequences for the whole community.

See the **Chapter 3 essay**, *Acid Rain: When Water Is Trouble*, in your Krogh textbook, page 44.

Among the most pH-sensitive organisms are those with calcareous shells such as molluscs (clams and snails), but crayfish as well as mayfly and stonefly larvae are also sensitive. Mollusc populations may begin to decline at a pH lower than 6.5. Populations of crayfish, mayfly larvae, and stonefly larvae begin declining when the pH drops below 6. Because of fluctuating weather patterns, the pH of precipitation can change from day to day. Therefore, long-term monitoring is needed to gain an accurate picture of the potential impact of acid precipitation in a given area.

Over the course of the semester, you will monitor the pH of precipitation and of a collecting site in your area. On average, the pH of the collecting site will probably be higher than the pH of precipitation in your area, because each body of water has a limited capacity to neutralize acids. This neutralizing capacity is determined by the nature of the soil and bedrock within the watershed *(Gorczyca, 1996)*.

A **watershed** includes a body of water, the area of rivers and streams draining into the body of water, and the land between them all.

1. To collect precipitation, place plastic cups or glass jars in undisturbed areas near your campus. At the beginning of each class period, using a pH meter or pH paper, record the pH of any precipitation that has been collected since the previous class. Allow snow to melt before measuring its pH. Record the date, form of precipitation (rain or snow), and pH in Table 27.3. When you complete your collecting season, record the average pH for all the precipitation samples in Table 27.3.

2. Collect water samples in clean jars from a collecting site in your area, especially the ones that you sampled in Exercise 27.1. Upon returning to the lab with a sample, measure the pH as you did in Step 1. Record your data in Table 27.3. When you complete your collecting "season," record the average pH for all the water samples in Table 27.3.

3. If your instructor also provides precipitation pH data from previous years, graph this data in Figure 27.5.

4. If your instructor also provides collecting site pH data from previous years, graph this data in Figure 27.6.

5. From Table 27.1, write the numbers of snails, crayfish, and mayfly naiads collected this year in Table 27.4. Record the relative abundance of *each species*.

6. Graph the relative abundance of each species in Figure 27.7. Using the symbols shown in the figure, graph one line for each species.

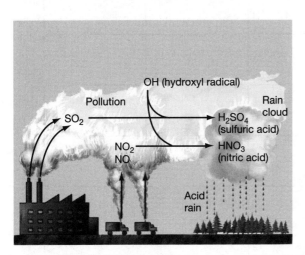

FIGURE 27.4   How Acid Rain Forms.

TABLE 27.3 ACID RAIN DATA				
pH of precipitation			**pH of collecting site**	
Date	**Form of precipitation**	**pH**	**Date**	**pH**
Average pH:			Average pH:	

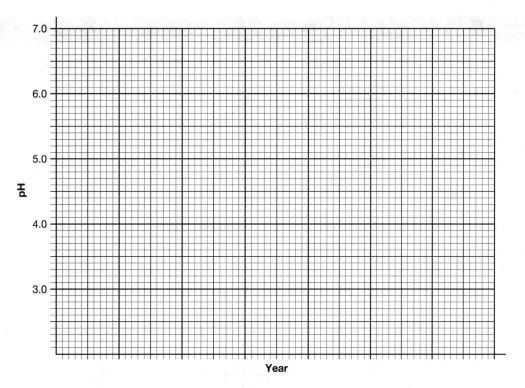

FIGURE 27.5   Graph of pH of Precipitation. Label the years on the x-axis.

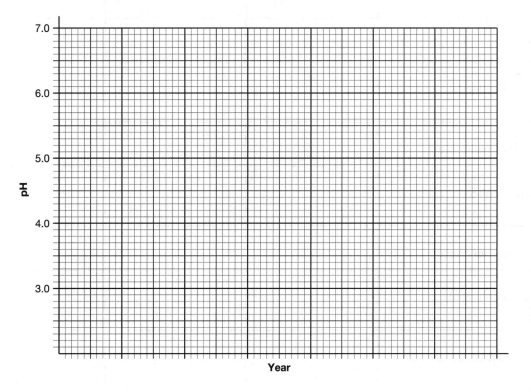

FIGURE 27.6   Graph of pH of Collecting Site. Label the years on the x-axis.

TABLE 27.4 RELATIVE ABUNDANCE OF pH-SENSITIVE SPECIES			
	Snails	**Crayfish**	**Mayflies**
Number collected this year			
Relative abundance			
Number collected year _____			
Relative abundance			
Number collected year _____			
Relative abundance			
Number collected year _____			
Relative abundance			
Number collected year _____			
Relative abundance			
Number collected year _____			
Relative abundance			
Number collected year _____			
Relative abundance			

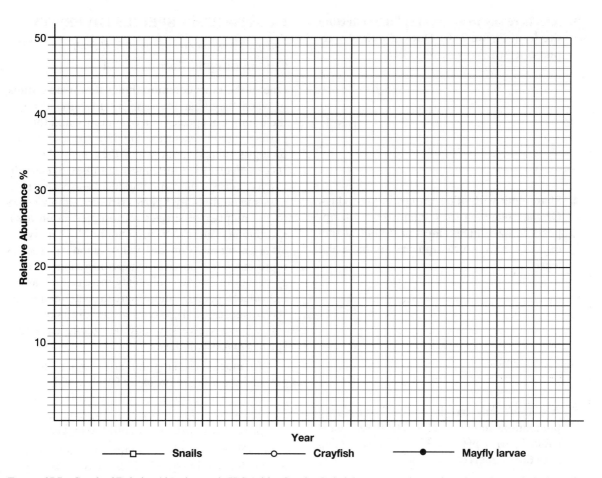

FIGURE 27.7   Graph of Relative Abundance of pH-Sensitive Species. Label the years on the x-axis, and use the symbols shown in the key.

Questions

1. What important information does the pH of precipitation give you that the pH of a collecting site may not?

2. Are there any trends in pH of precipitation on your graph in Figure 27.5? Does any trend warrant concern? Why or why not?

3. **Are there any trends in pH of the collecting site from Figure 27.6? Does any trend warrant concern? Why or why not?**

4. **Share data with other groups in your class. Of the collecting sites sampled, which appears to have the greatest acid-neutralizing capacity, and which appears to have the least?**

5. **Are there any trends in the invertebrate populations in Figure 27.7? Do these trends coincide with pH trends for the collecting sites the animals came from?**

6. **Are any pH-sensitive invertebrates absent from your data? Are there any present in your data that were not collected in previous years? What might this indicate?**

Exercise 27.4   SPECIES DIVERSITY

(Adapted from an exercise by B. B. Nickol)

Continuous with Exercise 27.1

One of the most common things that people think of when they hear the term *biodiversity* is simply the number of species in a community, and this is a perfectly good measure of biodiversity. But what a biologist thinks about when they hear the term *species diversity* is a little more complex. Species diversity consists of two components—species richness and equitability. **Species richness** simply refers to the number of species represented in an area. **Equitability** refers to the manner in which individuals of a community are distributed among the species represented. As species richness increases, so does species diversity. As equitability increases (or as the individuals become more spread out among the species represented), so does species diversity.

Species diversity may be quantified with a **species diversity index**.

An index is used to summarize large bodies of data. You are probably familiar with the use of indices from sociology, business, or economics classes, such as the consumer price index or the Dow-Jones Average.

Diversity indices summarize information about the numbers and kinds of organisms present in a community. Several diversity indices have been proposed. The Shannon-Wiener index, named for its innovators, is one of the best known. You will use a modified Shannon-Weiner index:

$$d = C \div N \times (N \log N - \Sigma n_i \log n_i)$$

where

$$d = \text{index of diversity}$$
$$C = 3.322$$

(*C* converts log bases from base 2 to base 10 logs).

n = # of individuals of the ith species

N = # of individuals in the entire community

For example, communities A and B in Table 27.5 are each composed of 4 species with a total of 20 individuals. In community A, each of the 4 species is represented by 5 individuals. In community B, 1 species is represented by 17 individuals and the other 3 are represented by only 1 individual each.

TABLE 27.5 SPECIES DIVERSITY CALCULATIONS								
Community A			**Community B**			**Community C**		
	n_i	$n_i \log n_i$		n_i	$n_i \log n_i$		n_i	$n_i \log n_i$
n_1	5	3.495	n_1	17	20.918	n_1	7	5.916
n_2	5	3.495	n_2	1	0	n_2	6	4.669
n_3	5	3.495	n_3	1	0	n_3	4	2.408
n_4	5	3.495	n_4	1	0	n_4	3	1.431
$N = 20$		$\Sigma = 13.980$	$N = 20$		$\Sigma = 20.918$	$N = 20$		$\Sigma =$
$N \log N = 26.021$			$N \log N = 26.021$			$N \log N =$		
$d = 1.999$			$d = 0.847$			$d =$		
$R =$			$R =$			$R =$		

You immediately notice that community A is much more equitable than community B. What difference will this make in the diversity index?

For community A, begin by taking the log of n_1 and multiplying that by n_1. There are 5 individuals in species n_1, so first determine the log of 5. For most calculators, this is done by keying in the number and then simply pressing the "log" button.

$$\log 5 = 0.699$$
$$0.699 \times 5 = 3.495$$

Repeat this procedure for each species. Calculate $\Sigma n_i \log n_i$ by summing the column: 13.980. The total number of individuals in the community is N. Since N = 20; then

$$20 \log 20 = 26.021$$

Now just plug everything into the equation and solve for d:

$$d = 3.322 \div 20(26.021 - 13.980)$$
$$d = 0.166 (12.041)$$
$$d = 1.999$$

When you repeat these calculations for Community B, you should get a d value of 0.847.

But what do these numbers mean? Species diversity values are relative and have meaning only within the context of comparisons. When two communities are sampled in *equivalent ways* and species diversity indices are calculated, the community with the higher d value is more diverse. Any community of a given number of species and individuals will have a maximum diversity index possible (d_{max}) and a minimum diversity index possible (d_{min}), which are determined by species richness and equitability. These numbers are hypothetical, because in nature, the trophic structuring of communities requires that there are fewer consumers than producers **[Krogh section 32.3]**. You would never expect to observe d_{min} or d_{max}, but these numbers are important because we can compare our diversity values to the maximum or minimum diversity that is possible.

The condition in which all individuals are spread out evenly among the species represented is represented by d_{max}, and is represented by Community A in the preceding example. Community B represents (d_{min}), because all species are represented by a single individual, except for one species that contains all the rest.

Low equitability indicates that one or a few species in a community is predominant. The degree of predominance in a community is expressed by an **index of redundancy** (R). High R values indicate a predominance of certain species (or poor equitability), whereas low R values indicate a more even distribution of individuals among the species represented in a community (or high equitability). The index of redundancy is calculated as follows:

$$R = (d_{max} - d) \div (d_{max} - d_{min})$$

Assume we have community C comprised of 4 species and 20 individuals, distributed as in Table 27.5. That gives a d value of 1.925. The d values of communities A and B are d_{max} and d_{min} respectively. Thus, the R value for Community C is

$$R = (1.999 - 1.925) \div (1.999 - 0.847)$$
$$R = 0.064$$

1. To investigate the effects of equitability on species diversity, calculate d for the samples in Table 27.6. Note that community size (N) and species richness is the same for each community. Only the number of individuals (n_i) per species is different.

2. What effect does community size (N) have on species diversity? Calculate d for the community listed in Table 27.7.

3. To investigate the effects of species richness on species diversity, calculate d for the samples in Table 27.8.

4. Calculate d and R values for the communities you sampled in Exercise 27.1. Show all your work on a separate paper.

5. Compare the d value calculated in Step 4 to the d value from previous years of the same community, or, compare the d value calculated in Step 4 to that of another community.

TABLE 27.6	EFFECT OF EQUITABILITY ON SPECIES DIVERSITY							
	Community A		**Community B**		**Community C**		**Community D**	
	n_i	$n_i \log n_i$	n_i	$n_i \log n_i$	n_i	$n_i \log n_i$	n_i	$n_i \log n_i$
n_1	20		40		90		96	
n_2	20		30		4		1	
n_3	20		15		3		1	
n_4	20		10		2		1	
n_5	20		5		1		1	
$N = 100$		$\Sigma =$	$N = 100$	$\Sigma =$	$N = 100$	$\Sigma =$	$N = 100$	$\Sigma =$
$N \log N =$			$N \log N =$		$N \log N =$		$N \log N =$	
$d =$			$d =$		$d =$		$d =$	
$R =$			$R =$		$R =$		$R =$	

TABLE 27.7 EFFECT OF COMMUNITY SIZE ON SPECIES DIVERSITY					
	Community I			**Community II**	
	n_i	$n_i \log n_i$		n_i	$n_i \log n_i$
n_1	40		n_1	196	
n_2	40		n_2	1	
n_3	40		n_3	1	
n_4	40		n_4	1	
n_5	40		n_5	1	
	$N = 200$	$\Sigma =$		$N = 200$	$\Sigma =$
$N \log N =$			$N \log N =$		
$d =$			$d =$		
$R =$			$R =$		

Questions

1. What effect does equitability have on diversity? Compare the communities in Table 27.6.

2. What effect does doubling the number of individuals (N) have on diversity? Compare d of communities I and II (Table 27.7) to communities A and D (Table 27.6).

3. How does species richness affect diversity? Compare d (d_{max} and d_{min}) of communities A and B (Table 27.8) to communities I and II (Table 27.7) respectively.

4. From the diversity value calculated for the community you sampled, how is the diversity of the community different from previous years? (And/or, how is it different from another community sampled by your classmates?) Which is more diverse? Why? (Hint: Compare species richness and equitability between each.)

TABLE 27.8 EFFECT OF SPECIES RICHNESS ON SPECIES DIVERSITY					
	Community A			**Community B**	
	n_i	$n_i \log n_i$		n_i	$n_i \log n_i$
n_1	20		n_1	191	
n_2	20		n_2	1	
n_3	20		n_3	1	
n_4	20		n_4	1	
n_5	20		n_5	1	
n_6	20		n_6	1	
n_7	20		n_7	1	
n_8	20		n_8	1	
n_9	20		n_9	1	
n_{10}	20		n_{10}	1	
	$N = 200$	$\Sigma =$		$N = 200$	$\Sigma =$
$N \log N =$			$N \log N =$		
$d =$			$d =$		
$R =$			$R =$		

Exercise 27.5 DIVERSITY OF INSTITUTIONS

The type of analysis that you conducted in the previous exercises may be used for more than just biological communities. You often hear of universities, cities, towns, high schools, churches, and other organizations making claims of being diverse communities. Are they really so?

Diversity refers to the condition of being different or having differences. How are they measuring diversity?

1. Obtain demographic data from three organizations in the same category, such as from three churches, three universities, or three cities. The demographic data you gather may include religion, ethnicity, gender, or education level, just to name a few.

Questions

1. What are the sources of your data? If you gathered the data yourself, describe how you did it.

2. Compare data from all three communities. What is their species (demographic) richness and equitability?

3. Which community is most diverse?

4. What characteristics of the community affect its diversity?

5. Compare data from all three communities. What is their species (demographic) richness and equitability?

6. Which community is most diverse? Explain.

7. What characteristics of the community affect its diversity?

8. Cite all references (see Appendix 1 for proper format).

FIGURE 27.8   Two Major Physiographic Regions of Arkansas.

Exercise 27.6   ECOSYSTEMS

Web exercise

An **ecosystem** refers to a community of organisms and the environment with which they interact. Thus, an ecosystem consists of biotic (living) and abiotic (nonliving) components. The abiotic features of an ecosystem include such things as climate, topography, and elevation. These abiotic components influence the conditions under which the biotic components of the ecosystem must exist. A **biome** is an ecosystem dominated by a large vegetation formation, whose boundaries are largely determined by climate. Biomes include tundra, taiga, temperate deciduous forest, grassland, chaparral, desert, and rainforest [**Krogh section 32.5**].

However, you do not have to resort to a global view to see remarkable differences in vegetation types. Such differences exist on a much smaller scale within the confines of a single state or province. It's great to learn about ecology on a global scale, but we shouldn't neglect our own back yard. Any given geographic area may be divided into various numbers of physiographic regions.

The term *physiographic* refers to physical geography or makeup of the land.

For instance, the state of Arkansas is divided into two major physiographic regions, the interior highlands and the gulf coastal plain (Figure 27.8). These two regions are very distinct, each being characterized by unique vegetation and to an extent, distinct fauna. Beyond the major physiographic regions, the state of Arkansas may be further subdivided into natural divisions (Figure 27.9).

1. Conduct research into the physiography of your state or province. You should identify the major physiographic regions as well as "minor" natural divisions included within each physiographic area. You may use library or Internet resources.

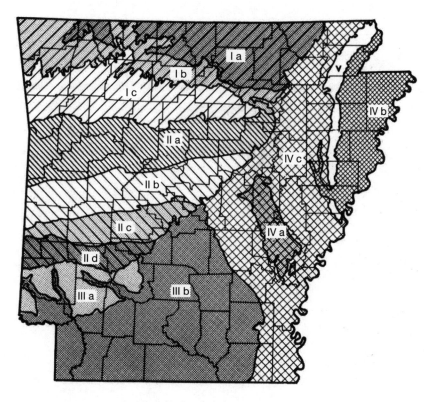

FIGURE 27.9 Natural Divisions of Arkansas. I. OZARK MOUNTAINS: a. Salem Plateau b. Springfield Plateau c. Boston Mountains. II. OUACHITA MOUNTAINS: a. Arkansas River Valley b. Fourche Mountains c. Central Ouachita Mountains d. Athens Piedmont Plateau. III. WEST GULF COASTAL PLAIN: a. Southwestern Arkansas b. South-central Arkansas. IV. MISSISSIPPI ALLUVIAL PLAIN (DELTA): a. Grand Prairie b. Northeastern Arkansas c. Eastern Alluvial Plain V. CROWLEY'S RIDGE.

Often, state agricultural extension services, state geological departments, and United States Geological Survey publications and Web sites are helpful (*www.usgs.gov*). Additionally, many university libraries have "local" sections. Also, general books on the physical geography and natural history are available for all states and regions. These may be easily found in your institutional or local library.

2. Print out a map for each area, and provide information on climate and vegetation (major growth structure), and identify at least two animals that are unique to each region.

3. Cite all references (see Appendix 3).

28

Predation

... Nature, red in tooth and claw ...
Alfred Lord Tennyson, 1850

Predation refers to the feeding of **predators** upon **prey**. Predator–prey interactions may play an important role in population cycles (Figure 28.1). Both prey availability and the abundance of predators serve as density-dependent factors that contribute to the carrying capacity (K) for a population (Figure 26.4). Predator–prey interactions have a profound impact not only upon population dynamics, but upon evolution as well. Predators often serve as an agent of natural selection by feeding on the weaker or "less fit" individuals in the prey population **[Krogh section 31.6]**.

Exercise 28.1 PREDATOR–PREY INTERACTIONS AND POPULATION DYNAMICS

Extended 2- to 3-week exercise

In this exercise, you will design an experiment investigating predator–prey interactions and population dynamics of two genera of ciliates, *Paramecium* and *Didinium*. *Didinium* is a voracious predator of *Paramecium*. In fact, *Paramecium* is all that *Didinium* eats. When *Paramecium* becomes scarce, *Didinium* enters an inactive cyst stage. When prey becomes abundant once again, *Didinium* excysts and begins to feed and reproduce again.

On your lab bench, you will find a culture of *Didinium*, a culture of *Paramecium*, several empty culture jars, pipettes, a container of culture medium, some cotton fibers, and graph paper.

1. Examine each culture at high power with a dissecting microscope.

2. Make separate temporary wet mounts of the *Paramecium* and *Didinium* cultures. Observe the slides with a compound microscope on low and

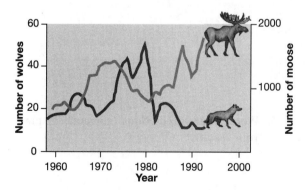

FIGURE 28.1   Predator and Prey Populations. The population dynamics of a predator species (the wolf) and its prey (the moose) in Michigan's Royale National Park over several decades. (Redrawn and adapted from an illustration in science, August 27, 1993, p. 1115.)

339

high power, and sketch the organisms in Question 1. Prepare a temporary wet mount of *Didinium*, but before you add a cover slip, add a drop of concentrated *Paramecium*. You may witness the trials of life right under your nose.

3. Design an experiment to test the predator–prey interactions of these organisms, using the materials available. Devise a way to count the organisms representing each species. There are many ways to accomplish this; be creative! Some ways to do this are by estimating the number of each species per drop with a compound microscope, or by using a gridded petri dish and a dissecting microscope, or by counting individuals of each species for a given amount of time. Keep track of the ciliate populations. Answer Question 2.

4. Count your organisms at least twice per week for the duration of the experiment; ask your instructor when you can do this.

Questions

1. **Sketch and label *Paramecium* and *Didinium* specimens. Note which is larger.**

2. **Write the components of your experimental design:**

   Hypothesis:

   Prediction:

Experimental group:

Control group:

Independent variable:

Dependent variable:

Controlled variables:

3. **Were your predictions correct? Explain your results.**

Exercise 28.2   OWL PELLETS

Mourn not for the Owl, nor his gloomy plight;
The Owl hath his share of good;
If a prisoner he be in the broad daylight,
He is lord in the dark greenwood!

Bryan Waller Procter

Barn owls (Figure 28.2) occur around the world and have become closely associated with humans, probably because of our intimate association with rodents. Owls are nocturnal, feeding at night on rodents, insectivores, birds, and the occasional bat, crayfish, or insect. They have a very interesting habit—a few hours after feeding, the owl "hacks up" the indigestible contents of a meal (hair and bones) in the form of dense hairballs called pellets. Then, a biologist happily appears on the scene to dig through the contents of the pellet. Leave it to those biologists.

Believe it or not, examining these pellets tells us not only what the bird had for dinner, but also a lot about the ecosystem in which the bird is living. In this exercise, you are going to examine owl pellets from a population of owls, attempt to identify and count its prey items, and conduct some basic ecological exercises that will enable you to make inferences about the environment in which the owls live [**Krogh section 32.3**].

1. Obtain a pellet from a population of owls. The pellet has been autoclaved to kill all microorganisms associated with the pellet. Don't worry—nothing will crawl out of it and bite you. Remove the wrapping and thoroughly soak the pellet in water for a few minutes. Carefully tease each pellet apart with dissecting needles and forceps, removing all bones. You will use only the skulls and lower jaws in an attempt to identify prey items.

2. Identify and count prey items from each pellet, using the taxonomic keys, diagrams, and comparative skull set provided. Skull anatomy is summarized in Figure 28.3. Identify the skulls using the materials provided by your instructor. You may also try these Web sites:

   http://www.pelletsinc.com/resources/skulls.html

   http://www.howe.k12.ok.us/~jimaskew/bpellet.htm

   Record your data in Table 28.1. Ask your instructor if you should dissect another pellet.

3. Obtain data from all other student groups in your class, and list all prey items found in Table 28.1.

4. Calculate the percentage that each prey item contributes to the diet of an owl. For example, suppose that the owl pellets dissected during your laboratory period yield 30 prey items, 12 of which are *Microtus*. The percentage that *Microtus* contributes to the diet of owls in your population is

   12 Microtus ÷ 30 total prey items × 100 = 40%

5. Determine the contribution of each prey item in terms of biomass, and record it in Table 28.1. Biomass refers to material from living or recently consumed organisms. The estimated biomass of each type of prey item is given in Table 28.2. For example, each *Microtus* contributes 40 grams of biomass:

   12 Microtus × 40 g per Microtus = 480 g

6. Determine the percentage contribution that each prey item makes to the owls' diet in terms of biomass. Assume from the example previously listed that the total biomass contributed by all prey items is 900 g. Then, the percent biomass contributed by *Microtus* is

   480 g Microtus ÷ 900 g × 100 = 53%

FIGURE 28.2   The Barn Owl, *Tyto alba*.

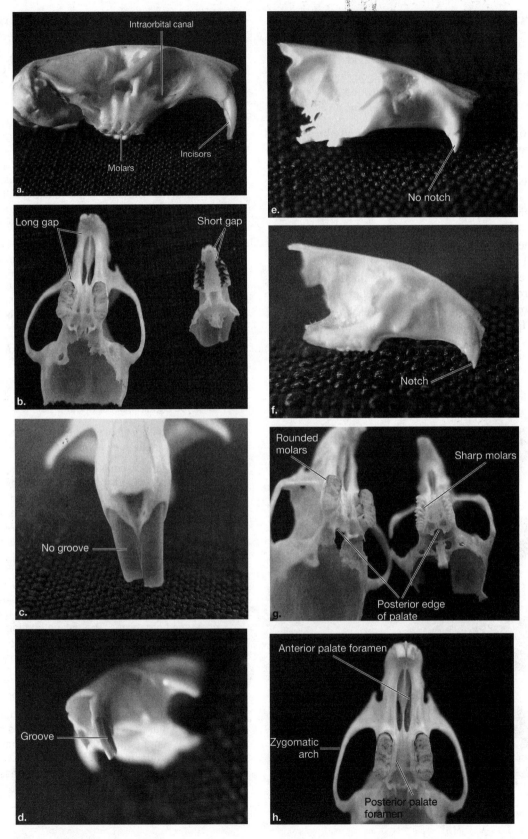

FIGURE 28.3   Skull Anatomy.

TABLE 28.1   OWL PELLET PREY ITEM DATA

Your data

Pellet	Name of prey items		Number of each	
#1				
#2				

Class data

Name of prey items	Number of each prey item found	% of total prey items	Biomass in pellets	% biomass in diet
Total:		100 %		100 %

TABLE 28.2   BIOMASS AND DIET OF PREY ITEMS

Prey item	Biomass (grams)	Diet
Mice: *Mus* (house mouse) *Perognathus* (pocket mouse) *Peromyscus* (deer mice and white-footed mice) *Reithrodontomys* (harvest mouse)	 14 27 18 10	 Seeds, rarely insects Seeds, some vegetation Seeds, berries, fruit, rarely insects Seeds, vegetation, rarely insects
Moles: *Scalopus*	 67	 Earthworms, insects, other invertebrates, rarely plants
Pocket gopher: *Thomomys*	 110	 Vegetation
Rats: *Oryzomys* (rice rat) *Rattus* (Norway rat) *Sigmodon* (cotton rat)	 50 215 100	 Vegetation, fruits, crabs, insects, snails, fungi Vegetation, seeds, insects, small animals Vegetation, insects, small animals
Shrews: *Blarina* *Cryptotis* *Sorex*	 15 6 3	 Insects, earthworms, snails, small animals, occasionally berries and vegetation Insects, earthworms, snails, small animals, occasionally berries and vegetation Insects, earthworms, snails, small animals, occasionally berries and vegetation
Vole *Microtus*	 40	 Grass, roots, seeds
Other: Bats Birds Reptiles	 – – –	 Insects Insects, seeds, grains Insects

7. Based on prey items found, construct a food web for the ecosystem that the owls live in, similar to the food web shown in **Figure 32.12** of your Krogh textbook. Feeding habits of the prey items are summarized in Table 28.2. Draw the food web in Question 2 or on a separate page.

Recall that the vast majority of terrestrial insects are herbivorous, and thus are primary consumers.

8. For extra credit: Your instructor may have you collect data from two owl populations. If so, calculate a species diversity index for each population of prey items, as in Exercise 27.4. Which community is more diverse, and why? Defend your answer and include a comparison of species richness and equitability (thus you will have to calculate R values) between the communities.

Questions

1. On the basis of the pooled class data, which prey item is most important? How do you determine this?

2. Construct a food web showing the trophic structure of the owls' community.

3. Read section 32.3 in your Krogh textbook. In terms of conservation of solar energy, which pathway of energy flow through the trophic levels is most efficient?

4. If an owl lives in an area with a few high-biomass prey items and many low-biomass prey items, which do you think the owl will consume most often?

5. What would happen to the ecosystem if the most important prey item disappeared?

6. What might happen to the ecosystem if the owls would disappear?

7. Extra credit: If you are comparing pellets from two different populations, which region is more diverse, and why? Provide your diversity calculations on a separate piece of paper.

29

Symbiosis

*. . . the play is the tragedy, 'Man,'
And its hero the Conqueror Worm.*
Edgar Allan Poe, 1843

While predation is a very intimate association in which one organism depends on another, it tends to be short-lived. Not so with **symbiosis**, which is a long-lasting association between two or more different species of organisms, with one symbiont living in or on the other **[Krogh section 31.6]**. There are many types of symbioses, but most may be divided into one of three categories. **Commensalism** is a relationship in which one symbiont benefits, but the other neither benefits nor is harmed. Most species of bacteria that inhabit the alimentary canal of humans are commensals. These organisms depend upon humans for nutrients and habitat, but do nothing to help or harm their human host. **Mutualism** is a relationship in which both symbionts benefit. The lichens you studied in Chapter 16 are mutualistic associations between algae and fungi. **Parasitism** is a one-sided relationship in which one of the symbionts (the parasite) lives in or on the other (the host), is physiologically dependent on the host, and imparts some degree of harm to the host. Some bacteria, protists, fungi, and plants are parasites—as are many animals, such as tapeworms.

Why study symbiosis so closely? Well, for one thing, symbiotic species outnumber free-living species. Another good reason—the eukaryotic cell may have a symbiotic origin. Our own cells probably resulted from a symbiosis between protists and bacteria. Imagine our ancestors as flagellated or ciliated single cells, carrying along with them a bacterium that couldn't survive without them, and after a period of time, the protist couldn't survive without the bacterium.

See Krogh's **Chapter 4 essay**, *The Stranger within: Endosymbiosis*, on page 84.

Exercise 29.1 HUMAN MITES

Terms such as *mutualism, parasitism,* and *commensalism* provide convenient, "pigeon-hole" categories, but the exact nature of a symbiotic relationship is often hard to define. Indeed, even though most species of bacteria in the human gut are commensals, other bacteria species in our gut provide protection from other pathogens by outcompeting them—and so can be considered mutualistic. Further, many symbiotic relationships are dynamic and may oscillate among the different categories! For instance, the follicle mite *Demodex folliculorum* (Figure 29.1) is almost always non-pathogenic, although in *rare* instances infestation with this mite may lead to dermatitis, acne, loss of eye-lashes, and hardened blemishes on the skin.

FIGURE 29.1   Human Follicle Mite, *Demodex folliculorum.*

A species of the same genus, *Demodex canis*, is responsible for red mange in dogs, a condition that is very serious and sometimes even fatal (to the dog).

Demodex folliculorum lives in the sebaceous (sweat) glands and hair follicles of humans, especially around the nose and eyelids. This microscopic mite is less than $\frac{1}{2}$ mm in length. Its prevalence is fairly high, ranging from about 20% in 20 year-olds to near 100% in the elderly population *(Roberts and Janovy, 2000).*

The **prevalence** of a symbiotic organism refers to the percentage of host in a population that is infected. For instance, if 1000 people are examined for a parasite and 150 are found to be infected, the prevalence of the parasite would be $150 \div 1000 = 15\%$.

1. Wash your hands before and after this step. Determine whether you have **Demodex** by examining facial oil from the follicles of the nose and/or region of the eyebrows. Using your fingernails, gently express a small amount of oil from the pores on the side of your nose. Don't be shy. After all, you've probably done it before. You may also wish to scrape along the eyebrows.

2. Transfer the oil to a microscope slide. Place a drop of mineral oil onto the material, place a coverslip over it, and examine the preparation using a compound microscope. If you find mites, examine them carefully and notice their movement.

3. Determine the prevalence of infection of this mite in your class. It's okay to admit that you have them—they are simply low-maintenance symbionts.

4. In case your class is unfortunate enough not to find living mites, your instructor may have prepared slides available or may have a wet mount of his or her own to show you. Observe them with a compound microscope.

Questions

1. **What type(s) of symbiotic relationship do humans and follicle mites share?**

2. **Sketch the mites you find in the wet mount or on the preserved slide.**

3. **What is the prevalence of *Demodex* infection in your class?**

4. **Would you expect this prevalence to increase or decrease as you and your classmates grow older? Why?**

Exercise 29.2 TERMITES AND PROTOZOANS

Many termites eat wood, which is composed primarily of cellulose. Digestion of cellulose requires the enzyme *cellulase*, which few animals produce; however, flagellate protozoans that do produce cellulase live in the termite gut (Figure 29.2). Thus, the termites are able to survive by eating wood, and in turn the flagellates have a home. Flagellates may make up 30 to 50% of the total weight of a wood-eating termite, which is impressive when you consider that protozoans are single-celled. This symbiosis is so well developed that the termites cannot live without the flagellates, and some genera of flagellates are found only within the gut of termites. Similar symbioses are common in many herbivorous mammals, like cattle and deer.

1. Place a drop of saline solution onto a microscope slide. Using forceps, obtain a termite from the container, and hold it in the solution. Using insect pins, remove the termite's head, and the light brown intestine should follow. Remove all parts of the exoskeleton, so that only soft parts are left on the slide.

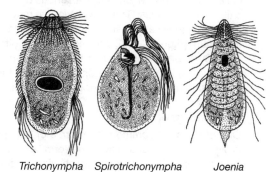

Trichonympha   *Spirotrichonympha*   *Joenia*

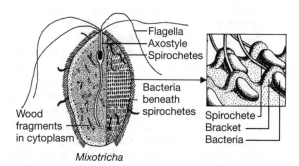

Mixotricha

FIGURE 29.2 Flagellates of Termites. The symbiotic bacteria of *Mixotricha* resemble flagella.

2. Using 2 insect pins, thoroughly mix the contents of the slide. Place a cover slip over the mixture and examine microscopically with low- and high-power objectives. Watch the movement of the flagella. Notice the large number of flagella, unlike the flagellates you examined earlier in the year that had only a single flagellum (such as *Euglena* and *Trypanosoma*). In some instances, what appear to be flagella are actually symbiotic bacteria that are associated with the protozoans. This is an example of **hypersymbiosis**.

Questions

1. **Sketch the different types of flagellates that you see. Do they appear to match any of those shown in Figure 29.2?**

2. **What type of symbiotic relationship do the termites and flagellates have?**

3. **If we humans cannot digest cellulose (fiber), why is it so important in our diet?**

Exercise 29.3   HUMAN HELMINTHS

... the study of life is, for the most part, parasitology. *(Carl Zimmer, 2000)*

Parasitology is the scientific study of parasites. In any study of biodiversity, parasites should be seriously considered because they are an important part of the overall picture of life. There are many more parasitic species than free-living species; virtually all animal species and many plant species serve as hosts to one or more species of parasites!

For example, a parasite survey revealed 23 species of worms (helminths) living inside 30 raccoons, and this number doesn't even include their parasitic bacteria, protozoa, fungi, or animals such as ticks or fleas *(Richardson et al., 1992)*.

Animals tend to carry their parasite burden very well; after all, a parasite shouldn't kill its host, or it will die too. All of the worms shown in Figure 29.3 were removed from the small intestine of a single skunk.

In Chapter 15, you were exposed to protozoa infecting humans—now we will get up close and personal with worms infecting humans. Review the life cycle of each parasite represented as you view the specimens.

1.   Parasites may exhibit rather complex life cycles requiring 2 or more species of hosts for completion. Such a life cycle is called an **indirect life cycle**.

The host in which the parasite attains sexual maturity is the **definitive host**. Hosts containing larval or juvenile stages are the **intermediate hosts**. Tapeworms have an indirect life cycle; 2 tapeworm species of humans are *Taenia solium*, the pork tapeworm, and *Taenia saginata*, the beef tapeworm. Observe the preserved tapeworms on display. Humans become infected by ingesting larval tapeworms that are in raw or poorly cooked pork or beef. Then adult worms grow in the human small intestine. Worldwide, about 77 million people are infected with the pork tapeworm and 10 million are infected with the beef tapeworm.

Figure 29.4 shows the life cycle of *Taenia solium*. Pigs are the intermediate host and humans are the definitive host. The life cycle of *T. saginata* is essentially identical to that of *T. solium*, except the intermediate host of *T. saginata* is a cow. However, there is a unique feature that renders *T. solium* potentially deadly; in addition to serving as definitive host, humans may also serve as

FIGURE 29.3   Roundworms collected from the Small Intestine of One Skunk.

intermediate host for *T. solium* (Figure 29.5). In this instance, humans swallow the eggs, becoming infected with the larvae of the tapeworm, called **cysticerci**, which cause a serious disease called cysticercosis. The larval worms may be found in virtually any tissue in the body, including the brain (Figure 29.6); this condition is often fatal. Cysticercosis is a major health problem in central Mexico and is becoming more important in the United States—it is now the leading cause of epilepsy in Los Angeles County, California *(Despommier et al., 2000)*.

One of the most common tapeworms of humans is *Hymenolepis nana*. This tapeworm is also a common parasite of rodents, which, like humans, may serve as definitive host. It infects about 4 of every 1000 people in the United States; and about 75 million worldwide, mostly children. A person gets infected either by ingesting eggs in grain or cereal, from petting fur that is contaminated with rodent feces, or by ingesting the intermediate host, which is a grain beetle. While most of us are not known as beetle eaters, a toddler or two has been known to eat them. Interestingly, Duclos and Richardson (2000) found that 75% of pet stores in south-central Connecticut were selling rodents infected with this tapeworm. It is possible that some pet owners are infected, but they would probably never know it because symptoms of infection (nausea, loss of appetite) are vague or absent altogether. **Zoonotic diseases**, or

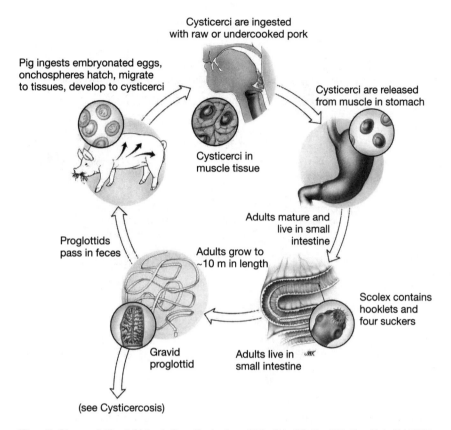

Cysticerci are ingested
with raw or undercooked pork

Pig ingests embryonated eggs,
onchospheres hatch, migrate
to tissues, develop to cysticerci

Cysticerci are released
from muscle in stomach

Cysticerci in
muscle tissue

Adults mature and
live in small
intestine

Proglottids
pass in feces

Adults grow to
~10 m in length

Scolex contains
hooklets and
four suckers

Gravid
proglottid

Adults live in
small intestine

(see Cysticercosis)

"Parasitic Diseases" 4th ed. (c) Apple Trees Productions, LLC., Pub. P.O. Box 280, New York, NY 10032

FIGURE 29.4   Life Cycle of *Taenia solium*, the Pork Tapeworm. The life cycle of the beef tapeworm is very similar, except the intermediate host is a cow.

zoonoses, are human diseases caused by parasites that normally live in nonhuman animals. It is estimated that 80% of all human infections are zoonotic in nature; zoonotic diseases comprise over half of the emerging diseases that the World Health Organization considers either new, volatile, or as posing an important public health threat *(Gauthier, et al., 2003)*.

2. Another important group of parasites is the **trematodes**, also called **flukes**. Examine the preserved trematode specimens on display. Figure 29.7 shows the life cycle of the sheep liver fluke, *Fasciola hepatica*. These parasites live as adults in the bile ducts of sheep, cattle, and sometimes humans. Worldwide about 2.4 million people are infected. The most important trematodes infecting humans are the blood flukes, or schistosomes (Figure 29.8). Adult schistosomes live in the veins that drain the intestine or urinary bladder, depending on the species. Instead of encysting on vegetation, like *Fasciola hepatica*,

schistosome cercariae enter the definitive host by penetrating the skin.

When schistosomes that normally infect birds burrow into the skin of a person , the rash called "swimmer's itch" appears.

Worldwide about 200 million people are infected with schistosomes, which cause 20,000 deaths per year. In addition, schistosomiasis often leads to decreased absorption of nutrients in the intestine, leading to reduced physical and mental development.

The suffix–*iasis* means "infected with."

3. Your instructor may have freshwater snails available in small containers. If the snails are infected and are "passing" cercariae, you may be able to see the tiny cercaria swimming when you hold the jar in front of a dark background. Examine all snail containers for the presence of cercariae, using a dissecting microscope.

Wear gloves when working with cercariae!

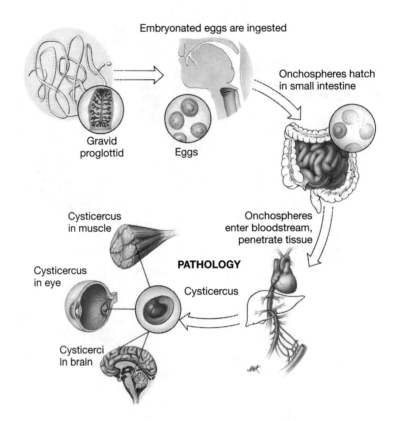

Embryonated eggs are ingested

Onchospheres hatch
in small intestine

Gravid
proglottid

Eggs

Cysticercus
in muscle

Onchospheres
enter bloodstream,
penetrate tissue

Cysticercus
in eye

PATHOLOGY

Cysticercus

Cysticerci
In brain

Parasitic Diseases" 4th ed. (c) Apple Trees Productions, LLC., Pub. P.O. Box 280, New York, NY 10032

FIGURE 29.5   Cysticercosis.

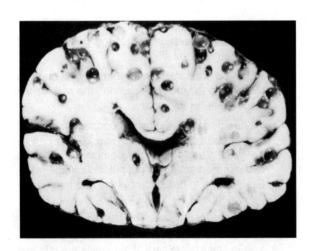

FIGURE 29.6   Cysticerci of *Taenia solium* in a Human Brain.

Observe the cercariae with a dissecting micro-scope, and then make a temporary wet mount of the cercaria and examine them with a compound microscope. There are many different types of cercariae. Can you identify the type of snails and cercariae that you have, using Figure 29.9? Sketch them (see Question 3).

4. Among the most important worms infecting hu-mans are the nematodes, or roundworms. Most nematodes exhibit a **direct life cycle** in which the parasite is passed directly from one host to the next (of the same host species). A good ex-ample of a direct life cycle is that of the human pinworm, *Enterobius vermicularis*. Obtain a pre-pared slide of a human pinworm; note the point-ed end of the female. About 400 million people are infected with pinworms at any given time, and the vast majority of these are children. Pinworms are easily transmitted from child to child in family settings, day care centers, and elementary schools—mostly because of unhygienic habits of children. Fortunately, infection is usually not seri-ous, and one-third of the infections are fully asymptomatic (*Roberts and Janovy, 2000*). Pin-worm infection is easily and effectively treatable, but reinfection is common because the eggs can float in the air and so are hard to remove from an

Total Parasite: Kills Host

Partial Parasite: Only harms host

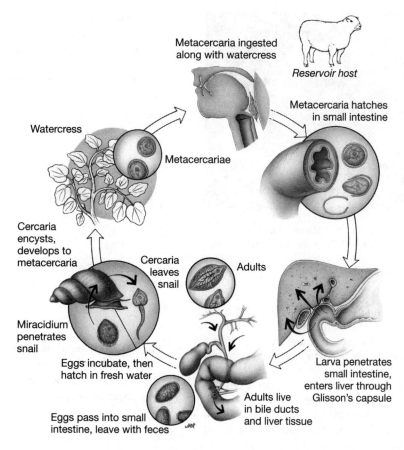

Metacercaria ingested
along with watercress

Reservoir host

Metacercaria hatches
in small intestine

Watercress

Metacercariae

Cercaria
encysts,
develops to
metacercaria

Cercaria
leaves
snail

Adults

Miracidium
penetrates
snail

Larva penetrates
small intestine,
enters liver through
Glisson's capsule

Eggs incubate, then
hatch in fresh water

Adults live
in bile ducts
and liver tissue

Eggs pass into small
intestine, leave with feces

"Parasitic Diseases" 4th ed. (c) Apple Trees Productions, LLC., Pub. P.O. Box 280, New York, NY 10032

FIGURE 29.7 Life Cycle of *Fasciola hepatica*. Adult worms mate in the bile ducts and eggs are released into the small intestine, then they are passed out in the feces. The intermediate host in this life cycle is a snail. Aquatic snails play the role of intermediate host in most trematode life cycles. The larval worms reproduce asexually in the snail, and each egg yields many larvae, called cercariae. The cercariae leave the snail and encyst on aquatic vegetation, transforming into metacercariae. In order for the life cycle to be completed, a cow, sheep, or human must ingest a metacercaria while eating vegetation. Humans usually become involved in the life cycle by eating watercress, a popular salad vegetable.

infected area. The pinworm life cycle is shown in Figure 29.10.

5. Of the parasitic worms infecting humans, the large human roundworm, *Ascaris lumbricoides*, is the most prevalent worldwide. One in every 4 people is infected with this nematode (almost 1.5 billion people worldwide). Each year, between 60,000 and 100,000 people, mostly children, die as a direct result of infection with *A. lumbricoides*. Examine the preserved specimens on display. This nematode has a direct life cycle in which the definitive host becomes infected by ingesting the eggs in fecally contaminated food or water (Figure 29.11). Adult worms live in the small intestine, where they feed on digested food materi-

als and epithelial cells. Figure 29.12a shows a child that is infected with *A. lumbricoides*. Her abdomen is distended because her small intestine is packed with worms. Figure 29.12b shows the worms that were expelled by the same girl after treatment with an anthelmintic (a drug used to kill helminth parasites).

6. Hookworms are another very important group of roundworms infecting people. Observe the hookworms on display. These also have a direct life cycle, but instead of entering by an oral route, the larvae penetrate directly through the skin of the host. Infection usually occurs when bare feet come in contact with fecally contaminated soil. After penetrating the skin, the larvae gain entry

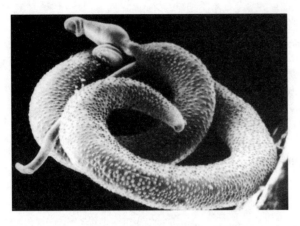

FIGURE 29.8   A Scanning Electron Micrograph of
Schistosoma mansoni. Flukes of this genus are unique in that
they are dioecious (have separate sexes). The female spends
most of her life within the gynocophoral groove of the male.

into the circulatory system. They undergo a migration through the lungs much like *A. lumbricoides*, and the adults eventually end up in the
small intestine. Hookworms feed on blood by
severing intestinal blood vessels with their cutting plates or teeth (Figure 29.13), depending on
the species. Hookworm infection often leads to
severe anemia. Worldwide, 1.3 billion people are
infected with hookworms, leading directly to
65,000 deaths per year.

Questions

1. How does a person become infected with the
   pork tapeworm? The beef tapeworm? Which
   of these infections is more serious? Why?

2. What simple step can a person take to avoid
   becoming infected with a pork or beef tapeworm? (Not including adopting a vegetarian
   diet.)

3. Sketch the trematode cercariae, and describe their movement. What type of cercariae are they? What type of snail was its
   host?

4. How does a person become infected with a
   pinworm?

5. Worldwide, what percentage of people is infected with *Ascaris lumbricoides*, the giant
   human roundworm? How does a person become infected?

6. Worldwide, what percentage of people is infected with hookworms? How does a person
   become infected?

7. Is infection with giant human roundworms
   or hookworms more serious? Why?

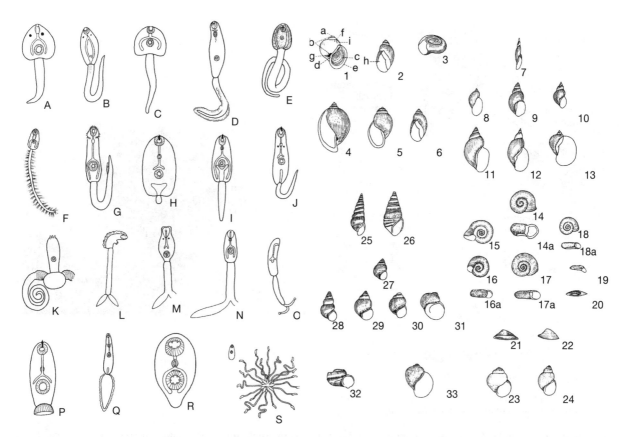

FIGURE 29.9   **Left: Types of Cercaria Found in Snails** A. amphistome B. monostome C. gymnocephalus D. gymnocephalus/pleurolophocercous E. cystophorous F. trichocercous G. echinostome H. microcercous I. xiphidiocercous J. ophthalmoxiphiodiocercous K. gasterostome furcocercous L. lophocercous furcocercous M. apharyngeate furcocercous N. pharyngeate furcocercous O. apharyngeate Monostome Furcocercous Cercaria, without Oral Sucker P. cotylocercous Q. rhopalocercous R. cercariaea S. rat-king cercaria. **Right: Common Freshwater Snails 1**. Diagram of spired operculate snail with dextral whorls, a. apex b. spire c. operculum d. columella e. lip f. suture g. umbilicus h. aperture i. body whorl. 2. Diagram of spired nonoperculate snail with sinistral whorls. 3. Diagram of discoidal snail with sinistral whorls. 4. *Physa sayi* 5. *Physa integra* 6. *Physa gyrina* 7. *Lymnaea haldemani* 8. *Pseudosuccinea columella* 9. *Fossaria modicella* 10. *Stagnicola caperata* 11. *Bulimnea megasoma* 12. *Lymnaea stagnalis* 13. *Lymnaea auricularia* 14-19. Planorbid snails 14-14a. *Helisoma* 15. *Planorbula* 16. *Segmentina* 17-17a. *Planorbus* 18-18a. *Gyraulus circumstriatus* 19. *Gyraulus deflectus* 20. *Menetus exacuous* 21-22. Limpets 21. *Ferrisia* 22. *Ancylus* 23. *Viviparus* 24. *Campeloma* 25. *Pleurocerca* 26. *Goniobasis* 27. *Cincinnatia* 28. *Pomatiopsis* 29. *Bithinia* 30. *Amnicola* 31. *Flumnicola* 32. *Valvata* 33. *Ampullaria*.

Exercise 29.4   JAR OF WORMS

> ...*Am I my brother's keeper?*
>
> *Genesis 4:9*

By this point, you may understand that parasitic worms and protozoans have a profound impact on all of humanity. It is estimated that of the 6 billion people on Earth, 4.5 billion are infected with worms, leading directly to 1 million deaths per year. This is just with the worms, and doesn't include the protozoans. When you also consider the toll taken by protozoan parasites, the numbers are truly staggering.

There is virtually no "hard" data on the overall prevalence of most parasitic diseases in the United States (*Hotez, 2002*). All estimates place the numbers of infected U.S. children in the millions (*Roberts and Janovy, 2000*); nevertheless, the vast majority of infected children live in developing nations that lie in tropical latitudes. Poverty is the premiere breeding ground for many diseases; poor socioeconomic conditions set the stage for optimal transmission of many diseases, including those caused by parasites.

An appreciation of the true magnitude of this problem requires you to alter your worldview;

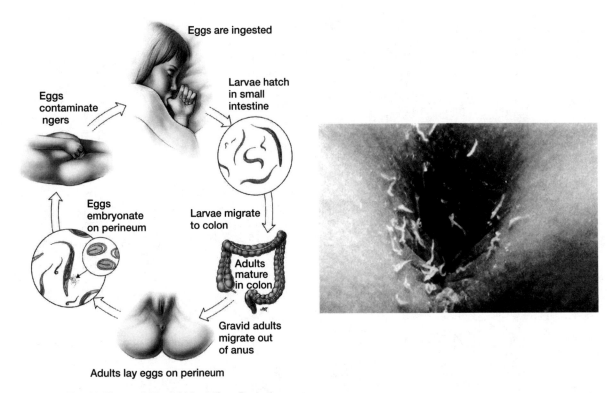

Eggs are ingested

Larvae hatch in small intestine

Eggs contaminate ngers

Larvae migrate to colon

Eggs embryonate on perineum

Adults mature in colon

Gravid adults migrate out of anus

Adults lay eggs on perineum

"Parasitic Diseases" 4th ed. (c) Apple Trees Productions, LLC., Pub. P.O. Box 280, New York, NY 10032

FIGURE 29.10   Life cycle of the Human Pinworm, *Enterobius vermicularis*. Adults reside in the lower intestine, where they copulate. The females migrate down the intestine to the anal opening where they deposit their eggs in the perianal region, thus evoking the primary symptom of pinworm infection: anal pruritus (itching). Of course, when a toddler has an itch, he will scratch, spreading pinworm eggs to the hands, which can then contaminate any toy or other item that is touched. Another child becomes infected as the pinworm eggs are ingested, then hatch in the small intestine. The worms then move to the lower intestine where they mature into adults. b. Female pinworms on perianal region of a five-year-old child. *New England Journal of Medicine* 328:927. © 1993, Massachusetts Medical Society. All Rights Reserved.

try to view the world as a child living in an overpopulated and impoverished third-world nation. In other words, in a global context, try to view the world through the eyes of an *average* person. This might be difficult for you to do. Approximately 80% of the world's population lives in developing nations. Yet, the 20% of the population living in developed nations (including you) consumes twice as much grain and fish, 3 times as much meat, 9 times as much paper, and 11 times as much gasoline as the 80% living in developing nations *(Mock, 2000)*. "For every person in the world to reach present levels of U.S. consumption…would require four more planet Earths." *(Wilson, 2002)*. For most of our sisters and brothers in this global community, things that we take for granted—running water, indoor toilets, and electricity—are luxury items.

Humans exacerbate many of their own diseases because of high population densities and subsequent environmental pollution. Most human parasites are transmitted through food and water contaminated with human feces (see Figure 29.15). *Only 10 to 15% of the world's population has adequate water and sewer systems.* As populations shift from rural to urban areas in developing countries, the sewer and water systems become overloaded, if they exist at all.

In addition to deaths directly attributed to parasite infection, millions more die because of intestinal infection in combination with malnutrition. Infection with intestinal parasites makes the victim more susceptible to other health problems.

When children are sick and hungry, how can they aspire to academic success? Ascariasis

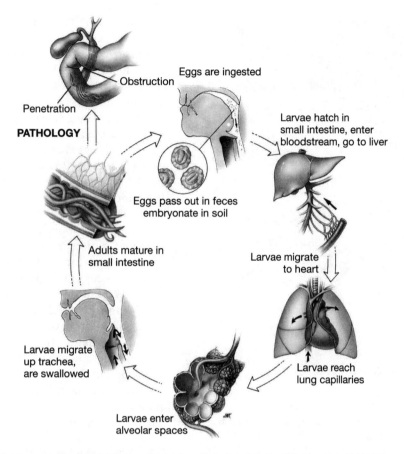

"Parasitic Diseases" 4th ed. (c) Apple Trees Productions, LLC., Pub. P.O. Box 280, New York, NY 10032

FIGURE 29.11   Life Cycle of *Ascaris lumbricoides*. This nematode has a direct life cycle in which the definitive host becomes infected by ingesting the eggs in fecally contaminated food or water. Once in the small intestine, the eggs hatch, releasing larvae that undergo a migration through the lungs. The larvae are eventually coughed up and swallowed, then mature to adults in the small intestine. It's a strange circle of events, but it works quite well.

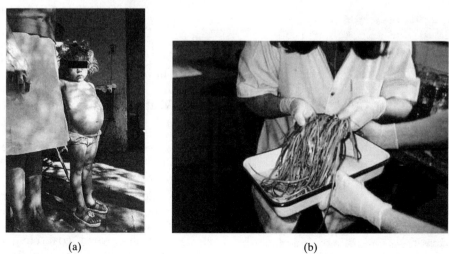

(a)   (b)

FIGURE 29.12   a. Child with Distended Abdomen due to Infection with *Ascaris lumbricoides*. b. Worms expelled by the child after treatment with an anthelmintic drug.

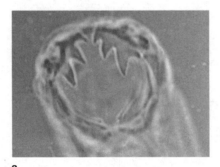

a

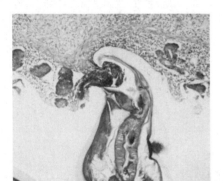

b

FIGURE 29.13   a. Hookworm Photomicrograph Showing Teeth in its Oral Cavity. b. Histological section of a hookworm attached to the mucosa of a heavily infected 12-week-old infant, who died of severe anemia.

FIGURE 29.14   *Ascaris lumbricoides*

and related helminth infections are a major adversary of childhood education. This is alarming, because 75% of the world's population is infected with helminths. What are *we* going to do about it? The United States spends less than 1/10th of 1% of its gross domestic product on foreign health assistance (*Cowley, 2003*). We are neglecting the *average* child, in a global context. Not only that, what about the 12 million U.S. children living in poverty? What about the 9 million without health insurance coverage? No child left behind, indeed.

The "average American" is beginning to realize that economic and political issues in developing nations may have very real and direct ramifications for us. One of the surest ways to promote economic and political instability is to pack a country full of sick and hungry people. See **Krogh Figure 32.9**. Treating parasitic infections in children would only be a beginning; it would be treating the "symptoms" but wouldn't change the underlying political and socioeconomic issues. Until education, sanitation, and socioeconomic conditions are improved, these parasitic infections and everything that goes with them will remain.

1. Study Figures 29.12, 29.14, and 29.15. Then write a 1–2 page essay of the impression that these worms invoke. Explain your answers to the following questions in your essay.

    How would you cope with poverty, poor sanitation systems, and intestinal infection?

    What can be done to alleviate ascariasis?

    Do you think the parasite can be eradicated? If so, how? If not, why?

    Do you, as a citizen of a highly developed nation, have a moral imperative to address this problem?

    What do the following two points have in common?

    1. List the five countries that you think are most successful. Look at a world map and find their latitude.

    2. Consider the impact of parasitic diseases in developing nations. Often, these nations are richly laden with resources (oil, mining, wood, coal, etc.), yet the overwhelming majority of their people are impoverished.

FIGURE 29.15   Overpopulation Contributes to the Transmission of Diseases. This photograph shows a young boy at a privy over water in Dhaka, Bangladesh. While public faucets provide drinking water, residents often rely on the water shown for bathing, laundry, and recreation.

Exercise 29.5   WATER

Web exercise

As you have gathered by now, much of the focus on ecology emphasizes the importance of unpolluted and potable (drinkable) water supplies. As an educated, relatively wealthy person, can you imagine not having water flowing from a tap in your home? What would your life be like without that luxury?

> Americans use 270 billion gallons of water on their 50 million lawns. This is enough water to give every person a shower for four days in a row. *(Burdick, 2003).*

As pointed out earlier, less than 15% of the world's population has adequate water supplies. The problem is particularly serious in Africa. In coming decades, water may become the most scarce and coveted natural resource, serving as an impetus for international hostility. The next world war may not be about religion or oil, but about water.

Issues about clean water aren't just a concern of developing nations. The United States Office of Homeland Security has noted that our water supplies may be vulnerable to terrorist acts. But many other threats exist, including overuse. On the bright side, water conservation efforts are finally beginning to take place, but are they happening fast enough?

Choose one of the options A–D listed below. Use at least two reliable websites for your report; see Appendix 1 for proper citation format.

A. Investigate the effects of pollution on water supplies and watersheds in the United States and Canada. What or who is causing the pollution?

What can be done to remedy this pollution, and how costly would it be?

B. OR, investigate the threat of terrorism on our water supplies. What potential danger does this have on the health and safety of those who consume the water? How is the water "cleaned up" after it is polluted?

C. OR, investigate the effects of water conservation, and on the flip side, the effects of careless water use on our water supplies. For example, what methods do water suppliers use to ensure the longevity of our water supply? What practical steps can consumers take to conserve water? Discuss the habits or hobbies of typical Americans that have an impact on our water supply.

D. OR, describe the difficulty of life in Africa because of lack of good water. Choose a particular region to study, and determine its water source. Describe how people living in this region use the water and how their lives are shaped by the water source.
    These Web sites may be helpful.
    http://water.usgs.gov/nawqa/
    http://water.usgs.gov/owq/dwi/index.html
    http://www.who.int/water_sanitation_health/en/
    http://www.epa.gov/water/index.html
    http://www.epa.gov/OW/you/intro.html
    *http://www.peacecorps.gov/wws/water/Africa
    /index.html*

30

Animal Behavior

Animal behavior is a fascinating area of biology. Stop for a moment and ponder your own behavior. How do you act when you know someone is observing you? How about when a person you are attracted to walks into the room? Do you behave differently when you are alone than when you are with another human? (Do you tell things to your dog that you would never tell another human?) Animals react in specific ways to members of their own species and to other species. Animal behaviors are quite complex because they may have genetic, environmental, hormonal, and/or social influences. Nevertheless, patterns in behavior are gradually becoming understood **[Krogh Chapter 33]**.

As we delve into the exercises in this chapter, keep in mind how the behavior of an animal can affect its success. Fleeing from a predator, of course, yields obvious and immediate benefits. But how does the animal know to flee? Does it flee in a reflex-like manner, or does it have to stop and think before deciding what to do? And what controls its response? Evolutionary benefits arising from animal behavior may not be so obvious, but they have profound implications on the success of the organism.

Exercise 30.1   TERRESTRIAL TAXIS
[Krogh section 33.3]

On a basic level, you already know that organisms exhibit movement in response to stimuli, called **taxis**. A *positive taxis* is movement toward a stimulus, and a *negative taxis* is movement away from a stimulus.

> The movement of plants in response to stimuli is called *tropism*. See Exercise 18.5.

To observe the phenomenon of taxis, we will use terrestrial sow bugs, which are ideal organisms to observe because they are easy to handle, and the stimuli they respond to are easy to apply in the laboratory. Sow bugs, also called **wood lice** (Figure 30.1), live in damp areas, such as under leaf litter, in various objects, or in moist soil. They are dorsoventrally flattened (flat on the dorsal and ventral sides) arthropods with many jointed appendages, and they belong to the subclass Malacostraca, which means *soft shell*.

1. For this experiment, you will test the effects of three variables on the taxis of sow bugs—amount of light, amount of moisture, and temperature. Predict whether the organisms will show a

FIGURE 30.1   The Sow Bug, a Terrestrial Isopod.

positive or negative taxis to each stimulus. Answer Questions 1 to 3.

2. Design your experiment, and use at least six sow bugs for each test. The "walking" surface for the sow bugs should have a consistent texture, and it should be level. Run each test separately, and run each for the same length of time. Don't forget to use a control group. Record your data in Table 30.1. Compare your findings with those of other groups.

Here is a list of materials that will be available to you:

Baking pans or trays

Cardboard

Tape

Paper towels

Water

Hot plate

Ice bath

Sow bugs

Timer

TABLE 30.1 SOW BUG TAXIS	
Stimulus	**+ or − taxis?**
Light	
Moisture	
Heat	

Questions

1. **Write the components for the light test.**

Hypothesis:

Prediction:

Experimental group:

Control group:

Independent variable:

Dependent variable:

Independent variable:

2. **Write the components for the moisture test:**
   Hypothesis:

Dependent variable:

   Prediction:

3. **Write the components for the temperature test:**
   Hypothesis:

   Prediction:

   Experimental group:

   Control group:

   Experimental group:

Control group:

Independent variable:

Dependent variable:

4. **Were your data consistent with your predictions? Explain any unexpected results.**

5. **Why is it important to maintain a level surface with consistent texture?**

6. **Why do you think the sow bugs behave as they do? How might this behavior affect the success of the organism?**

Exercise 30.2   AQUATIC TAXIS

This exercise uses an aquatic leech of the genus *Placobdella*.

See Exercise 19.7 for a discussion about annelids.

Leeches have a typical invertebrate nervous system of a double ventral nerve cord, with two anterior ganglia. They do not have "eyes" to sense their prey, but they do possess several types of specialized sensory cells in their epidermis, enabling them to exhibit an array of elaborate responses to external stimuli. This is not surprising, because leeches must react quickly in order to attach to a rapid-moving host *(Mann, 1961)*.

Leeches of the genus *Placobdella* are blood feeders but will occasionally feed on snails. Barnes (1980) pointed out that *Placobdella* species will feed on almost any species of turtle and alligators, and only rarely attack amphibians or mammals, including humans. Many leech species will also feed on humans, if given the opportunity. In field studies conducted by the authors, leeches of the genus *Placobdella* were quite eager to feed on a human host. Indeed, leeches can be acquired in generous numbers for laboratory studies simply by wading through a pond and allowing them to attach to the skin. You don't have to collect them this way—don't worry.

1. Conduct experiments to test and quantify the responses (taxis) of leeches to light, heat, and vibrations. Again, conduct only one test at a time. Predict whether the organisms will have a positive or negative taxis to each stimulus. Answer Questions 1 to 3.

2. Design your experiment, using at least six leeches for each test. Conduct each test separately, and run each test for the same length of time. Don't forget to use a control group. Record your data in Table 30.2. Compare your data with data from other groups.

TABLE 30.2 LEECH TAXIS	
Stimulus	**+ or − taxis?**
Light	
Vibration	

Here is a list of materials that will be available to you:

Baking pans or trays

Cardboard

Tape

Stirring straw

Leech chamber

Aged tap water

Leeches

Timer

Questions

1. **Write the components for the light test:**

   Hypothesis:

   Prediction:

   Experimental group:

Control group:

Independent variable:

Dependent variable:

2. **Write the components for the heat test:**

   Hypothesis:

   Prediction:

Experimental group:

Control group:

Independent variable:

Dependent variable:

3. **Write the components for the vibration test:**

   Hypothesis:

   Prediction:

Experimental group:

Control group:

Independent variable:

Dependent variable:

4. **Were your data consistent with your predictions? Explain any unexpected results.**

5. **Which stimulus did leeches respond to most dramatically? Explain how the behavior exhibited by your leeches may be important as a host-finding mechanism.**

Exercise 30.3 SOCIAL BEHAVIOR OF BETTA FISH

(Adapted from Morgan and Carter, 2002)

In this exercise you will investigate the agonistic behavior exhibited by male betta fish, *Betta splendens*. **Agonistic behavior** refers to aggressive or defensive behavior among individuals of the same species. Male bettas are highly aggressive and exceedingly territorial. They will even respond aggressively to their own image in a mirror. Bettas are endemic to Southeast Asia, including the northern Malay Archipelago, Thailand, Cambodia, and Vietnam, where they are commonly found in areas of heavy vegetation in irrigation ditches and flooded rice paddies.

In nature, agonistic behavior exhibited by bettas is used to establish territories. Males then defend their small territories from other males and even from females that are not visiting to spawn (lay eggs). Males prepare a "bubble nest" by releasing air bubbles on the surface of still water. When it is time to spawn, a female visits and swims in a circular motion with the male, which helps to ensure that the eggs will be fertilized by the male right after they are released by the female. After spawning, the female either leaves or is driven away by the male. The male gathers the eggs into his mouth, deposits them into the bubble nest, and then guards the nest until the eggs hatch.

Animals respond to stimuli in **action patterns**, behaving with a pattern of responses that have different functions. If a male betta perceives another fish as a threat, he will exhibit different types of agonistic behavior in response. However, like some other animal species, male bettas may also construct a **dominance hierarchy** with other "neighbor" male fish it already recognizes, with a dominant male controlling the behavior of other males in the group. Two male bettas should never be placed in the same tank! In nature, the "losing" fish can retreat, but in a confined area bettas may fight and injure one another.

1. Observe an isolated male betta, and identify its external anatomy (Figure 30.2). Observe its normal behavior when it is not exposed to other fish. Describe it (see Question 1).

2. Place a mirror in front of the tank, and observe the fish's agonistic responses. Describe and name the behaviors (create descriptive names) in the space provided, answering Questions 2 to 4.

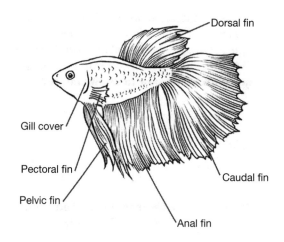

FIGURE 30.2 The Siamese Fighting Fish, *Betta splendens*.

3. Now conduct your own experiment to investigate the agonistic behavior exhibited by male bettas. Consider what might happen when you allow two males to see one another. Will they exhibit agonistic behavior, even if they are in separate tanks? What position(s) of agonistic behavior will they exhibit? Will they behave differently toward one another if they already know one another (conditioned) than when they are unfamiliar with one another (unconditioned)? Answer Question 5.

Here is the list of materials at your disposal:

Two conditioned bettas in separate tanks (familiar with one another)

Two unconditioned bettas in separate tanks (unfamiliar with one another)

Mirror

Cardboard to place between tanks

Questions

1. **Describe the normal behavior of male bettas. Sketch and indicate the colors of the fins, gills, and body of the fish as they appear during normal behavior.**

2. **Describe the different positions of agonistic behavior of male bettas. Draw the fish as it appears in each position of agonistic behavior. What are the different colors of the fins, gills, and body in response to its own image?**

Prediction:

Experimental group:

3. **What might be the functions and advantages of each type of agonistic behavior?**

Control group:

4. **Why does it react in such a manner to its own image?**

Independent variable:

Dependent variable:

5. **Write the components of your experimental design:**

    Hypothesis:

6. **Were your predictions correct? Explain your results.**

Appendix 1

HOW TO CITE REFERENCES

When your biology instructor assigns research papers or other written assignments, it is important that you cite your resources properly. Unless you are completing a Web exercise, choose hardcopy sources (books, magazines, journals) as your primary references, and then supplement these with secondary references (hardcopy, or even Internet sources, if your instructor allows them). As you accumulate information to complete your assignment, keep careful notes on your resources. A good *References* section (sometimes called *Literature Cited*) enables you or your instructor to easily access the sources in the future, if necessary.

There are many different formats for citing references in your text and in the references section of your manuscript. Listed below are examples from the *Council of Biology Editors* preferred methods for citing references (Knisely, 2002). One of these two methods is recommended:

> **Name–Year System:** Lists references in alphabetical order by author's last name.
> **Citation-Sequence System:** Lists references in the order they are cited in your text.

Notice that there are two main differences in these systems: the order in which the references are listed, and where the publication year is placed in the citation. Of course, you may refer to the *References and Suggested Readings* section of this laboratory manual for endless examples. Your instructor may choose one of the following formats, but if your instructor doesn't specify which type to use, choose the system you prefer.

For a Book:

Name–Year:

The citation in your text should read

> *...Plant stem growth is affected by the amount of previous exposure to light (Darwin, 1880).*

And in your References section, the citation should look like this:

> Darwin C. 1880. The power of movement in plants. London: John Murray. 592 p.

Citation-Sequence:

The citation in your text should read

> *...Plant stem growth is affected by the amount of previous exposure to light (1).*

And in your References section, the citation should look like this:

> 1. Darwin C. The power of movement in plants. London: John Murray. 1880. 592 p.

For an Article in a Journal:

Name–Year:
Bundy DAP, Chan MS, Savioli L. 1995. Hookworm infection in pregnancy. Transactions of the Royal Society of Tropical Medicine and Hygiene 89:521–522.

Citation-Sequence:
1. Bundy DAP Chan MS Savioli L. Hookworm infection in pregnancy. Transactions of the Royal Society of Tropical Medicine and Hygiene 1995; 89:521–522.

For a Chapter in a Book:

Name–Year:
Greene EC. 1949. Gross Anatomy. In: Farris EJ, Griffith JQ, editors. The Rat in Laboratory Investigation. New York: Hafner Publishing Co. pp 24–50.

Citation-Sequence:
1. Greene EC. Gross Anatomy. In: Farris EJ, Griffith JQ, editors. The Rat in Laboratory Investigation. New York: Hafner Publishing Co. 1949. pp 24–50.

For Internet Articles or Websites:

Name–Year:
Centers for Disease Control and Prevention.
2003. Summary of Notifiable Diseases—
United States, 2001. Morbidity and Mortality
Weekly Reports 50(53);1–108. Retrieved August 28, 2003 from the World Wide Web:
http://www2.cdc.gov/mmwr/summary.html

Citation-Sequence:
Centers for Disease Control and Prevention.
Summary of Notifiable Diseases—United
States, 2001. Morbidity and Mortality Weekly
Reports 2003; 50(53);1–108. Retrieved August
28, 2003 from the World Wide Web:
http://www2.cdc.gov/mmwr/summary.html

Sometimes, the authors are not listed on the website articles; if this happens, cite the reference title and year in your text (if you're using the Name–Year system) and begin the citation with "[Anonymous]." Then continue with the date the article was published, as shown above.

Be careful when using Internet resources for your assignments! Avoid using websites that are not published by recognizable scientific agencies or institutions, unless your assignment or instructor advises you to do so. Websites should not be used as primary sources of information (unless, of course, the exercise in entirely Web-based) but might be allowed as secondary sources of information, to provide more support for your paper. There is little accountability for what is published on the World Wide Web; Internet material does not have the same reliability as peer-reviewed books or journal articles (Knisely, 2002). What does "peer-reviewed" mean? It refers to the process that editors (who are usually well-known 'experts' in their fields) use to make sure that submitted manuscripts are checked for accuracy and validity. This process includes sending a submitted manuscript to a number of "experts" in that particular topic, who critique the manuscript for scientific content, brevity, relevance, as well as general format, grammar, and sentence structure. In other words, publishing in the scientific arena isn't an easy task—many rules have to be followed. All a person has to do to publish on the Web, however, is know how to build a website.

If you need more guidance, refer to the publications listed in the references section for this appendix.

Appendix 2

DARWIN: *THE POWER OF MOVEMENT IN PLANTS*

Excerpts from CHAPTER IX: Sensitiveness of Plants to Light: Its Transmitted Effects.

...A pot with seedlings of *Phalaris Canariensis*, which had been raised in darkness, was placed in a completely darkened room, at 12 feet from a very small lamp. After 3 h. the cotyledons were doubtfully curved towards the light, and after 7 h. 40 m. from the first exposure, they were all plainly, thought slightly, curved towards the lamp. Now, at this distance of 12 feet, the light was so obscure that we could not see the seedlings themselves, nor read the large Roman figures on the white face of a watch, nor see a pencil line on paper, but could just distinguish a line made with Indian ink. It is a more surprising fact that no visible shadow was cast by a pencil held upright on a white card; the seedlings, therefore, were acted on by a difference in the illumination of their two sides, which the human eye could not distinguish. On another occasion, even a less degree of light acted, for some cotyledons of Phalaris became slightly curved towards the same lamp at a distance of 20 feet; at this distance we could not see a circular dot 2.29 mm. (.09 inch) in diameter made with Indian ink on white paper, though we could just see a dot 3.56 mm. (.14 inch) in diameter; yet a dot of the former size appears large when seen in the light.

We next tried how small a beam of light would act; as this bears on light serving as a guide to seedlings whilst they emerge through fissured or incumbered ground. A pot with seedlings of Phalaris was covered by a tin-vessel, having on one side a circular hole 1.23 mm. in diameter (i.e. a little less than the 1/20th of an inch); and the box was placed in front of a paraffin lamp and on another occasion in front of a window; and both times the seedlings were manifestly bent after a few hours towards the little hole.

(p. 457)...

The cotyledons of Phalaris bend much more slowly towards a very obscure light than towards a bright one. Thus, in the experiments with seedlings placed in a dark room at 12 feet from a very small lamp, they were just perceptibly and doubtfully curved towards it after 3 h., and only slightly, yet certainly, after 4 h. After 8 h. 40 m. the chords of their arcs were deflected from the perpendicular by an average angle of only 16°. Had the light been bright, they would have become much more curved in between 1 and 2 h. Several trials were made with seedlings placed at various distances from a small lamp in a dark room; but we will give only one trial. Six pots were placed at distances of 2, 4, 8, 12, 16, and 20 feet from the lamp, before which they were left for 4 h. As light decreases in a geometrical ratio, the seedlings in the 2nd pot received 1/4th, those in the 3rd pot 1/16th, those in the 4th 1/36th, those in the 5th 1/64th, and those in the 6th 1/100th of the light received by the seedlings in the first or nearest pot. Therefore it might have been expected that there would have been an immense difference in the degree of their heliotropic curvature in the several pots; and there was a well-marked difference between those which stood nearest and furthest from the lamp, but the difference in each successive pair of pots was extremely small. In order to avoid prejudice, we asked three persons, who knew nothing about the experiment, to arrange the pots in order according to the degree of curvature of the cotyledons. The first person arranged them in proper order, but doubted long between the 12 feet and 16 feet pots; yet these two received light in the proportion of 36 to 64. The second person also arranged them properly, but doubted between the 8 feet and 12 feet pots, which received light in the proportion of 16 to 36. The third person arranged them in wrong order, and doubted about four of the pots. This evidence shows conclusively how little the curvature of the seedlings differed in the successive pots, in comparison with the great difference in the amount of light which they received; and it should be noted that there was no

371

excess of superfluous light, for the cotyledons become but little and slowly curved even in the nearest pot. Close to the 6th pot, at the distance of 20 feet from the lamp, the light allowed us just to distinguish a dot 3.56 mm (.14 inch) in diameter, made with Indian ink on white paper, but not a dot 2.29 mm (.09 inch) in diameter.

The degree of curvature of the cotyledons of Phalaris within a given time, depends not merely on the amount of lateral light which they may then receive, but on that which they have previously received from above and on all sides. Analogous facts have been given with respect to the nyctitropic and periodic movements of plants. Of two pots containing seedlings of Phalaris which had geminated in darkness, one was still kept in the dark, and the other was exposed (Sept. 26th) to the light in a greenhouse during a cloudy day and on the following bright morning. On this morning (27th), at 10:30 A.M., both pots were placed in a box, blackened within and open in front, before a north-east window, protected by a linen and muslin blind and by a towel, so that but little light was admitted, though the sky was bright. Whenever the pots were looked at, this was done as quickly as possible, and the cotyledons were then held transversely with respect to the light, so that their curvature could not have been thus increased or diminished. After 50 m. the seedlings which had previously been kept in darkness, where perhaps, and after 70 m. were certainly, curved, though very slightly, towards the window. After 85 m. some of the seedlings, which had previously been illuminated, were perhaps a little affected, and after 100 m. some of the younger ones were certainly a little curved towards the light. At this time (i.e. after 100 m.) there was a plain difference in the curvature of the seedlings in the two pots. After 2 h. 12 m. the chords of the arcs of four of the most strongly curved seedlings in each pot were measured, and the mean angle from the perpendicular of those which had previously been kept in darkness was 19°, and of those which had previously been illuminated only 7°. Nor did this difference diminish during two additional hours. As a check, the seedlings in both pots were then placed in complete darkness for two hours, in order that apogeotropism should act on them; and those in the one pot which were little curved became in this time almost completely upright, whilst the more curved ones in the other pot still remained plainly curved.

Two days afterwards the experiment was repeated, with the sole difference that even less light was admitted through a window, as it was protected by a linen and muslin blind and by two towels; the sky, moreover, was somewhat less bright. The result was the same as before, excepting that everything occurred rather slower. The seedlings which had been previously kept in darkness were not in the least curved after 54 m., but were so after 70 m. Those which had previously been illuminated were not at all affected until 130 m. had elapsed, and then only slightly. After 145 m. some of the seedlings in this latter pot were certainly curved towards the light; and there was now a plain difference between the two pots. After 3 h. 45 m. the chords of the arcs of 3 seedlings in each pot were measured, and the mean angle from the perpendicular was 16° for those in the pot which had previously been kept in darkness, and only 5° for those which had previously been illuminated.

The curvature of the cotyledons of Phalaris towards a lateral light is therefore certainly influenced by the degree to which they have been previously illuminated. We shall presently see that the influence of light on their bending continues for a short time after the light has been extinguished. These facts, as well as that of the curvature not increasing or decreasing in nearly the same ratio with that of the amount of light which they receive, as shown in the trials with the plants before the lamp, all indicate that light acts on them as a stimulus, in somewhat the same manner as on the nervous system of animals, and not in a direct manner on the cells or cell-walls which by their contraction or expansion cause the curvature.

It has already been incidentally shown how slowly the cotyledons of Phalaris bend towards a very dim light; but when they were placed before a bright paraffin lamp their tips were all curved rectangularly towards it in 2 h. 20 m. The hypocotyls of *Solanum lycopersicum* had bent in the morning at right angles towards a north-east window. At 1 P.M. (Oct. 21st) the pot was turned round, so that the seedlings now pointed from the light, but by 5 P.M. they had reversed their curvature and again pointed to the light. They had thus passed through 180° in 4 h., having in the morning previously passed through about 90°. But the reversal of the first half of the curvature will have been aided by apogeotropism.

Similar cases were observed with other seedlings, for instance, with those of *Sinapis alba*...

(p. 465)...

In our various experiments we were often struck with the accuracy with which seedlings pointed to a light although of small size. To test this, many seedlings of Phalaris, which had germinated in darkness in a very narrow box several feet in length, were placed in a darkened room near to and in front of a lamp having a small cylindrical wick. The cotyledons at the two ends and in the central part of the box, would therefore have to bend in widely different directions in order to point to the light. After they had become rectangularly bent, a long white thread was stretched by two persons, close over and parallel, first to one and then to another cotyledon; and the thread was found in almost every case actually to intersect the small circular wick of the now extinguished lamp. The deviation from accuracy never exceeded, as far as we could judge, a degree or two. This extreme accuracy seems at first surprising, but is not really so, for an upright cylindrical stem, whatever its position may be with respect to the light, would have exactly half its circumference illuminated and half in shadow; and the difference in illumination of the two sides is the exciting cause of heliotropism, a cylinder would naturally bend with much accuracy towards the light. The cotyledons, however, of Phalaris are not cylindrical, but oval in section; and the longer axis was the shorter axis (in the one which was measured) as 100 to 70. Nevertheless, no difference could be detected in the accuracy of their bending, whether they stood with their broad or narrow sides facing the light, or in any intermediate position; and so it was with the cotyledons of *Avena sativa*, which are likewise oval in section. Now, a little reflection will show that in whatever position the cotyledons may stand, there will be a line of greatest illumination, exactly fronting the light, and on each side of this line an equal amount of light will be received; but if the oval stands obliquely with respect to the light, this will be diffused over a wider surface on one side of the central line than on the other. We may therefore infer that the same amount of light, whether diffused over a wider surface or concentrated on a smaller surface, produces exactly the same effect; for the cotyledons in the long narrow box stood in all sorts of positions with reference to the light yet all pointed truly towards it.

That the bending of the cotyledons to the light depends on the illumination of one whole side or on the obscuration of the whole opposite side, and not on a narrow longitudinal zone in the line of the light being affected, was shown by the effects of painting longitudinally with Indian ink one side of five cotyledons of Phalaris. These were then placed on a table near to a south-west window, and the painted half was directed either to the right or left. The result was that instead of bending in a direct line towards the window, they were deflected from the window and towards the unpainted side, by the following angles, 35°, 83°, 31°, 43°, and 39°. It should be remarked that it was hardly possible to paint one-half accurately, or to place all the seedlings which are oval in section in quite the same position relatively to the light; and this will account for the differences in angles. Five cotyledons of Avena were also painted in the same manner, but with greater care; and they were laterally deflected from the line of the window, towards the unpainted side, by the following angles, 44°, 44°, 55°, 51°, and 57°. This deflection of the cotyledons from the window is intelligible, for the whole unpainted side must have received some light, whereas the opposite and painted side received none; but a narrow zone on the unpainted side directly in front of the window will have received most light, and all the hinder parts (half an oval in section) less and less light in varying degrees; and we may conclude that the angle of deflection is the resultant of the action of the light over the whole of the unpainted side.

It should have been premised that painting with Indian ink does not injure plants, at least within several hours; and it could injure them only by stopping respiration. To ascertain whether injury was thus soon caused, the upper halves of 8 cotyledons of Avena where thickly coated with transparent matter,—4 with gum, and 4 with gelatine; they were placed in the morning before a window, and by the evening they were normally bowed towards the light, although the coatings now consisted of dry crusts of gum and gelatine. Moreover, if the seedlings which were painted longitudinally with Indian ink had been injured on the painted side, the opposite side would have gone on growing, and they would consequently have become bowed towards the painted side; whereas the curvature was always, as we have seen, in the opposite direction, or towards the unpainted side which was exposed to the light. We witnessed the effects of injuring longitudinally one side of the cotyledons of Avena and Phalaris; for before we knew that grease was highly injurious to them, several were painted down one side with a mixture of oil and lamp-black, and where then exposed before a window; others similarly treated were afterwards tried in darkness. These cotyledons soon became plainly bowed towards the blackened side, evidently owing to the grease on this side having checked their growth, whilst growth continued on the opposite side. But it deserves notice that the curvature differed from that caused by light, which ultimately becomes abrupt near the ground. These seedlings did not afterwards die, but were much injured and grew badly...

CONCLUDING REMARKS AND SUMMARY OF CHAPTER

(p. 486)...

Light exerts a powerful influence on most vegetable tissues, and there can be no doubt that it generally tends to check their growth. But when the two sides of a plant are illuminated in a slightly different degree, it does not necessarily follow that the bending towards the illuminated side is caused by changes in the tissues of the same nature as those which lead to increased growth in darkness. We know at least that a part may bend from the light, and yet its growth may not be favoured by light. This is the case with the radicles of *Sinapis alba*, which are plainly apheliotropic; nevertheless, they grow quicker in darkness than in light. So it is with many aërial roots, according to Wiesner; but there are other opposed cases. It appears therefore, that light does not determine the growth of apheliotropic parts in any uniform manner.

We should bear in mind that the power of bending to the light is highly beneficial to most plants. There is therefore no improbability in this power having been specially acquired. In several respects light seems to act on plants in nearly the same manner as it does on animals by means of the nervous system. With seedlings the effect, as we have just seen, is transmitted from one part to another. An animal may be excited to move by a very small amount of light; and it has been shown that a difference in the illumination of the two sides of the cotyledons of Phalaris, which could not be distinguished by the human eye, sufficed to cause them to bend. It has also been shown that there is no close parallelism between the amount of light which acts on a plant and its degree of curvature; it was indeed hardly possible to perceive any difference in the curvature of some seedlings in Phalaris exposed to a light, which, though dim, was very much brighter than that to which others had been exposed. The retina, after being stimulated by a bright light, feels the effect for some time; and Phalaris continued to bend for nearly half an hour towards the side which had been illuminated. The retina cannot perceive a dim light after it has been exposed to a bright one; and plants which had been kept in the daylight during the previous day and morning, did not move so soon towards an obscure lateral light as did others which had been kept in complete darkness.

Even if light does act in such a manner on the growing parts of plants as always to excite in them a tendency to bend towards the more illuminated side—a supposition contradicted by the foregoing experiments on seedlings and by all apheliotropic organs—yet the tendency differs greatly in different species, and is variable in degree in the individuals of the same species, as may be seen in almost any pot of seedlings of a long cultivated plant. There is therefore a basis for the modification of this tendency to almost any beneficial extent. That it has been modified, we see in many cases: thus, it is of more importance for insectivorous plants to place their leaves in the best position for catching insects than to turn their leaves to the light, and they have no such power. If the stems of twining plants were to bend towards the light, they would often be drawn away from their supports; and as we have seen they do not thus bend...

GLOSSARY

Heliotropism: the bending of an organ to light.
Apheliotropism: bending away from light.
Geotropism: bending toward the center of the earth, with the pull of gravity.
Apogeotropism: bending away from the center of the earth, opposite the pull of gravity.
Indian ink: black waterproof ink.
Lamp-black: fine soot from the burning of lamp oil.
Nyctitropic: closure in response to darkness.

Appendix 3

FRESHWATER MACROINVERTEBRATES OF THE UNITED STATES

This appendix provides photographs and information concerning the natural history of most common freshwater macroinvertebrates that you may encounter in the field. This information is not designed to replace taxonomic keys; rather, it may supplement them and serve as a quick reference. The photographs and drawings in this guide will enable you to recognize the basic *types* of common aquatic invertebrates. It will also provide you with information on their natural history, which is needed for the exercises in Chapter 27. For more specific identification of aquatic invertebrates, have Pennak (1959) and at least one of these references on hand: Borrer and White (1970), Lehmkuhl (1970), Eddy and Hodson (1982), Reid (1987), Thorp and Covich (1991), and Merrit and Cummins (1996).

PHYLUM PLATYHELMINTHES (FLATWORMS)

Class Turbellaria (Planarians)

Free-living flatworms (Figure 1) glide along the substrate of most bodies of fresh water. The gliding motion is a result of ciliary action in a thin coat of mucus that is secreted by the worms. They feed on a wide range of material with a preference for small invertebrates. They occasionally feed on larger dead animals, algae, or detritus. Although

turbellarians superficially resemble leeches, they may easily be distinguished from the latter because of the head bearing two conspicuous eyespots and sensory structures called auricles that resemble "ears." Also, planarians are not segmented; leeches are. Planarians are monoecious and reproduce both sexually and asexually. They are particularly noted for their tremendous ability to regenerate lost parts.

PHYLUM ANNELIDA (SEGMENTED WORMS)

Class Hirudinea (Leeches)

Leeches (Figure 2) are best known for their sanguineous (L. bloodthirsty) nature and segmented bodies, although not all leeches are parasitic blood feeders; many types of leeches are predators of macroinvertebrates (such as snails). There are many kinds of leeches, and they exhibit great diversity in body size, shape, and coloration. Leeches are easily recognized by the presence of an anterior and posterior sucker, which is lacking in planarians and earthworms.

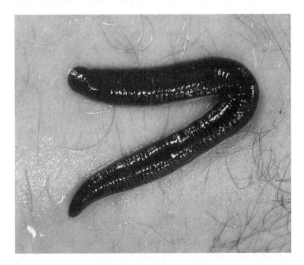

FIGURE 2   A Leech.

FIGURE 1   Planaria (*Dugesia tigrina*).

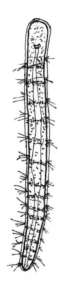

FIGURE 3   A Freshwater Oligochaete, *Aeolosoma* sp. (after Pennak, 1953).

Class Oligochaeta

Freshwater oligochaetes are very similar to their terrestrial cousins, the earthworms; however, they are usually smaller (1–30 mm) and exhibit prominent seatae (Figure 3). These little detritivores are common in the benthos (mud and detritus on the bottom) of most freshwater habitats.

PHYLUM MOLLUSCA

Class Bivalvia

This group includes the mussels and clams (see Exercise 19.6), which are highly variable in size. Bivalves typically burrow their stout muscular foot into the substrate. Water containing food and oxygen is carried into the mollusc through the incurrent siphon and water carrying carbon dioxide and metabolic wastes exits through the excurrent siphon. Food particles, caught in mucus on the soft body of the mollusc are directed into the mouth by cilia. Molluscs are filter feeders, feeding on zooplankton, phytoplankton, and detritus.

Class Gastropoda (Snails)

Snails, characterized by their distinct coiled shell, are important players in freshwater aquatic ecosystems. Using their scraping radula, snails feed primarily on algae covering the substrate, and

FIGURE 4   *Stagnicola* sp., a Typical Spiraled Snail.

FIGURE 5   *Helisoma* sp., a Typical Planorbid Snail.

occasionally feed on detritus. The shell may be a spiral (Figure 4) or discoidal (planorbid) type (Figure 5). Two of the most common genera of spiral-shelled snails in ponds and streams are *Physa* with a sinistral shell and *Stagnicola* with a dextral shell. Two examples of common planorbid snails are the tiny *Gyraulus* and the large *Helisoma*.

PHYLUM ARTHROPODA

Class Arachnida
Order Hydracarina (Water Mites)

Water mites are small, often brightly colored, active swimmers found in freshwater, lentic habitats. They

FIGURE 6   A Stilt Spider (*Tetragnatha* sp.)

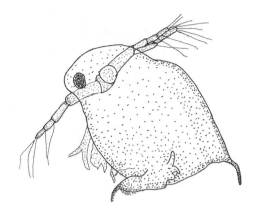

FIGURE 7   *Daphnia* sp., a Typical Cladoceran.

are globular to ovoid and are between 0.5 to 3 mm long. Most water mites are carnivores, feeding on tiny insects, worms, and other small invertebrates. Some are parasitic.

Order Araneae (Stilt Spiders)

Stilt spiders (Figure 6), also known as pond spiders, stretch spiders, and long-jawed spiders, are commonly found carnivores along shores of ponds and streams, where they spin webs in vegetation to capture small flying insects such as midges. Stilt spiders are true spiders (Order Araneae), although they resemble the harvestman (Order Opiliones; see Figure 20.3b). Stilt spiders are characterized by their unusually large chelicerae.

Class Crustacea
Orders Cladocera, Copepoda, and Ostracoda (Microcrustaceans)

Microscopic crustaceans usually make up the majority of zooplankton in a body of water. All of these are detritivores and play very important roles in the foodweb of most freshwater ecosystems. The microcrustaceans are among the most important primary consumers in many communities.

Cladocerans (Figure 7), also known as water fleas or sea monkeys, are 2 to 3 mm long and may occur in large numbers. They may be seen swimming through the water. They are fairly easy to recognize because of their jerky movement. Highly modified legs (that appear as fans) direct debris from the substrate toward the cladoceran's mouthparts.

Copepods (Figure 8) feed on bacteria, algae, and detritus. A few species are parasitic on fish and other aquatic organisms. Copepods are dorsoventrally flattened and may be easily recognized by the characteristic pair of posterior extensions called caudal rami.

Ostracods (Figure 9), commonly known as seed shrimps, are laterally flattened and possess a bivalve carapace making them appear superficially as tiny clams. When the "shell" is open, the antennae extend outward, and the tiny mouthparts and legs may be seen moving rapidly to create a current of water that flows between the valves of the shell. An ostracod eats bacteria, algae, and detritus from the current.

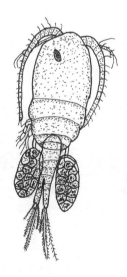

FIGURE 8   *Cyclops* sp., a Typical Copepod. Note the conspicuous egg-containing ovisacs attached to the female in this drawing.

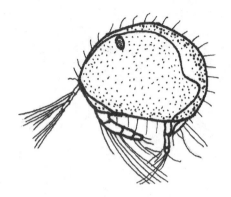

FIGURE 9   A typical ostracod.

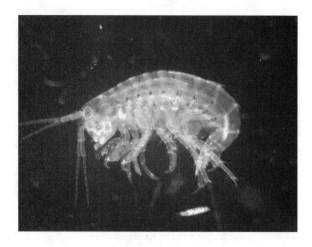

FIGURE 11   *Hyalella azteca*, a Common Amphipod.

Order Isopoda

Aquatic isopods (Figure 10) are very similar to their terrestrial cousins, the pill bugs and sow bugs, and they play the same ecological role as detritivores. Isopods are dorsoventrally flattened with total length ranging from 5–20 mm. Although several genera of isopods inhabit North American waters, *Asellus* is the most commonly encountered genus. Fertilized eggs develop in the female's marsupium (pouch on the ventral surface). After hatching, the young remain in the marsupium for several days before being released.

Order Amphipoda

Amphipods (Figure 11), sometimes called **sideswimmers** or **scuds**, are common **detritivores** in

FIGURE 10   *Asellus* sp., a Typical Freshwater Isopod.

most aquatic ecosystems, and play very important roles as primary consumers in most aquatic foodchains. Amphiods are easily distinguished from isopods because amphipods are compressed laterally, while isopods are dorsoventrally flattened. Amphipods exhibit interesting reproductive behaviors. The male and female will remain in pairs from 1 to 7 days before mating, while the male carries the females on his back. They separate while the female molts. They then pair up again and copulation occurs within 24 hours. Amphipods possess a marsupium that contains the eggs and the young hatchlings. The most common genera of amphipods are *Gammarus* and *Hyalella* (the one North American species is *Hyalella azeteca*). The two genera may easily be distinguished because amphipods of the genus *Gammarus* tend to be larger and possess an accessory flagellum that originates around the third segment of the first antenna (Figure 12).

Order Decapoda (Crayfish)

Crayfish (Figure 13) are fascinating critters that closely resemble their marine cousins, the lobsters, although crayfish are smaller. (See Exercise 20.3 for a detailed account of crayfish anatomy.) Crayfish (also called crawfish or crawdads) are easily recognized by the presence of large chelipeds (pencers), which are used for food handling and defense (and even offense). Crayfish feed on a wide variety of food items and are usually categorized as scavengers. Although crayfish will eat dead animal

(a)

(b)

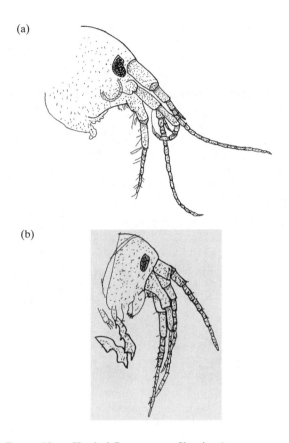

FIGURE 12   a. Head of *Gammarus* sp. Showing Accessory Flagellum Originating Around the 3rd Segment of the 1st Antenna. b. Head of *Hyalella azteca*. Note that the accessory flagellum is lacking.

FIGURE 13   A Crayfish (*Orconectes* sp.).

material, they are seldom predaceous. They prefer plant material when it is abundant. Females lay eggs, which are "stuck" to her ventral side on her swimmerets. After hatching, young remain attached to the swimmerets for several weeks. The lifespan of crayfish is usually less than 20 months, although crayfish may live for over 2 years. It is the opinion of one of the authors (DJR) that boiled crayfish is true ambrosia (food of the gods).

Class Insecta
Order Plecoptera (The Stoneflies)

Adult stoneflies are terrestrial and are normally found near streams where the aquatic naiads occur, preferring swift-moving water. Stonefly naiads (Figure 14) are readily recognized by the presence of two prominent segmented cerci extending from the posterior end (for comparison, mayflies have three cerci) and two claws on each foot. Indistinct external filamentous gills may or may not be present. Feeding habits vary by family. Representatives of families Perlidae and Perlodidae are carnivorous, feeding on other insects such as mayfly naiads and midge larvae. Others feed on algae and detritus. The Perlidae possess branched filamentous gills on the lateral and ventral surface of the thorax while the other stonefly naiads typically do not. Representatives of the family Perlidae are the stonefly naiads most frequently collected. Stoneflies are important as indicators of water quality because they are sensitive to many types of organic and chemical pollution. Adult stoneflies (Figure 15) have delicate wings and long antennae and retain the segmented cerci. They usually emerge in the fall to early winter and live for only a few weeks. Adults of some species feed on algae and plants but adults of other species do not feed at all. Stonefly naiads appear to be a preferred prey item of trout, and often serve as models for fly fishermen who tie their own flies.

FIGURE 14   Stonefly Naiad.

FIGURE 15   Adult Stonefly.

Naiads are most readily collected from the bottom of stones in fast-running water.

Order Ephemeroptera (The Mayflies)

Mayfly naiads (Figure 16) are very common in stream and pond environments and may be distinguished from stonefly naiads in that mayflies usually possess prominent, paired tracheal gills on the lateral and dorsal aspects of the abdomen. Also, they usually possess three, rarely two, distinct caudal filaments (as opposed to two possessed by stoneflies). Mayfly naiads usually live only about a year before emergence of adults but the naiads of some species may persist two or three years. Mayfly naiads are herbivorous, feeding on algae and plants. Adults are short-lived (from a few hours to about a week) and do not feed. They have highly-veined, nearly transparent wings and two or three caudal filaments.

Mayflies may be extremely abundant and often play an important role in the ecology of freshwater ecosystems. They are a favorite food of many fish, including trout; fly-tiers often try to mimic mayflies with their lures. Mayflies are another important indicator species in regard to water quality, as they are sensitive to organic pollution, acidity, and many types of chemical pollution.

Mayflies often emerge in exceedingly large numbers, forming mating swarms in the spring and summer. These mating swarms are composed almost entirely of males. Some cities near large lakes have experienced such intense swarms of

(a)

(b)

FIGURE 16   a. Mayfly Naiad (Family Baetidae). b. Mayfly Naiad (Family Hepatgeniidae).

mayflies that dead mayflies make roadways slippery and impassable. Many kinds of mayflies are attracted to light, and their bodies are found in piles several inches deep near exterior lights, and may have to be removed with snow shovels.

Order Odonata (Dragonflies and Damselflies)

The odonates, dragonflies and damselflies, constitute one of the most well known and widely appreciated groups of aquatic organisms. They are commonly depicted in popular art, and have been throughout the ages. One of the reasons for their broad appeal is their beauty. Another is probably their ubiquity. Palmer (1949) said of dragonflies, "As mosquito destroyers should be encouraged, and as insects of beauty are to be admired." Both dragonflies and damselflies exhibit a lifecycle with an incomplete metamorphosis. The naiads of both are entirely aquatic while the adults of both are common flying insects around ponds, lakes, and streams. Naiads of both damselflies and dragonflies are voracious carnivores and consume an array of aquatic insects, crustaceans, annelids, and even snails. They will essentially eat any small invertebrate. They capture their prey with a specialized labium that looks and functions rather like a large spoon (Figure 17).

Damselfly naiads (Figure 18) may be distinguished from dragonfly naiads (Figure 19) in the possession of three prominent caudal gills. In

(a)

(b)

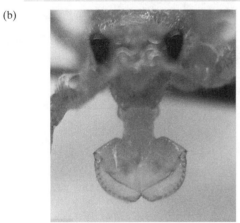

FIGURE 17   a. Dragonfly Naiad with Labium Extended.
b. Close-up of Labium.

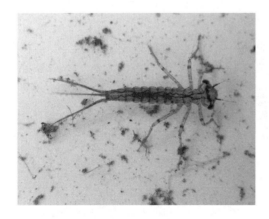

FIGURE 18   Dameselfly Naiad.

FIGURE 19   Dragonfly Naiad.

dragonfly naiads, the gills are internal and are found in a rectal chamber. Contraction of muscles of the rectal walls causes the rectum to expand and contract, pumping oxygen rich water in, and carbon dioxide laden water out through the anal opening. Also, dragonflies tend to be much thicker-bodied than damselflies.

Adult dragonflies (Figure 20) and damselflies may live for several weeks (usually 2–3 weeks). They feed in flight on small insects such as midges and mosquitos. Because of the thousands of small units, called ommatidia, comprising their large compound eyes, they have remarkable eyesight. Damselflies are normally found near water, but dragonflies are strong fliers and may be seen several miles away from water.

The damselfly mating ritual is a lesson in acrobatics. Using a pair of claspers on the distal tip of the abdomen, the male grasps the female just behind the neck. After being grabbed by the male, the female loops the tip of her abdomen up toward

FIGURE 20   Adult Dragonflies. Left to right a clubtail (Family Gomphidae); a darner (Family Aeshnidae); a skimmer (Family Libellulidae).

the male's genitalia located on his second abdominal segment. Their attached bodies are a leggy heart-shaped entanglement. Females lay eggs into plant material using the sharp ovipositor on the posterior end. Dragonflies may be distinguished from damsel flies fairly easily on the basis of a few characteristics. Dragonflies tend to be larger and stronger fliers than damselflies. When at rest, damselflies hold their wings folded upward (Figure 21) whereas, dragonflies hold their wings unfolded horizontally (Figure 20). Finally, the hind wings of dragonflies differ from the front wings in that they are broader at the base; whereas, the front and hind wings of damselflies are alike and are more narrow at the base than in distal regions (far end) of the wings.

Identifying naiads of the various families of dragonflies and damselflies provides something of a challenge for the beginning taxonomist, although the adults of each may be readily identified to the family level.

One name commonly applied to all adult dragonflies is *snake doctor*. Despite their large size and foreboding appearance, adult dragonflies are harmless. The largest dragonfly, with its chewing mouthparts, can only give a slightly painful pinch when held. Dragonflies do not sting! Some common types of dragonflies include clubtails (Gomphidae), darners (Aeshnidae), and skimmers (Libellulidae) (Figure 20).

Damselflies are among the most beautiful of all the invertebrates. There are 3 common families of damselflies in North America, Coenagrionidae and Lestidae (the narrow-winged damsel flies) and Calopterygidae (broad-winged damselflies). The families are easily differentiated based on wing veination. Wings of narrow-winged damselflies (Figure 22) are stalked (more narrow where they attach to the body) and possess far fewer antenodal cross veins than do broad-winged damselflies (Figure 23). The two families of narrow-winged damselflies may be differentiated by the location of the M3 vein, which arises nearer the arculus in the family Lestidae and nearer the nodus in the family Coenagrionidae.

The two most common species of narrow-winged damselflies (Coenagrionidae) are the civil bluet, *Enallagma civile,* and the forktail, *Ischnura verticalis*. These damselflies are commonly found inhabiting vegetation around marshes, ponds, and lakes. The male civil bluet is a pale blue, while the female is greenish-blue. Males and females are

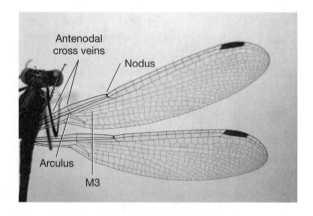

FIGURE 22   Wing of Narrow-Winged Damselfly (Lestidae).

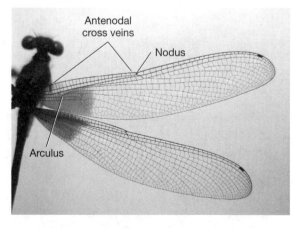

FIGURE 23   Wing of Broad-Winged Damselfly (Calopterygidae).

FIGURE 21   Adult Male American Rubyspot Damselfly (*Hetaerina americana*) Resting on a Blade of Grass.

FIGURE 24   Male Civil Bluet (*Enallagma civile*).

FIGURE 26   Male and Female *Calopteryx maculata*.

characterized by black markings or bands along the abdomen (Figure 24). Representatives of *Ischnura verticalis* are dark-colored with green stripes on the thorax. Representatives of the genus *Lestes*, the most common genus in the family Lestidae, are weak fliers found around marshes and ponds.

The two most common species in the family Calopterygidae, *Hetaerina americana* (the American ruby-spot) (Figure 25) and the ebony jewel-wing or black-winged damselfly, *Calopteryx maculata* (Figure 26) are common around streams and rivers. The American ruby-spot is reddish with a bright red spot at the base of the wings. Representatives of

Calopteryx maculata exhibit a variety of dark colors and the female has a distinctive white spot near the tip of the wings. Male Calopterygidae are exceedingly territorial, especially *Hetaerina americana*. The male "stakes out" a territory and will aggressively fight off invading males.

Damselflies are often confused with adult ant lions (order Neuroptera; family Myrmeleontidae). See Figure 27. Ant lions are easily distinguished from damselflies by the presence of their knobbed antennae. Myrmeleontids are entirely terrestrial. They get the common name *ant lion* from the larvae, which dig pits (about 5 cm in diameter) in sand. They wait for ants or other small insects to fall into their pit before seizing them with their large jaws. The larvae are also sometimes called *doodlebugs*.

FIGURE 25   Male American Ruby-Spot (*Hetaerina americana*).

FIGURE 27   Adult Ant Lion (Order Neuroptera; Family Myrmeleontidae).

Order Hemiptera (Bugs)

Many invertebrates are often referred to as *bugs*, but technically speaking bugs are insects belonging to the order Hemiptera. Bugs play important roles in aquatic ecosystems. Of the common aquatic bugs, all are carnivores (and so they occupy higher levels in the food web) with the exception of the water boatman (Corixidae), which serves as an important primary consumer.

Families Gerridae and Veliidae (Water Striders and Ripple Bugs)

Water striders (Gerridae) and ripple bugs (Veliidae) are voracious carnivores found in most bodies of fresh water. They skim along the surface of the water and feed on other small aquatic insects and crustaceans (Figure 28). They actually run, not swim, along the surface of the water. They feed by piercing prey with their sharp beak and then sucking out the juices. There is a lot of variation in size, depending on their species. Representatives of the two common families may be differentiated by examination of the hind legs. The hind femur in Gerridae extends well beyond the tip of the abdomen, whereas in Veliidae, the hind femur does not extend beyond the tip of the abdomen.

Family Notonectidae (Back Swimmers)

Back swimmers (Figure 29) commonly occur in ponds, lakes, and backwaters of streams. They are carnivorous, feeding mainly on insects with their piercing sucking mouthparts, although occasionally they attack tadpoles and small fish. Their typical

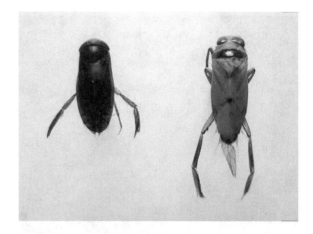

FIGURE 29   Left Water Boatman (Family Corixidae) Right Backswimmer (Family Notonectidae).

means of capturing prey is to drift up underneath the prey item. Notonectids are bullet-shaped with a keel running down the length of the abdomen. As the name implies, they swim on their back. They are light colored on top, and dark on the bottom. This acts as camouflage because the dark surface blends with their aquatic background, making the bug inconspicuous to predators that look down from above. Likewise, the light colored dorsal surface blends in with the lighter colored sky, such that predators viewing the bug from below cannot see them easily. The two common genera of notonectids, *Notonecta* and *Buenoa*, are most easily differentiated based on size. Representatives of the genus *Notonecta* are larger and broader and are 8 to 17 mm long, whereas the smaller, narrower representatives of *Buenoa* are 5 to 9 mm. When being handled, these bugs are capable of delivering a painful "sting" with their beak, although they seldom do so.

Family Nepidae (Water Scorpions)

Water scorpions (Figure 30) are also carnivorous bugs that feed on other insects with their piercing sucking mouthparts. They are found in marshes, ponds, and slowly moving streams. Water scorpions are usually brown and superficially resemble walking sticks. They breathe through a long respiratory tube on the posterior end. The tube is exposed above the surface of the water occasionally to permit the entry of oxygen into the bug's tracheal system. If handled carelessly, they can inflict a painful "sting" with their beak. The most common and widely distributed genus is *Ranatra*.

FIGURE 28   Water Strider (Family Gerridae).

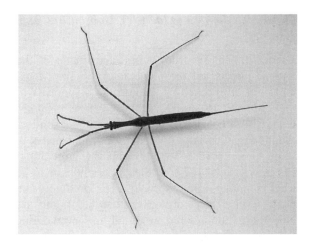

FIGURE 30   Water Scorpions (Family Nepidae). a. *Ranatra* sp. b. *Nepa apiculata*, another North American water scorpion.

Family Belostomatidae (Giant Water Bugs or Toe Biters)

The belostomatids (Figure 31) are the largest of the aquatic bugs, typically ranging from 20 to 70 mm in length (Figure 32). They are found in ponds and small lakes where they feed on a wide range of prey including insects, tadpoles, small frogs, and fish. Females of two genera, *Belastoma* and *Abedus*, lay their eggs on the back of the male, where they remain until they hatch. Belostomatids can inflict a painful "sting" with their beak (Figure 33) if handled carelessly.

Family Corixidae (Water Boatman)

Water boatman (Figure 34) are often the most abundant aquatic insects in ponds, lakes, and streams. Unlike most of the aquatic bugs, water boatmen feed primarily on detritus and algae. The front tarsi are broad and scoop-like, and are used to gather food with a sweeping motion. The hind legs are highly modified for swimming; they are

FIGURE 31   Giant Water Bug (Family Belostomatidae), a.k.a. Toe Biters.

FIGURE 32   Belostomatids Exhibit Great Diversity in Size.

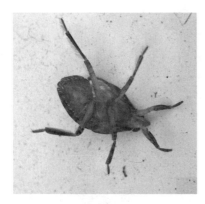

FIGURE 33   Ventral View of Giant Water Bug Showing Piercing and Sucking "Beak."

Order Megaloptera (Alderflies, Dobsonflies, and Fishflies)

Among the most striking and charismatic of the aquatic insects are the megalopterans. The Megaloptera is considered by some to be a suborder of the order Neuroptera. The most common family, Corydalidae, consists of the fishflies and dobsonflies. Adults are very large (6–7 cm long) and males of some species possess huge mandibles (Figure 35). It has been speculated that these mandibles may be used in mating or may serve to deter predators like bats. Despite their ominous appearance, adults are harmless to humans and are usually observed during the spring and early summer near streams, ponds, and lakes.

Megaloptera larvae are known as hellgrammites (Figure 36). These large, voracious carnivores

FIGURE 34   Water Boatman.

FIGURE 35   Male and Female Dobsonflies (Order Megaloptera).

flattened and fringed, and appear and function like tiny oars. Corixids are most easily confused with back swimmers (Figure 29); however, corixids have a light-colored ventral surface and dark-colored dorsal surface, lack a keel on the ventral side, and are more broad (not bullet-shaped). Corixids may be extremely abundant in lakes and ponds and have been (and continue to be) used as a source of human food in some cultures. They are high in protein and quite palatable, at least, according to one of the authors (DJR).

FIGURE 36   A Hellgrammite (dobsonfly larva Order Megaloptera).

may attain a length of up to 90 mm when mature (larval stages persist 2–3 years). They feed primarily on insects, but will eat any suitably sized animal. Hellgramites may inflict a painful bite if handled carelessly. Larvae of alderflies may be distinguished from dobsonflies and fishflies by the presence of a pronounced, single caudal filament that is lacking in dobsonflies.

Order Trichoptera (Caddis Flies)

Adult caddis flies (Figure 37) are drab, delicate, inconspicuous, moth-like insects that live only a few days. Adults are good fliers and possess four highly veined membranous wings. Adults of some species feed on nectar and it is likely that adults of some other species do not feed at all. The larvae exhibit a wide array of life styles and feeding habits, and are thus found in a wide range of aquatic habitats ranging from marshes and ponds to fast-flowing streams and rivers. Most larvae (Figure 38) construct a tube-like case; depending on species, the case may be constructed from silk, leaves, sand, twigs, or detritus. The cases are so distinctive that the species of larva can often be identified based on observation of the case alone. Depending on species, larvae may be detritivores, herbivores, carnivores, or some combination thereof, with the majority being omnivorous.

Order Coleoptera (The Beetles)

The order Coleoptera constitutes one of the most extensive and diverse orders in the animal kingdom. Like their terrestrial counterparts, the aquatic beetles exhibit a great extent of diversity regarding the roles they play in aquatic ecosystems. Distinguishing

(a)

(b)

FIGURE 38   a&b Caddis Fly Larvae.

between the various types of adult beetles can be very difficult; this is especially true of the larvae. Some basic taxonomic information follows, along with photographs to make identification easier. Nevertheless, using the keys listed in the introduction to this appendix will be necessary to identify some individuals to the proper family.

Family Haliplidae (Crawling Water Beetles)

The family Haliplidae (Figure 39) is comprised of a group of small herbivorous beetles that are often

FIGURE 37   An adult caddis fly (Order Trichoptera).

FIGURE 39   Crawling Water Beetle (Family Haliplidae).

found crawling around in mats of algae and other submerged aquatic vegetation. Adults are 2 to 5 mm long and are oval, with the widest point near the anterior edge of the elytra. The elytra is the hard outer covering on beetles, formed by the first (outer) pair of wings. The thread-like antennae consist of 10 segments. Longitudinal rows of indentations or punctures are usually present on the elytra. These beetles are usually a brownish color with black spots. Slender and elongate haliplid larvae may be distinguished by the presence of a single tarsal claw (all other aquatic beetle larvae have two tarsal claws).

Family Dytiscidae (Predaceous Diving Beetles)

As the name implies, predaceous diving beetles (Figure 40) are voracious carnivores feeding on an array of other invertebrates, with the larger ones feeding occasionally on tadpoles, frogs, and small fish. Most are shiny and black, but they may be brown or even yellowish and often have light markings. Dytiscids are highly variable in size, ranging from 1.5 to 35 mm. The antennae are long, thread-like, and have 11 segments. The scutellum, a triangular region formed at the juncture of the elytra with the pronotum, is usually visible. The hind legs are flattened and fringed with hairs.

Legs of the larvae are five segmented with two tarsal claws. The abdomen of dytiscid larvae have eight visible segments with hooks lacking on the terminal abdominal segment although paired cerci are often present. Larvae, like the adults, are predaceous and have pronounced, curved mandibles. Because of their voraciousness, these larvae are sometimes referred to as water tigers.

Family Gyrinidae (Whirligig Beetles)

Whirligig beetles (Figure 41) get their common name from their swimming habits. They are usually seen in groups swimming erratically, often in circles or irregular curves, on the surface of lentic bodies of water. This erratic swimming behavior is unique to the gyrinids and is diagnostic for the family. Most gyrinids are carnivorous predators, although their tendency to occasionally feed on dead insects and vegetation has led some entomologists to categorize them as scavengers. Gyrinids are usually black, shiny beetles ranging in size from 3–15 mm. The antennae are short (3 segmented) and clubbed (the 3rd segment is enlarged). The scutellum is usually not visible. The most distinctive morphological characteristic unique to whirligigs is their two pair of compound eyes with one pair being found on the dorsal side of the head and one pair being found on the ventral side (Figure 42). The eyes are divided by the insertion of the antennae, so that only one pair is visible when

FIGURE 40 Predaceous Diving Beetles (Family Dytiscidae).

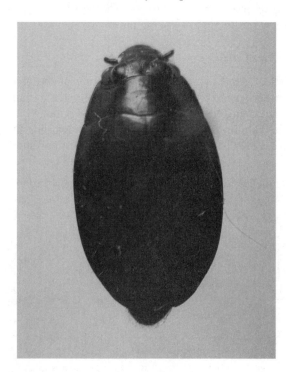

FIGURE 41 Whirligig Beetle (Family Gyrinidae).

FIGURE 42   Lateral View of Whirligig Beetle Showing Compound Eyes Divided at the Edge of the Head by the Insertion of the Antennae.

FIGURE 43   Water Scavenger Beetles (Family Hydrophilidae).

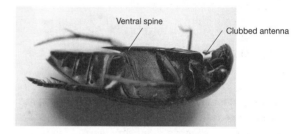

FIGURE 44   Ventral Surface of Water Scavenger Beetle Showing Sharp Spine Projecting Posteriorly Between the Hind Legs and Clubbed Antennae.

viewing the dorsal surface of the beetle and only one pair is visible when viewing the ventral surface. It has been suggested that this arrangement enables them to watch for potential predators from both below and above the water's surface. When disturbed, adult gyrinids often secrete a substance that smells much like pineapple.

Gyrinid larvae are voracious carnivores that superficially resemble the house centipede because of the presence of long respiratory filaments on the abdominal segments. They also have four (two pair) of terminal hooks on the last abdominal segment. Gyrinids have large mandibles.

Family Hydrophilidae (Water Scavenger Beetles)

Water scavenger beetles are common inhabitants of lentic environments that have a lot of vegetation and/or algae. They exhibit a high degree of variability in size and color (Figure 43). Hydrophilids are easily confused with dytiscids. They differ from the dytiscids by their short-clubbed antennae, which are usually hidden beneath the head. The antennae have 7 to 9 segments, and the last 3 segments are clubbed. The scutellum is usually visible. A sharp ventral spine is often exhibited projecting posteriorly from between the hind legs

(Figure 44). They have pronounced maxillary palps that are often longer than the antennae. Adults feed primarily on algae and detritus.

Hydrophyllid larvae are voracious predators and have pronounced mandibles. Larvae are highly variable in form, making them particularly difficult to identify. The abdomen is usually "wrinkled" with more than eight abdominal segments being visible. Long lateral filaments (gills) may be present, similar to those of gyrinid larvae; however, if abdominal filaments are present, hooks are lacking on the terminal abdominal segment.

Family Psephenidae (Riffle Beetles or Water-Penny Beetles)

Riffle beetles are found associated with shallow, swift (fast-flowing) water in streams. Such stretches of streams characterized by shallow swift water are called riffles. This is the only beetle that lives primarily on or near riffles. Adults are usually called riffle beetles and the larvae are called water pennies. Adults are not well equipped for swimming and are

FIGURE 45   Larva of the Water-Penny Beetle, *Psephenus herricki* (Family Psephenidae) a. Dorsal surface.

usually found on stones and logs in or near water. Both adults and larvae feed primarily on algae. The most common species, *Psephenus herricki* is widely distributed throughout the eastern United States. Adults are black with oval, flattened bodies and relatively long legs with large, simple tarsal claws. They range from 3.5 to 6 mm in length. There are six or seven ventral segments on the abdomen. A projection on the ventral surface extends from the pronotum and fits into a groove in the pronotum. The entire body is pubescent (covered with small hairs); the dorsal surface is sparsely pubescent and the ventral surface is densely pubescent.

The larvae, called water-pennies (Figure 45), are extremely unusual and barely resemble beetles at all. They have a concave oval body. The entire body acts like a sucker enabling them to firmly attach to stones in riffles where they feed on algae. Legs are not visible when viewed from the dorsal surface. The ventral surface bears five pair of delicate, white gills. The entire life cycle takes two years to complete but only the larvae survive the winter. The presence of water pennies indicates year-round flow and well-aerated water.

Order Diptera (The True Flies)

This group contains some of the ecologically most important aquatic insects as well as some of the most annoying (mosquitos and biting midges).

Family Tipulidae (Crane Flies)

Crane flies (Figure 46) are often mistaken for large mosquitos. Thankfully, they are not mosquitos, since they range from 10–25 mm in length. They do not bite or injure humans in any way. Crane flies are found around streams, marshes, and lakes, where they occur singly or in swarms. They are strongly attracted to light and are often found in houses where they fly awkwardly near windows. In

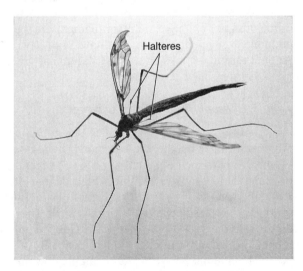

FIGURE 46   Adult Crane Fly (Family Tipulidae). Note conspicuous halteres, a defining characteristic of the order Diptera.

FIGURE 47   Crane Fly Larva.

all dipterans, the hind wings are replaced by haltares, knobbed structures that are probably used for balance. Crane flies possess unusually pronounced haltares. Adults of some species of crane flies feed on nectar, while others do not feed.

Crane fly larvae (Figure 47) occur in a wide range of habitats and only a few are truly aquatic. Larvae range from brown to white, are 10 to 50 mm long, and are characterized by the presence of a large lobed spiracular disc at the posterior end. The spiracular disc is usually surrounded by hairs and delicate anal gills. Tipulidae larvae lack legs and have a head capsule that may withdraw partially or completely into the thorax. Depending on species, larvae may be herbivorous, carnivorous, or feed on detritus.

Family Culicidae (Mosquitos)

These dipterans are real pests! Mosquitos are small flies with piercing/sucking mouthparts. Most male mosquitos feed on nectar and juices of ripe fruit, whereas females, in addition to feeding on nectar, usually require a blood meal to facilitate egg production. Mosquitos are found near lentic bodies of water. Females lay their eggs in water and larvae develop there. The four most common genera of mosquitos in the United States are *Anopheles*, *Aedes*, *Culex*, and *Psorophora*. Mosquitos have been the focus of a phenomenal amount of research because of the role they play as vectors of human diseases, including malaria (*Anopehles*), west Nile virus (*Anopheles, Aedes, Culex, Psorophora*, and others), yellow fever (*Aedes*), dengue (*Aedes*), filariasis (*Anopheles, Aedes, Culex*), and several kinds of encephalitis (*Anopheles, Aedes, Culex*, and *Psorophora*). Additionally, canine heartworm disease is vectored by mosquitos.

Mosquitos feeding on humans are likely to belong to one of these four genera. Representatives of the genus *Anopheles* are characterized by long palps (nearly as long as the proboscis) (Figure 48). Among the remaining three genera of mosquitos with short palps, *Culex* spp. have short palps and a blunt abdomen (Figure 49). Representatives of the genera *Aedes* and *Psorophora* also have short palps and may be differentiated by the arrangement of white ornamentation on the dorsal aspect of the abdominal segments (Figures 50 and 51).

Mosquito larvae may be differentiated from other dipteran larvae in that the head region is greatly enlarged and is the broadest part of the body (Figure 52). Mosquito larvae breathe with a tube or pair of spiracles that extend, almost like a snorkel tube, to the water surface where they feed on algae, zooplankton, and detritus. When disturbed, they swim downward in a characteristic wriggling motion; their common name is *wrigglers*.

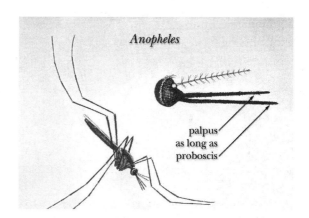

FIGURE 48   Key Characteristics of *Anopheles*.

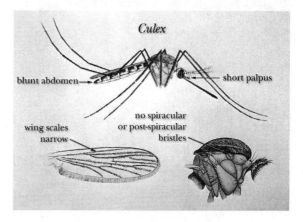

FIGURE 49   Key Characteristics of *Culex*.

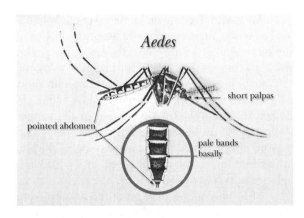

FIGURE 50   Key Characteristics of *Aedes*.

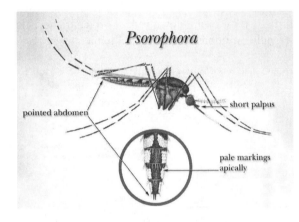

FIGURE 51   Key Characteristics of *Psorophora*.

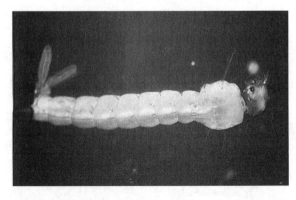

FIGURE 52   Mosquito larvae.

Despite the horrendous attributes of mosquitos, they do have redeeming qualities, at least from an ecological perspective. They play important roles in both aquatic and terrestrial foodwebs; adult mosquitos, particularly males, are important as pollinators of several types of wildflowers.

Family Simuliidae (Black Flies)

Black flies are another nuisance group, except from the viewpoint of fish like smallmouth bass and trout. Like mosquitos, the tiny adult female black flies are aggressive feeders on warm-blooded hosts, and can inflict a very painful bite for their size. Females also feed on nectar, while males feed exclusively on nectar. Black flies are found near swift-flowing rivers and streams, where they lay their eggs on vegetation, rocks, and logs. Larvae are characterized by a flat disk on the posterior end armed with many hooks that are used to attach to the substrate in riffles. The posterior end is greatly enlarged making these larvae fairly easy to distinguish from other dipteran larvae (Figure 53). Furthermore, the head of the larva bears two prominent fan-like structures that strain plankton and detritus from the water for food. In the United States, black flies are primarily just a nuisance to humans, although swarms of black flies have been known to attack and kill entire herds of livestock. In other parts of the world, black flies serve as intermediate host to the parasitic worm *Onchocerca volvulus*, which lives in Africa, Arabia, Mexico, and parts of Central America. Infection with this parasite may cause a grossly deforming condition known as elephantiasis. In parts of Africa, infection with this parasite leads to a condition known as river blindness. It is estimated that as many as 30 million Africans may currently be infected with this parasite and that between 1 and 2 million people are living who have lost

FIGURE 53   Black Fly Larva (Family Simuliidae).

their eyesight as a result of this parasite being transmitted by the bite of a black fly.

Family Chironomidae (Midges)

The chironomid midges, from an ecological standpoint, often play a very important role in aquatic ecosystems because of their sheer abundance. Adult midges are a very prominent feature around most ponds, marshes, and lakes. Once adults emerge, they do not feed and live only 2 to 10 days. Males initiate massive swarms at night, in order to attract females for mating. Midges are strongly attracted to light, and sometimes become unbearably dense. Midges will travel up to about 1/3 mile away from a body of water. Adult chironomid midges are small (no more than 7.5 mm long), and soft bodied with long legs and antennae. Sexes are easily differentiated; the antennae of male midges are plumose but those of females are not (Figure 54). Chironomid midges do not bite.

Chironomid larvae often constitute the predominant benthic organisms in lentic environments.

(a)

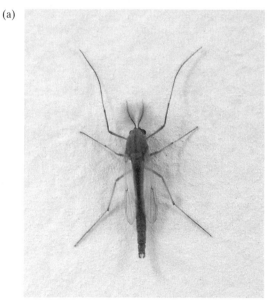

(b)

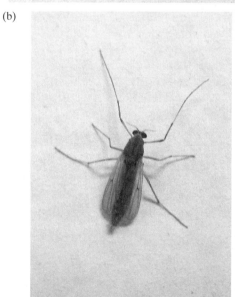

FIGURE 54   a. Male Chironomid Midge. Note plumose antennae and claspers. b. Female Chironomid Midge.

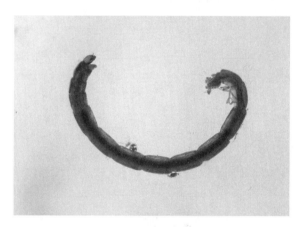

FIGURE 55   A Bloodworm (chironomid midge larva).

FIGURE 56   Larva of Biting Midge (*Probezzia* sp.; Family Ceratopogonidae).

Larvae usually possess red hemoglobin-rich blood, which give them an intense red color (Figure 55) and thus, their common name is *blood worms*. Chironomid larvae are 12-segmented with a pair of prolegs on the first thoracic and last abdominal segments. Gill filaments are often present near the posterior end.

Family Ceratopogonidae (Biting Midges)

Of the more than 60 genera of biting midges, only 4 feed on mammals; however, these 4 types are well known for their biting capability. As with other blood-feeding dipterans, only the females take a blood meal. Most of the other genera feed on insects. The common name of these midges, *no-see-ums*, eludes to their small size (often less than 1 mm long). Representatives of many species can readily pass through window screens. The larvae (Figure 56) are found in algal mats and may be herbivorous, carnivorous, or omnivorous, depending on the species. They range from 3 to 12 mm in length and are rather nondescript, for the most part lacking seatae and totally lacking prolegs. They have a distinct head capsule and retractable gill filaments on the posterior-most segment. Although ceratopogonids are often not present or overlooked, when they do occur they may be present in very high numbers.

References and Suggested Readings

CHAPTER 1

Allen GE, Baker JJW. 2001. Biology: Scientific process and social issues. Bethesda, MD: Fitzgerald Science Press, Inc. 221 p.

Buffaloe ND. 1962. Principles of biology. Englewood Cliffs, NJ: Prentice Hall, Inc. 365 p.

Carolina Biological Supply Company. 1998. Plant mineral requirements sets: Instructions. Burlington, NC: Carolina Biological Supply Company. 6 p.

Chalmers AF. 1976. What is this thing called science? London: University of Queensland Press, 179 p.

Huxley TH. 1854. On the educational value of the natural history sciences. In: Science and Education (1898). New York: D. Appleton and Company. 451 p.

Huxley TH. 1909. Autobiography and selected essays by Thomas Henry Huxley. In: Snell, ALF, editor. Boston: Houghton Mifflin Co. 138 p.

Janovy J. 1985. On becoming a biologist. New York: Harper & Row, Publishers, Inc. 159 p.

Krogh D. 2002. Biology: A guide to the natural world. 2nd ed. Upper Saddle River, NJ: Prentice Hall. 760 p.

Lorenz K. 1973. Civilized man's eight deadly sins. Wilson, MK, translator. New York: Harcourt Brace Jovanovich, Inc. 107 p.

Magner LN. 1979. A history of the life sciences. New York: Marcel Dekker, Inc. 489 p.

Morgan JG, Carter MEB. 1996. Investigating biology, 2nd ed. Menlo Park, CA: Benjamin/Cummings Publishing Company, Inc. 601 p.

Morholt E, Brandwein PF. 1986. A sourcebook for the biological sciences, 3rd ed. San Diego: Harcourt, Brace, Jovanovich, Publishers. 813 p.

Overton WR. 1982. Judgement, injunction, and memorandum opinion in the case of McLean v. Arkansas Board of Education. Science 215: 934–943.

Wilson EO. 1998. Consilience: the unity of knowledge. New York: Alfred A. Knopff. 332 p.

CHAPTER 2

Appleman P, editor. 2001. Darwin, 3rd ed. New York: W. W. Norton & Company. 695 p.

Barber L. 1980. The heyday of natural history. Garden City, NY: Doubleday & Company, Inc. 320 p.

Buffaloe ND. 1962. Principles of biology. Englewood Cliffs, NJ: Prentice Hall, Inc. 365 p.

Burroughs J. 1901. Songs of nature. New York: Grosset & Dunlap, Publishers. 359 p.

Carson R. 1956. The sense of wonder. New York: Harper & Row, Publishers, Inc. (Reprinted by The Nature Company, Berkely, California). 72 p.

Darwin CR. 1859. On the origin of species. London: John Murray and Sons, 698 p.

Desmond A. 1994. Huxley: From devil's disciple to evolution's high priest. Reading, MA: Helix Books/Perseus Books. 820 p.

Flower WH, Lydekker R. 1891. An introduction to the study of mammals: Living and extinct. London: Adam and Charles Black. 763 p.

Gamwell L. 2003. Beyond the visible—microscopy, nature, and art. Science 299: 49–50.

Goodrich SG. 1859a. Illustrated natural history of the animal kingdom, being a systematic and popular description of the habits, structure, and classification of animals from the highest to the lowest forms, with their relations to agriculture, commerce, manufactures, and the arts, Vol. I. New York: Derby & Jackson. 680 p.

Goodrich SG. 1859b. Illustrated natural history of the animal kingdom, being a systematic and popular description of the habits, structure, and classification of animals from the highest to the lowest forms, with their relations to agriculture, commerce, manufactures, and the arts, Vol. II. New York: Derby & Jackson. 680 p.

Hobbes T. 1651. Leviathan. London: J.M. Dent and Sons, Ltd., (1914). 392 p.

Huxley L. 1900. Life and letters of Thomas Henry Huxley, Vol. I. New York: D. Appleton & Company. 539 p.

Huxley L. 1901. Life and letters of Thomas Henry Huxley, Vol. II. New York: D. Appleton & Company. 541 p.

Huxley TH. 1876. On the study of biology. In: Science and Education (Huxley, 1898). New York: D. Appleton and Company. pp. 262–293.

Huxley TH. 1882. On science and art in relation to education. In: Science and Education (Huxley, 1898). New York: D. Appleton and Company. pp. 160–188.

Janovy J. 1985. On becoming a biologist. New York: Harper & Row Publishers, Inc. 159 p.

Magner LN. 1979. A history of the life sciences. New York: Marcel Dekker, Inc. 489 p.

Morris D. 1962. The biology of art. London: Methuen & Co., Ltd. 176 p.

Paley W. 1802. Natural theology or evidences of the existence and attributes of the deity, collected from the appearances of nature. London: R. Faulder. 598 p.

Steele JD, Jenks JWP. 1887. Popular zoology. New York: Chautauqua Press. 319 p.

Wallace AR. 1880. The Malay Archipelago: The land of the orang-utan and the bird of paradise, a narrative of travel, with studies of man and nature, 7th ed. London: Macmillan and Co. 653 p.

Wilson EO. 1994. Naturalist. Washington, DC: Island Press. 380 p.

Wilson EO. 1998. Consilience: The unity of knowledge. New York: Alfred A. Knopff. 332 p.

CHAPTER 3

Krogh D. 2000. Biology: A guide to the natural world. Upper Saddle River, NJ: Prentice Hall, Inc. 678 p.

Mathis ML. 1996. Biology 1400 laboratory manual, Fall 1996 edition: Biology for general education. Plymouth, MI: Hayden-McNeil Publishing, Inc. 102 p.

CHAPTER 4

Hooke R. 1665. Micrographia or some physiological descriptions of minute bodies made by magnifying glasses with observations and inquiries thereupon. New York: Facsimile by Dover Publications, Inc. (1961). 270 p.

Magner LN. 1979. A history of the life sciences. New York: Marcel Dekker, Inc. 489 p.

Mathis ML. 1996. Biology 1400 laboratory manual, Fall 1996 edition: Biology for general education. Plymouth, MI: Hayden-McNeil Publishing, Inc. 102 p.

Morgan JG, Carter MEB. 1996. Investigating biology, 2nd ed. Menlo Park, CA: Benjamin/Cummings Publishing Company, Inc. 601 p.

Perry JW, Morton D. 1995. Laboratory manual for Starr & Taggart's biology: the unity and diversity of life and Starr's biology: concepts and applications. Belmont, CA: Wadsworth Publishing Company. 515 p.

CHAPTER 5

Mathis ML. 1996. Biology 1400 laboratory manual, Fall 1996 edition: Biology for general education. Plymouth, MI: Hayden-McNeil Publishing, Inc. 102 p.

Perry JW, Morton D. 1995. Laboratory manual for Starr & Taggart's biology: The unity and diversity of life and Starr's biology: concepts and applications. Belmont, CA: Wadsworth Publishing Company. 515 p.

CHAPTER 6

Alberts B, Bray D, Lewis J, Raff M, Roberts K, Watson JD. 1989. Molecular biology of the cell, 2nd ed. New York: Garland Publishing Co. 1218 p.

[Anonymous]. 1981. Small animal metabolism apparatus: instruction manual. Richmond, VA: Phipps and Bird Company. 6 p.

Carolina Biological Supply Company. 1974. Carolina enzyme: A teacher's manual. Burlington, NC: Carolina Biological Supply Company. 13 p.

Fox SI. 1993. Human physiology, 4th ed. Dubuque, IA: Wm. C. Brown. 671 p.

Glider W. 1993. Enzyme function. In: Biological sciences 101L general biology lab. University of Nebraska-Lincoln. 22 p.

Hickman CP, Roberts LS, Hickman FM. 1984. Integrated principles of zoology, 7th ed. St. Louis, MO: Times Mirror/Mosby College Publishing. 1065 p.

Stryer L. 1988. Biochemistry, 3rd ed. New York: W. H. Freeman and Co. 1089 p.

Tanabe MJ. 1991. Pineapple. World Book Encyclopedia 5: 476–477.

Tobin AJ, Dusheck J. 1998. Asking about life. New York: Saunders College Publishing. 962 p.

CHAPTER 7

Benjamin CL, Garman GR, Funston JH. 1997. Human biology. New York: McGraw-Hill Companies, Inc. 615 p.

Bres M, Weisshaar A. 1998. Thinking about biology: An introductory biology laboratory manual. Upper Saddle River, NJ: Prentice Hall. 376 p.

Hickman CP, Roberts LS, Hickman FM. 1984. Integrated principles of zoology, 7th ed. St. Louis, MO: Times Mirror/Mosby College Publishing. 1065 p.

Morholt E, Brandwein PF. 1986. A sourcebook for the biological sciences, 3rd ed. San Diego, CA: Harcourt Brace Jovanovich, Publishers. 813 p.

Perry JW, Morton D. 1995. Laboratory manual for Starr & Taggart's biology: The unity and diversity of life and Starr's biology: concepts and applications. Belmont, CA: Wadsworth Publishing Company. 515 p.

Raven PH, Evert RF, Eichhorn SE. 1992. Biology of plants, 5th ed. New York: Worth Publishers. 791 p.

Stryer L. 1988. Biochemistry, 3rd ed. New York: W. H. Freeman and Co. 1089 p.

CHAPTER 8

Hickman CP, Roberts LS, Hickman FM. 1984. Integrated principles of zoology, 7th ed. St. Louis, MO: Times Mirror/Mosby College Publishing. 1065 p.

Eiseley L. 1964. The hidden teacher. In: The Star Thrower (1978). New York: Times Books. pp. 82–93.

Mader SS. 2001. Biology, 7th ed. Boston: McGraw-Hill. 944 p.

Mathis ML. 1996. Biology 1400 laboratory manual, Fall 1996 edition: Biology for general education. Plymouth, MI: Hayden-McNeil Publishing, Inc. 102 p.

Perry JW, Morton D. 1995. Laboratory manual for Starr & Taggart's biology: The unity and diversity of life and Starr's biology: concepts and applications. Belmont, CA: Wadsworth Publishing Company. 515 p.

CHAPTER 9

Mathis ML. 1996. Biology 1400 laboratory manual, Fall 1996 edition: Biology for general education. Plymouth, MI: Hayden-McNeil Publishing, Inc. 102 p.

Starr C. 2000. Biology: Concepts and applications, 4th ed. Pacific Grove, CA: Brooks/Cole. 788 p.

CHAPTER 10

[Anonymous]. 1988. Carolina *Drosophila* manual. Burlington, NC: Carolina Biological Supply Company. 31 p.

Bliss CI. 1967. Statistics in biology. New York: McGraw-Hill. 558 p.

Fortey R. 2000. Trilobite: Eyewitness to evolution. New York: Vintage Books. 284 p.

Hickman CP, Roberts LS, Hickman FM. 1984. Integrated principles of zoology, 7th ed. St. Louis, MO: Times Mirror/Mosby College Publishing. 1065 p.

Magner LN. 1979. A history of the life sciences. New York: Marcel Dekker, Inc. 489 p.

Wagner RF, Hedrick PW. 1989. Genetics. Dubuque, IA: Wm. C. Brown Publishers. 569 p.

CHAPTER 11

[Anonymous]. 2003. Glowing fish to be first genetically changed pet. Cable News Network. Retrieved November 21, 2003 from the World Wide Web: http://www.cnn.com/2003/us/11/21/offbeat.glofish.reut/index.html

Bancroft H. 1957. Introduction to biostatistics. New York: Hoeber-Harper. 210 p.

Carolina Biological Supply Company. 1998. Outbreak! Fingerprinting virus DNA: teacher's manual. Burlington, NC: Carolina Biological Supply Company. 13 p.

Dennis DT, Hughes JM. Multidrug resistance in plague. New England Journal of Medicine 337: 702–704.

Galimand M, Guiyoule A, Gerbaud G, Rasoamanana B. Chanteau S, Carniel E, Caurvalin P. 1997. Multidrug resistance in *Yersinia pestis* mediated by a transferable plasmid. New England Journal of Medicine 337: 677–680.

Krogh D. 2000. Biology: A guide to the natural world, 1st ed. Upper Saddle River, NJ: Prentice Hall. 678 p.

National Centre for Biotechnology Education. 1997. The University of Reading, U.K.

Restriction analysis and electrophoresis of DNA. Burlington, NC: Carolina Biological Supply. 8 p.

Palladino MA. 2002. Understanding the human genome project. San Francisco: Benjamin Cummings. 36 p.

Rapoza M., Kreuzer H. 1999. Transformations: A teacher's manual. Burlington, NC: Carolina Biological Supply. 21 p.

CHAPTER 12

Armstrong K. 1993. A history of God: The 4000-year quest of Judaism, Christianity and Islam. New York: Alfred A. Knopf. 460 p.

Barber L. 1980. The heyday of natural history 1820-1870. Garden City, NY: Doubleday and Co, Inc. 320 p.

Brooker RJ. 1999. Genetics: Analysis and principles. Menlo Park, CA: Addison Wesley Longman, Inc. 772 p.

Browne J, Campbell NA, Mitchell LG, Reece JB. Biology: Concepts and connections. San Francisco, CA: Addison Wesley Longman, Inc. 809 p.

Darwin CR. 1858. On the tendency of species to form varieties; and on the perpetuation of varieties and species by natural means of selection. Journal of the Proceedings of the Linnean Society, Zoology 3. In: Appleman P, editor. 2001. Darwin, 3rd ed. New York: W.W. Norton and Company. 695 p.

Darwin C. 1859. The origin of species. New York: Reprinted by P.F. Collier and Son Company (Harvard Classics, Vol. 11, 1909). 552 p.

Darwin C. 1958. The autobiography of Charles Darwin (1809-1882). Edited with appendix and notes by his granddaughter N. Barlow. New York: W.W. Norton and Company, Inc. 253 p.

Dobzhansky T, Ayala FJ, Stebbins GL, Valentine JW. 1977. Evolution. San Francisco, CA: W. H. Freeman and Co. 572 p.

Eiseley L. 1958. Darwin's century: Evolution and the men who discovered it. New York: Anchor Books. 378 p.

Gibbons A. 2002. First member of human family uncovered. Science 297: 171–173.

Gore R. 2002. The first pioneer? A new find shakes the human family tree. National Geographic 202 (2)

Gunstream SE. 2001. Biological explorations: A human approach, 4th ed. Upper Saddle River, NJ: Prentice Hall. 436 p.

Hickman CP, Roberts LS, Hickman FM. 1984. Integrated principles of zoology, 7th ed. St. Louis, MO: Times Mirror/Mosby College Publishing. 1065 p.

Klein R. 2003. Whither the Neanderthals? Science 299: 1525-1527.

Krogh D. 2000. Biology: A guide to the natural world, 1st ed. Upper Saddle River, NJ: Prentice Hall. 678 p.

Lamarck JB. 1809. Zoological philosophy. Translated by H. Elliot, 1914. London: MacMillan Publishers. 410 p.

Lemonick M, Dorfman A. 2002. Father of us all? Time (July 22, 2002) 160: 40–47.

Magner LN. 1979. A history of the life sciences. New York: Marcel Dekker, Inc. 489 p.

Malthus TR. 1798. An essay on the principle of population. In: Population, Evolution, and Birth Control: A Collage of Controversial Ideas, 2nd ed. San Francisco: W.H. Freeman and Company, 386 p.

Marsh OC. 1877. (Letter Written to Charles Darwin.) New Haven, CT: Peabody Museum exhibit, Yale University.

Mayr E. 1976. Evolution and the diversity of life: Selected essays. Cambridge, MA: The Belknap Press of Harvard University Press. 721 p.

Miller KR. 1999. Finding Darwin's God: A scientist's search for common ground between God and evolution. New York: HarperCollins Publishers, Inc. 338 p.

Murray NP, Buffaloe ND. 1983. Creationsim and evolution: The real issues. In: Zetterberg, J.P, editor. Evolution versus Creationism: The Public Education Controversy. Phoenix, AZ: Oryx Press. pp. 454–476.

Pickrell J. 2002. Unified *Erectus*: fossil suggests single human ancestor. Science News (March 23, 2002) 161: 179.

Poirier FE. 1990. Understanding human evolution, 2nd ed. Englewood Cliffs, NJ: Prentice Hall. 356 p.

Relethford J. 1990. The human species: An introduction to biological anthropology. Mountain View, CA: Mayfield Publishing Co. 339 p.

Stringer C. 2003. Out of Ethiopia. Nature 423: 692–695.

Tobin AJ, Dusheck J. 1998. Asking about life. Fort Worth, TX: Saunders College Publishing. 964 p.

University of Indianapolis. Index of Biology 504. Retrieved March 15, 2002 from the World Wide Web: http://biology.uindy.edu/Biol504/HUMANSTRATEGY/19brainsize.htm

Wallace AR. 1858. On the tendency of varieties to depart indefinitely from the original type. Journal of the Proceedings of the Linnean Society, Zoology 3. In: Appleman, P, editor. Darwin, 3rd ed. (2001). New York: W.W. Norton and Company. p. 61–64.

White TD, Asfaw B, DeGusta D, Gilbert H, Richards GD, Suwa G, Howell FC. 2003. Pleistocene *Homo sapiens* from Middle Awash, Ethiopia. Nature 423: 742–747.

Wong K. 2003. An ancestor to call our own. Scientific American (January, 2003): 54–63.

Zetterberg JP. (editor). 1983. Evolution versus creationism: The public education controversy. Phoenix, AZ: Oryx Press. 516 p.

CHAPTER 13

Burns JM. 1968. A simple model illustrating problems of phylogeny and classification. Systematic Zoology 17: 170–173.

Harlow WM. 1957. Trees of the eastern and central United States and Canada. New York: Dover Publications, Inc. 288 p.

Hickman CP, Roberts LS, Hickman FM. 1984. Integrated principles of zoology, 7th ed. St. Louis, MO: Times Mirror/Mosby College Publishing. 1065 p.

Hickman CP, Roberts LS, Larson A. 1993. Integrated principles of zoology, 9th ed. St. Louis, MO: Mosby. 983 p.

International Union of Biological Sciences. 1985. International code of zoological nomenclature, 3rd ed. Berkeley, CA: University of California Press. 338 p.

Krogh D. 2000. Biology: A guide to the natural world. Upper Saddle River, NJ: Prentice Hall. 678 p.

Little EL. 1980. The Audubon Society field guide to North American trees. New York: Alfred A. Knopf, Inc. 714 p.

Magner LN. 1979. A history of the life sciences. New York: Marcel Dekker, Inc. 489 p.

Margulis L, Schwartz KV. 1982. Five kingdoms: An illustrated guide to the phyla of life on earth. San Francisco, CA: W. H. Freeman. 338 p.

Mathis ML. 1996. Biology 1400 laboratory manual, fall 1996 edition: Biology for general education. Plymouth, MI: Hayden-McNeil Publishing, Inc. 102 p.

Mayr E. 1942. Systematics and the Origin of Species. New York: Columbia University Press. 334p.

Mayr E. 1969. Introduction: the role of systematics in biology. In: National Research Council. Systematic Biology (Publication 1692). Washington, DC: National Academy Press. p. 4–15.

Mayr E. 1976. Evolution and the diversity of life: Selected essays. Cambridge, MA: The Belknap Press/Harvard University Press. 721 p.

Peattie DC. 1966. A natural history of trees of eastern and central North America. Boston: Houghton Mifflin Co. 606 p.

Raven PH, Evert RF, Eichhorn SE. 1992. Biology of plants, 5th ed. New York: Worth Publishers. 791 p.

Simpson GG. 1961. Principles of animal taxonomy. New York: Columbia University Press. 247 p.

Wiley EO. 1978. The evolutionary species concept reconsidered. Systematic Zoology 27: 17–26.

CHAPTER 14

[Anonymous]. 1997. Water quality kit: Teacher's guide. Goshen, IN: Micrology Laboratories, LLC. 4 p.

Ballard RD, Hively W. 2000. The eternal darkness: A personal history of deep-sea exploration. Princeton, NJ: Princeton University Press. 388 p.

Centers for Disease Control and Prevention. 2003. Summary of notifiable diseases—United States, 2001. Morbidity and Mortality Weekly Reports 50(53): 1-108. Retrieved August 28, 2003 from the World Wide Web: http://www2.cdc.gov/mmwr/summary.html.

Davis BD, Dulbecco R. Eisen HN, Ginsberg HS. 1990. Microbiology, 4th ed. Philadelphia: J. B. Lippincott Co. 1215 p.

Dennis DT, Hughes JM. Multidrug resistance in plague. The New England Journal of Medicine 337: 702–704.

Dennis DT, Gage KL, Gratz N, Poland JD, Tikhomirov E. 1999. Plague manual: Epidemiology, distribution, surveillance and control. Geneva, Switzerland: World Health Organization. 171 p.

Freeman BA. 1979. Burrows' textbook of microbiology, 21st ed. Philadelphia: W.B. Saunders Company. 1138 p.

Galimand M, Guiyoule A, Gerbaud G, Rasoamanana B, Chanteau S, Carniel E, Caurvalin P. 1997. Multidrug resistance in Yersinia pestis mediated by a transferable plasmid. The New England Journal of Medicine 337: 677–680.

Garrett L. 1994. The coming plague. New York: Farrar, Straus, and Giroux. 750 p.

Geldreich EE. 1997. Coliforms: A new beginning to an old problem. In: Kay, D, Fricker, C, (editors).

Coliforms and E. coli: Problem or solution? Cambridge, UK: The Royal Society of Chemistry. pp. 3–11.

Greenaway K. 1881. Mother Goose or the old nursery rhymes. London: Frederick Warne and Company, Ltd. 52 p.

Griffen DW, Lipp EK, McLaughlin MR, Rose JB. 2001. Marine recreation and public health microbiology: Quest for the ideal indicator. BioScience 51: 817–825.

Hooke R. 1665. Micrographia or some physiological descriptions of minute bodies made by magnifying glasses with observations and inquiries thereupon. New York: Facsimile by Dover Publications, Inc. (1961). 270 p.

Horrox R. 1994. The black death. New York: Manchester University Press. 364 p.

Jensen MM, Wright DN, Robinson RA. 1997. Microbiology for the health sciences, 4th ed. Upper Saddle River, NJ: Prentice Hall. 530 p.

Kay D, Fricker C, (editors). 1997. Coliforms and E. coli: Problem or solution? Cambridge, UK: The Royal Society of Chemistry. 267 p.

Krogh D. 2000. Biology: A guide to the natural world. Upper Saddle River, NJ: Prentice Hall. 678 p.

Link VB. 1955. A history of plague in the United States of America. Public Health Monographs 26: 1–120.

Mader SS. 2000. Laboratory manual, inquiry into life, 9th ed. Boston: McGraw-Hill. 488 p.

Maki DG, et al. 1976. Nationwide epidemic of septicemia caused by contaminated intravenous products: Epidemiological and clinical features. American Journal of Medicine 60: 471–485.

Nohl J. 1969. The black death: a chronicle of the plague. New York: Harper & Row. 284 p.

Pichichero ME, Green JL, Francis AB, Marsocci SM, Murphy ML. 2000. Outcomes after judicious antibiotic use for respiratory tract infections seen in a private pediatric practice. Pediatrics 105: 753–759.

Pratt HD, Stark HE. 1973. Fleas of public health importance and their control. Atlanta, GA: U. S. Dept. of Health and Human Services, Public Health Service, Centers for Disease Control, HHS Publication No. (CDC) 86-8396. 48 p.

Rapoza M, Kreuzer H. 1999. Transformations: a teacher's manual. Burlington, NC: Carolina Biological Supply. 21 p.

Ryan KJ, Falkow S. 1994. Corynebacteria, *Listeria, Bacillis,* and other aerobic and facultative gram-positive rods. In: Ryan KJ, editor. Sherris Medical Microbiology: An Introduction to Infectious Diseases, 3rd ed. Stamford, CT: Appleton and Lange. pp. 285–308.

State of Connecticut, Department of Environmental Protection. 1997. Water quality standards. Retrieved October 26, 2003 from the World Wide Web: http://www.dep.state.ct.us/wtr/wq/wqs.pdf

Tanner L. 2000. Antibiotic overload. The Associated Press. Retrieved September 12, 2003 from the World Wide Web: http://more.abcnews.go.com/sections/living/dailynews/antobioticskids_0000404.html

United States Environmental Protection Agency, Office of Water, Reports and References. 2003. Bacterial water quality standards for recreational waters (freshwater and marine waters). EPA 823-R-03-008. Retrieved September 12, 2003 from the World Wide Web:

http://www.epa.gov/waterscience/beaches/local/statrept.pdf

United States Environmental Protection Agency, Office of Research and Development. 2000. Membrane filter method for the simultaneous detection of total coliforms and *Escherichia coli* in drinking water. EPA 600-R-00-013.

Washington Virtual Classroom. 2003. Fecal coliform (from Mitchell and Stapp, 1992). Retrieved September 12, 2003 from the World Wide Web: http://www.wavcc.org/wvc/cadre/WaterQuality/fecal.htm

Wilson JB, Russell KE. 1964. Isolation of *Bacillus anthracis* from soil stored 60 years. Journal of Bacteriology 87: 237–238.

Wolf HW. 1972. The coliform count as a measure of water quality. In: Mitchell R, editor. Water Pollution Microbiology. New York: Wiley-Interscience. pp. 333–345.

CHAPTER 15

[Anonymous]. 2000. *An gorta mor*: The great hunger. Hamden, CT: Arnold Bernhard Library (pamphlet, The Lender Family Special Collection), Quinnipiac University. 13 p.

Black SJ, Seed JR. (editors). 2001. The African trypanosomes. Norwell, MA: Kluwer Academic Publishers. pp. vii–ix.

Bremen JD, Egan A, Keusch GT. 2001. The intolerable burden of malaria: A new look at the numbers. American Journal of Tropical Medicine and Hygiene 64: iv–vii.

Burkholder JM, Glasgow HB. 2001. History of toxic *Pfiesteria* in North Carolina estuaries from 1991 to the present. BioScience 51: 827–841.

Butler D. 1997. Time to put malaria control on the global agenda. Nature 386:535–538.

Carr A. 1995. Notes on the behavioral ecology of sea turtles. In: Bjorndal, KA, editor. Proceedings of the World Conference on Sea Turtle Conservation, Washington, D. C., 26-30 November 1979 with Contributions on Recent Advances in Sea Turtle Biology and Conservation, 1995. Washington, DC: Smithsonian Institution Press. pp. 19–26.

Centers for Disease Control. 2003. Local transmission of *Plasmodium vivax* malaria–Palm Beach County, Florida, 2003. Morbidity and Mortality Weekly Reports 52(38): 908–911.

Despommier DD, Gwadz RW, Hotez PJ, Knirsch CA. 2000. Parasitic diseases, 4th ed. New York: Apple Trees Productions, LLC. 345 p.

Glasgow HB, Burkholder JM, Schmechel DE, Fester PA, Rublee PA. 1995. Insidious effects of a toxic dinoflagellate on fish survival and human health. Journal of Toxicology and Environmental Health 46: 501–522.

Graham LE, Wilcox LW. 2000. Algae. Upper Saddle River, NJ: Prentice Hall. 640 p.

Grattan LM, Oldach D, Morris JG. 2001. Human health risks of exposure to *Pfiesteria piscicida*. BioScience 51: 853–857.

Hickman CP, Roberts LS, Hickman FM. 1984. Integrated principles of zoology, 7th ed, St. Louis, MO: Times Mirror/Mosby College Publishing. 1065 p.

Hotez PJ. 2002. Reducing the global burden of human parasitic diseases. Comparative Parasitology 69(2): 140–145.

Janovy J. 1991. Zoology lab: A laboratory manual for biological sciences 112 (University of

Nebraska-Lincoln). Edina, MN: Burgess International Group, Inc. 169 p.

Johnson GB. 1997. The living world. Dubuque, IA: William C. Brown Publishers. 658 p.

Magnien RE. 2001. The dynamics of science, perception, and policy during the outbreak of *Pfiesteria* in the Chesapeake Bay. BioScience 51: 843–852.

Markell EK, John DT, Krotoski WA. 1999. Markell and Voge's medical parasitology, 8th ed. Philadelphia, PA: W. B. Saunders, Co. 501 p.

Marquardt WC, Demaree RS, Grieve RB. 2000. Parasitology and vector biology, 2nd ed. San Diego, CA: Harcourt Academic Press. 702 p.

Pennak RW. 1953. Fresh-Water invertebrates of the United States. New York: The Ronald Press Co. 769 p.

Raven PH, Evert RF, Eichhorn SE. 1992. Biology of plants, 5th ed. New York: Worth Publishers. 791 p.

Roberts LS, Janovy J. 2000. Gerald D. Schmidt and Larry S. Robert's foundations of parasitology, 6th ed. Boston, MA: McGraw-Hill. 670 p.

CHAPTER 16

Davis BD, Dulbecco R, Eisen HN, Ginsberg HS. 1990. Microbiology, 4th ed. Philadelphia, PA: J. B. Lippincott Co. 1215 p.

Hale ME. 1979. How to know the lichens. 2nd ed. Boston: Wm. C. Brown/McGraw-Hill, 246 p.

Lincoff GH, Pacioni G. 1981. Simon & Schuster's Guide to Mushrooms. New York: Simon and Schuster, Inc. 512 p.

Mader SS. 2001. Biology, 7th ed. New York: McGraw-Hill Higher Education. 944 p.

Matossian MK. 1989. Poisons of the past: Molds, epidemics, and history. New Haven, CT: Yale University Press. 208 p.

Morholt E, Brandwein PF. 1986. A sourcebook for the biological sciences, 3rd ed. San Diego, CA: Harcourt Brace Jovanovich, Publishers. 813 p.

Raven PH, Evert RF, Eichhorn SE. 1992. Biology of plants, 5th ed. New York: Worth Publishers. 791 p.

Starr C. 1994. Biology: Concepts and applications, 2nd ed. Belmont, CA: Wadsworth Publishing Co. 645 p.

Wilson E. 2002. The future of life. New York: Alfred A. Knopf. 229 p.

CHAPTER 17

Hickman CP, Roberts LS, Hickman FM. 1984. Integrated principles of zoology, 7th ed. St. Louis, MO: Times Mirror/Mosby College Publishing. 1065 p.

Raven PH, Evert RF, Eichhorn SE. 1992. Biology of plants, 5th ed. New York: Worth Publishers. 791 p.

Tennyson A. 1869. Flower in the crannied wall. In: Buckley, JH, editor (1958). Poems of Tennyson. Cambridge, MA: The Riverside Press. 542 p.

Wilson E. 2002. The future of life. New York: Alfred A. Knopf. 229 p.

CHAPTER 18

Darwin C. 1880. The power of movement in plants. London: John Murray. 592 p.

Darwin C. 1881. The formation of vegetable mould through the action of worms with observations on their habits. London: John Murray. 326 p.

Darwin C. 1893. The movements and habits of climbing plants, 2nd ed. New York: D. Appleton and Co. 208 p.

Eberhard C. 1996. Saunders general biology laboratory manual, updated version. Fort Worth, TX. Saunders College Publishing. 584 p.

Eiseley L. 1957. How flowers changed the world. In: The Star Thrower (1978). New York: Times Books. 264 p.

Ferry JF, Ward HS. 1959. Fundamentals of plant physiology. New York: Macmillan Publishing Co. 288 p.

Gunstream SE. 1999. Explorations in basic biology, 8th ed. Upper Saddle River, NJ: Prentice Hall. 481 p.

Hopkins WG. 1995. Introduction to plant physiology. New York: John Wiley and Sons, Inc. 464 p.

Kaufman PB, Labovitch J, Anderson-Prouty A, Ghosheh NS. 1975. Laboratory experiments in plant physiology. New York: Macmillan Publishing Co., Inc, 262 p.

Krogh D. 2000. Biology: A guide to the natural world. Upper Saddle River, NJ: Prentice Hall. 678 p.

Mathis ML. 1996. Biology 1400 laboratory manual, Fall 1996 edition: Biology for general education. Plymouth MI: Hayden-McNeil Publishing, Inc. 102 p.

Raven PH, Evert RF, Eichhorn SE. 1992. Biology of plants, 5th ed. New York: Worth Publishers. 791 p.

CHAPTER 19

Belding DL. 1942. Textbook of clinical parasitology. New York: Appleton-Century-Crofts, Inc. 888 p.

Cobb NA. 1915. Nematodes and their relationships. United States Department of Agriculture Yearbook 1914: 457–490.

Darwin C. 1881. The formation of vegetable mould through the action of worms with observations on their habits. London: John Murray. 326 p.

Despommier DD, Gwadz RW, Hotez PJ, Knirsch CA. 2000. Parasitic diseases, 4th ed. New York: Apple Trees Productions, LLC. 345 p.

Gould SJ. 1989. Wonderful life: The Burgess Shale and the nature of history. New York: W. W. Norton and Co. 347 p.

Halton CM. 1989. Those amazing leeches. Minneapolis, MN: Dillon Press, Inc. 120 p.

Hickman CP, Hickman FM. 1993. Laboratory studies in integrated zoology, 8th ed. St. Louis, MO: Mosby. 420 p.

Hickman CP, Roberts LS, Hickman FM. 1984. Integrated principles of zoology, 7th ed. Times Mirror/Mosby College Publishing, St. Louis, Missouri, 1065 p.

Janovy J. 1991. Zoology lab: A laboratory manual for biological sciences 112. Edina, MN: Burgess International Group, Inc. 169 p.

Meinkoth NA. 1981. The Audubon Society field guide to North American seashore creatures. New York: Alfred A. Knopf. 813 p.

Pennak RW. 1953. Fresh-water invertebrates of the United States. New York: The Ronald Press Co. 769 p.

Richardson DJ, Nickol BB. 1995. The genus Centrorhynchus (Acanthocephala) in North America with description of Centrorhynchus robustus n. sp., redescription of Centrorhynchus

conspectus, and a key to species. Journal of Parasitology 81: 767–772.

Roberts LS, Janovy J. 1996. Gerald D. Schmidt & Larry S. Roberts' foundations of parasitology, 5th ed. Dubuque, IA: Wm. C. Brown Publishers. 659 p.

Wilson EO. 1987. The little things that run the world. Conservation Biology 1: 344–346.

Wilson EO. 2002. The future of life. New York: Alfred A. Knopf. 229 p.

CHAPTER 20

Duncanson HB. 1895. Suggestive lessons in elementary zoology. Lincoln, NE: Hunter Printing Co. 119 p.

Fortey R. 2000. Trilobite: Eyewitness to evolution. New York: Vintage Books. 284 p.

Goddard J. 1993. Physician's guide to arthropods of medical importance. Boca Raton, FL: CRC Press. 332 p.

Gould SJ. 1989. Wonderful life: The Burgess Shale and the nature of history. New York: W. W. Norton and Co. 347 p.

Hickman CP, Hickman, FM. 1993. Laboratory studies in integrated zoology, 8th ed. St. Louis, MO: Mosby. 420 p.

Hickman CP, Roberts LS, Hickman FM. 1984. Integrated principles of zoology, 7th ed. St. Louis, MO: Times Mirror/Mosby College Publishing. 1065 p.

Huxley TH. 1880. The crayfish: An introduction to the study of zoology. New York: D. Appleton and Company. 371 p.

Markell EK, John DT, Krotoski WA. 1999. Markell and Voge's medical parasitology, 8th ed. Philadelphia, PA: W. B. Saunders Co. 501 p.

Nauss RW. 1944. Medical parasitology and zoology. New York: Paul B. Hoeber, Inc., Harper and Brothers. 534 p.

Spielman A, Levi HW. 1970. Probable envenomation by Chiracanthium mildei: A spider found in houses. American Journal of Tropical Medicine and Hygiene 14: 729-732.

Wilson EO. 1991. In the company of ants. Bulletin of the American Academy of Arts and Sciences 45: 13–23.

Wilson E. 2002. The future of life. New York: Alfred A. Knopf. 229 p.

CHAPTER 21

Ackerman J. 1998. Dinosaurs take wing. National Geographic 194: 74 – 99.

Carrol RL. 1988. Vertebrate paleontology and evolution. New York: W.H. Freeman and Co. 698 p.

Chiarelli AB. 1973. Evolution of the primates: An introduction to the biology of man. London: Academic Press. 354 p.

Darwin C. 1880. The descent of man, and selection and relation to sex. New York: D. Appleton and Company. 688 p.

Gunstream SE. 1997. Biological explorations: A human approach, 3rd ed. Upper Saddle River, NJ: Prentice Hall. 426 p.

Hickman CP, Roberts LS. 1994. Biology of animals, 6th ed. Dubuque, IA: Wm. C. Brown Publishers. 764 p.

Hickman CP, Roberts LS, Larson A. 1993. Integrated principles of zoology, 9th ed. St. Louis, MO: Mosby. 983 p.

Janovy J. 1991. Zoology lab: A laboratory manual for biological sciences 112. Edina, MN: Burgess International Group, Inc. 169 p.

Krogh D. 2000. Biology: A guide to the natural world. Upper Saddle River, NJ: Prentice Hall. 678 p.

Poirier FE. 1990. Understanding human evolution, 2nd ed. Englewood Cliffs, NJ: Prentice Hall. 356 p.

Wilson E. 2002. The future of life. New York: Alfred A. Knopf. 229 p.

CHAPTER 22

Fox SI. 1993. Human physiology, 4th ed. Dubuque IA: Wm. C. Brown Publishers. 671 p.

CHAPTER 23

Fox SI. 1993. Human physiology, 4th ed. Dubuque IA: Wm. C. Brown Publishers. 671 p.

Gunstream SE. 2001. Biological explorations: A human approach, 4th ed. Upper Saddle River, NJ: Prentice Hall. 436 p.

Graham TM. 1998. Biology: Life features, 2nd ed. Needham Heights, MA: Simon and Schuster Custom Publishing. 464 p.

Lione A. 1979. Reaction time kit instructions. Burlington, NC: Carolina Biological Supply Company. 20 p.

CHAPTER 24

Fox SI. 1993. Human physiology, 4th ed. Dubuque IA: Wm. C. Brown Publishers. 671 p.

Gunstream SE. 2001. Biological explorations: A human approach, 4th ed. Upper Saddle River, NJ: Prentice Hall. 436 p.

CHAPTER 25

Bres M, Weisshaar A. 1998. Thinking about biology: An introductory biology laboratory manual. Upper Saddle River, NJ: Prentice Hall. 376 p.

Farris EJ, Griffith JQ. 1949. The rat in laboratory investigation. New York: Hafner Publishing Co. 542 p.

Fox SI. 1993. Human physiology, 4th ed. Dubuque IA: Wm. C. Brown Publishers. 671 p.

Graham TM. 1998. Biology: Life features, 2nd ed. Needham Heights, MA: Simon and Schuster Custom Publishing. 464 p.

Greene EC. 1949. Gross anatomy. In: Farris EJ, Griffith JQ, (editors). The Rat in Laboratory Investigation. New York: Hafner Publishing Co. p. 24-50.

Gunstream SE. 1997. Biological explorations: A human approach, 3rd ed. Upper Saddle River, NJ: Prentice Hall. 426 p.

Hickman CP, Hickman FM. 1993. Laboratory studies in integrated zoology, 8th ed. St. Louis, MO: Mosby. 420 p.

Krogh D. 2000. Biology: A guide to the natural world. Upper Saddle River, NJ: Prentice Hall. 678 p.

Sealander JA. 1979. A guide to Arkansas mammals. Conway, AR: River Road Press. 313 p.

Skavaril R, Finnen M, Lawton S. 1993. General biology lab manual: Investigations into life's phenomena. Fort Worth, TX: Saunders College Publishing. 385 p.

Van De Graaff KM. 1988. Human anatomy, 2nd ed. Dubuque, IA: Wm. C. Brown Publishers. 800 p.

Whitaker JO. 1980. The Audubon Society field guide to North American mammals. New York: Alfred A. Knopf. 745 p.

CHAPTER 26

[Anonymous]. 1988. Carolina *Drosophila* manual. Burlington, NC: Carolina Biological Supply Company. 31 p.

DeBuhr LE. 1994. Using *Lemna* to study geometric population growth. In: Moore R, editor. Biology Labs That Work: The Best of How-to-do-its. Reston, VA: National Association of Biology Teachers. 192 p.

Fisher RC, Ziebur AD. 1982. Integrated algebra, trigonometry, and analytic geometry, 4th ed. Englewood Cliffs, NJ: Prentice Hall. 500 p.

Hopkins WG. 1995. Introduction to plant physiology. New York: John Wiley and Sons, Inc. 464 p.

Lino M. 2002. Expenditures on children by families, 2001 annual report. U.S. Department of Agriculture, Center for Nutrition Policy and Promotion. Misc. Pub. No. 1528-2001, May 2002. Retrieved August 28, 2003 from the World Wide Web: http://www.usda.gov/cnpp/Crc/crc2001.pdf

Lorenz K. 1973. Civilized man's eight deadly sins. Translated by Marjorie Kerr Wilson. San Diego, CA: Harcourt Brace Jovanovich. 107 p.

Malthus TR. 1798. An essay on the principle of population. In: Population, Evolution, and Birth Control: A Collage of Controversial Ideas, 2nd ed. San Francisco: W.H. Freeman and Company. 386 p.

Morholt E, Brandwein PF. 1986. A sourcebook for the biological sciences, 3rd ed. San Diego: Harcourt Brace Jovanovich, Publishers. 813 p.

Palmer EL. 1949. Fieldbook of natural history. New York: McGraw-Hill. 664 p.

Raven PH, Evert RF, Eichhorn SE. 1992. Biology of plants, 5th ed. New York: Worth Publishers. 791 p.

Robinson WL, Bolen EG. 1984. Wildlife ecology and management. New York: MacMillan Publishing Company. 478 p.

Wilcox WF, editor. 1937. Natural and political observations made upon the Bills of Mortality by John Graunt (reprint of the 1st edition, 1662). Baltimore: Johns Hopkins Press. In: Fox JP, Hall CE, Elveback LR, 1970. Epidemiology: Man and disease. London: Collier-MacMillan Limited. 339 p.

Wilson E. 2002. The future of life. New York: Alfred A. Knopf. 229 p.

CHAPTER 27

Audesirk T, Audesirk G, Byers BE. 2002. Biology: life on earth, 6th ed. Upper Saddle River, NJ: Prentice Hall. 892 p.

Eberhard C. 1996. Saunders general biology laboratory manual: Updated version. Fort Worth, TX: Saunders College Publishing. 584 p.

Gorczyca T. 1996. Issues of the environment: acid rain. Retrieved September 12, 2003 from the World Wide Web: http://www.necc.mass.edu/MRVIS/MR1_6/start.htm

Leopold A. 1933. Game management. New York: Charles Scribner's Sons. 481 p.

Patten BC. 1962. Species diversity of net phytoplankton of Raritan Bay. Journal of Marine Research 20: 57–75.

Pennsylvania Fish and Boat Commission. 2000. The basics of water pollution in Pennsylvania. Retrieved September 12, 2003 from the World Wide Web: http://sites.state.pa.us/PA_Exec/Fish_Boat/jf2001/wpollbas.htm

Schreiber RK. Acidic deposition ("Acid rain"). National Biological Service. Retrieved September 12, 2003 from the World Wide Web: http://biology.usgs.gov/s+t/noframe/u204.htm

Sealander JA. 1979. A guide to Arkansas mammals. Conway, AR: River Road Press. 313 p.

Shannon CE. 1948. A mathematical theory of communication. Journal of Bell Systematic Technology 27: 379-423, 623–656.

United States Environmental Protection Agency, Office of Water. 2003. Volunteer stream monitoring: a methods manual. EPA 841-B-97-003. Retrieved September 12, 2003 from the World Wide Web: http://www.epa.gov/owow/monitoring/volunteer/stream/index.html

Wiener W. 1948. Cybernetics. New York: Wiley and Sons. 194 p.

Wilhm JL, Dorris TC. 1968. Biological parameters from water quality criteria. BioScience 18: 477–481.

CHAPTER 28

[Anonymous]. 1993. Investigation of NW v. SE owl pellets: teacher's guide. Rochester, NY: Ward's Natural Science Establishment, Inc. 5 p.

Bunn DS, Warburton AB, Wilson RDS. 1982. The barn owl. Berkhamsted, UK: T. and A. D. Poyser. 264 p.

Burton JA, Channell J. 1991. The pocket guide to mammals of North America. London: Parkgate Books. 192 p.

Collins HH. 1981. Harper & Row's complete field guide to North American wildlife, eastern ed. New York: Harper & Row, Publishers. 714 pp.

Craighead JJ, Craighead FC. 1969. Hawks, owls, and wildlife. New York: Dover Publications Inc. 443 p.

Gaussoin B, Lapsansky J. 1988. The barn owl and the pellet. Bellingham, WA: Pellets, Inc. 28 p.

Johnsgard PA. 1988. North American owls: biology and natural history. Washington, DC: Smithsonian Press. 295 p.

Procter BW. 1901. The owl. In: Burroughs J, editor. Songs of Nature. New York: Grosset and Dunlap Publishers. p. 31–32.

Sealander JA. 1979. A Guide to Arkansas mammals. Conway, AR: River Road Press. 313 p.

Skavaril R, Finnen M, Lawton S. 1993. General biology laboratory manual: investigations into life's phenomena. Fort Worth, TX: Saunders College Publishing. 385 p.

Tennyson A. 1850. In memoriam A.H.H. In: Poems of Tennyson. Cambridge, MA: The Riverside Press. pp. 178-259.

New York: Reprinted by Houghton Mifflin Company (1958).

Whitaker JO. 1980. The Audubon Society field guide to North American mammals. New York: Alfred A. Knopf. 745 p.

CHAPTER 29

Ahmadjian V, Paracer S. 1986. Symbiosis: an introduction to biological associations. Hanover, NH: University Press of New England, for Clark University. 212 p.

Belding DL. 1942. Textbook of clinical parasitology. New York: Appleton-Century-Crofts, Inc. 888 p.

Benjamin CL, Garman GR, Funston JH. 1997. Human biology. New York: The McGraw-Hill Companies, Inc. 615 p.

Brooks DR, Hoberg EP. 2000. Triage for the biosphere: the need and rationale for taxonomic inventories and phylogenetic studies of parasites. Comparative Parasitology 67: 1–25.

Bundy DAP, Chan MS, Savioli L. 1995. Hookworm infection in pregnancy. Transactions of the Royal Society of Tropical Medicine and Hygiene 89: 521-522.

Burdick A. 2003. A farewell to sprinklers. Discover 24:7. pp. 26-27.

Chandler AC. 1955. Introduction to parasitology with special reference to the parasites of man, 9th ed. New York: John Wiley and Sons, Inc. 799 p.

Cleveland LR, Grimstone AV. 1964. The fine structure of the flagellate *Myxotricha paradoxa* and its associated microorganisms. Proceedings of the Royal Society of London B. 159: 668–686.

Colville J (editor). 1991. Diagnostic parasitology for veterinary technicians. Goleta, CA: American Veterinary Publications, Inc. 266 p.

Costello E. 2003. Gut-sucking worms. SuperScience 14: 6–9.

Cowley, G. 2003. Where living is lethal. Newsweek (Sept. 22, 2003): 78–80.

Crompton DWT. 1988. The prevalence of ascariasis. Parasitology Today 4: 162–169.

Crompton DWT. 1999. How much helminthiasis is there in the world? Journal of Parasitology 85: 397–403.

Desowitz RS. 1981. New Guinea tapeworms and Jewish grandmothers. New York: W.W. Norton and Co. 224 p.

Despommier DD, Gwadz RW, Hotez PJ, Knirsch CA. 2000. Parasitic diseases, 4th ed. New York: Apple Trees Productions, LLC. 345 p.

Duclos LM, Richardson DJ. 2000. *Hymenolepis diminuta* in pet store rodents. Comparative Parasitology 67: 197–201.

Garrett L. 1994. The coming plague. New York: Farrar, Straus and Giroux. 750 p.

Gauthier, JL, Gupta A, Hotez P. 2003. Stealth parasites: the underappreciated burden of parasitic zoonoses in North America. In: Richardson DJ, Krause PJ, editors. North American Parasitic Zoonoses, Norwell, MA: Kluwer Academic Publishers. p. 1–21.

Gibbons A. 1992. Researchers fret over neglect of 600 million patients. Science 256: 1135.

Hopkins DR. 1992. Homing in on helminths. American Journal of Tropical Medicine and Hygiene 46: 626–634.

Hotez, PJ. 2002. Reducing the global burden of human parasitic diseases. Comparative Parasitology 69(2): 140–145.

Howard LM. 1971. The relevance of parasitology to the growth of nations. Journal of Parasitology 57: 143–147.

Laureti E. 1999. Fish and fishery products: World apparent consumption statistics based on food balance sheets. Rome: Food and Agriculture Organization of the United Nations, Fisheries Circular No. 821, Revision 5.

Le Riche WH. 1967. World incidence and prevalence of the major communicable diseases. In: Health of mankind. Boston: Little, Brown & Co. pp. 1–42.

May RM. 1988. How many species are there on Earth? Science 241: 1441–1449.

May RM. 1992. How many species inhabit the Earth? Scientific American 267: 42–48.

Mock G. 2000. How much do we consume? In: World Resources, 2000-2001: People and Ecosystems: The Fraying Web of Life. World Resources Institute. Retrieved September 12, 2003 from the World Wide Web: http://pubs.wri.org/pubs_pdf.cfm?PubID=3027

Nauss RW. 1944. Medical parasitology and zoology. New York: Paul B. Hoeber, Inc., 534 p.

Poe EA. 1985. The conqueror worm. In: Works of Edgar Allan Poe. New York: Gramercy Books. pp. 749–750.

Price PW. 1980. Evolutionary biology of parasites. Princeton, NJ: Princeton University Press. 237 p.

Richardson DJ. 2002. Intestinal tapeworm infections. In: Richardson DJ, Krase PJ, editors. North American Parasitic Zoonoses. Kluwer Academic Publishers, Norwell, MA: p. 73–83.

Roberts LS, Janovy J. 2000. Gerald D. Schmidt & Larry S. Robert's foundations of parasitology, 5th ed. Dubuque, IA: Wm. C. Brown Publishers. 659 p.

Schmidt GD, Roberts LS. 1989. Foundations of parasitology, 4th ed. St. Louis, MO: Times Mirror/Mosby College Publishing. 750 p.

Sorvillo FJ, Waterman SH, Richards FO, Schantz PM. 1992. Cysticercosis surveillance: locally acquired and travel-related infections and detection of intestinal tapeworm carriers in Los Angeles County. American Journal of Tropical Medicine and Hygiene 47: 365–371.

Stoll NR. 1947. This wormy world. Journal of Parasitology 33:1–18.

TRB. 1969. Sleep well. The New Republic (March 6) 160: 6.

Windsor DA. 1998. Most of the species on Earth are parasites. International Journal for Parasitology 26: 1127–1129.

World Development Report. 1993. Investing in health, world development indicators. New York: Oxford University Press.

Zimmer C. 2000. Parasite rex: inside the bizarre world of nature's most dangerous creatures. New York: Touchstone Books/Simon and Schuster. 298 p.

CHAPTER 30

[Anonymous]. 1999. Betta splendens: Siamese fighting fish. Retrieved September 13, 2003 from the World Wide Web: http://www.notcatfish.com/findex/fish/betta_splendens.htm

Barnes RD. 1980. Invertebrate zoology, 4th ed. Philadelphia: Saunders College Publishing. p.

Mann KH. 1961. Leeches (Hirudinea): their structure, physiology, ecology, and embryology. International Series of Monographs on Pure and Applied Biology: Vol. 11. New York: Pergamon Press. 201 p.

Morgan JG, Carter MEB. 2002. Investigating biology, 4th ed. San Francisco: Benjamin Cummings. 776 p.

Moser WE. 1991. Leeches (Annelida:Hirudinea) in central and western Nebraska. Transactions of the Nebraska Academy of Sciences 18: 87–91.

Moser WE, Willis MS. 1994. Predation on gastropods by Placobdella spp. (Clitellata: Rhynchobdellida). American Midland Naturalist 132: 399–400.

Nico L. 1999. Nonindigenous aquatic species: Betta splendens Regan 1910. United States Geological Survey. Retrieved September 13, 2003 from the World Wide Web: http://nas.er.usgs.gov/queries/SpFactSheet.asp?speciesID=326

Pennak RW. 1953. Freshwater invertebrates of the United States. New York: The Ronald Press Co. 769 p.

APPENDIX 1

Knisely K. 2002. A student handbook for writing in biology. Massachusetts: Sinauer Associates, Inc. 205 p.

Style Manual Committee, Council of Biology Editors. 1994. Scientific style and format: The CBE manual for authors, editors, and publishers. 6th ed. Cambridge, MA: Cambridge University Press. 825 p.

APPENDIX 2

Darwin C. 1880. Sensitiveness of Plants to Light: Its Transmitted Effects. In: The Power of Movement in Plants. London: John Murray. 592 p.

APPENDIX 3

Bland RG, Jaques HE. 1978. How to know the insects, 3rd ed. Dubuque, IA: William C. Brown. 409 p.

Borror DJ, White RE. 1970. A field guide to insects: America north of Mexico. Boston: Houghton Mifflin Co. 404 p.

Centers for Disease Control and Prevention, Public Health Service, U.S. Department of Health and Human Services. 1960. Mosquitos of public health importance and their control. HHS Publication No. 86-8396.

Delettre Y, Morvan N. 2000. Dispersal of adult aquatic Chironomidae (Diptera) in agricultural landscapes. Freshwater Biology 44: 399–411.

Eddy S, Hodson AC. 1982. Taxonomic keys to the common animals of the north central states. Minneapolis, MN: Burgess Publishing Co. 205 p.

Huggins DG, Liechti PM, Ferrington LC. 1985. Guide to the freshwater invertebrates of the Midwest. University of Kansas, Technical Publication Number 11.

Lehmkuhl DM. 1979. How to know the aquatic insects. Dubuque, IA: William C. Brown. 168 p.

Marquardt WC, Demaree RS, Grieve RB. 2000. Parasitology and vector biology, 2nd ed. San Diego, CA: Harcourt Academic Press. 702 p.

Merritt RW, Cummins KW. 1996. An introduction to the aquatic insects of North America, 3rd ed. Dubuque, IA: Kendall/Hunt Publishing. 862 p.

Palmer EL. 1949. Fieldbook of natural history. New York: McGraw-Hill Book Co, Inc. 664 p.

Pennak RW. 1953. Freshwater invertebrates of the United States. New York: The Ronald Press Co. 769 p.

Reid GK. 1987. Pond life. New York: Golden Books. 160 p.

Roberts LS, Janovy J. 2000. Gerald D. Schmidt and Larry S. Roberts' foundations of parasitology, 6th ed. Boston: McGraw-Hill Higher Education. 670 p.

Thorp JH, Covich AP. 1991. Ecology and classification of North American freshwater invertebrates. San Diego, CA: Academic Press, Inc. 1056 p.

Turell MJ, Sardelis MR, Dohm DJ, O'Guinn ML. 2001. Potential North American vectors of West Nile virus. Annals of the New York Academy of Science 951: 317–324.

White DJ, Kramer LD, Backenson PB, Lukacik G, Johnson G, Oliver J, Howard JJ, Means RG, Eidson M, Gotham I, Kulasekera V, Campbell S. 2001. Mosquito surveillance and polymerase chain reaction detection of West Nile virus, New York State. Emerging Infectious Diseases 7: 643–649.

Art Sources

Ahmadjian V, Paracer S. 1986. Symbiosis: an introduction to biological associations. Hanover, NH: University Press of New England, for Clark University. 212 p.

[Anonymous]. 1988. Carolina *Drosophila* manual. Burlington, NC: Carolina Biological Supply Company. 31 p.

Audesirk T, Audesirk G, Byers BE. 2002. Biology: life on earth, 6th ed. Upper Saddle River, NJ: Prentice Hall. 892 p.

Bres M, Weisshaar A. 1998. Thinking about biology: an introductory biology laboratory manual. Upper Saddle River, NJ: Prentice Hall. 376 p.

Buffaloe ND. 1962. Englewood Cliffs, NJ: Principles of biology. Prentice Hall, Inc. 365 p.

Despommier DD, Gwadz RW, Hotez PJ, Knirsch CA. 2000. Parasitic diseases, 4th ed. New York: Apple Trees Productions, LLC. 345 p.

Fulton JF, Wilson LG. 1966. Selected readings in the history of physiology, 2nd ed. Springfield, IL: Charles C. Thomas Publishers, Ltd. 492 p.

Goodrich SG. 1859a. Illustrated natural history of the animal kingdom, being a systematic and popular description of the habits, structure, and classification of animals from the highest to the lowest forms, with their relations to agriculture, commerce, manufactures, and the arts, Vol. I. New York: Derby & Jackson. 680 p.

Goodrich SG. 1859b. Illustrated natural history of the animal kingdom, being a systematic and popular description of the habits, structure, and classification of animals from the highest to the lowest forms, with their relations to agriculture, commerce, manufactures, and the arts, Vol. II. New York: Derby & Jackson. 680 p.

Greenaway K. 1881. Mother Goose or the old nursery rhymes. London: Frederick Warne and Company, Ltd. 52 p.

Gunstream SE. 1999. Explorations in basic biology, 8th ed. Upper Saddle River, NJ: Prentice Hall. 481 p.

Gunstream SE. 2001. Biological explorations: a human approach, 4th ed. Upper Saddle River, NJ: Prentice Hall. 436 p.

Harter J. 1979. Animals: 1419 Copyright-free illustrations of mammals, birds, fish, insects, etc. A pictorial archive from nineteenth-century sources. New York: Dover Publications, Inc. 282 p.

Harter J. 1988. Plants: 2400 Copyright-free illustrations of flowers, trees, fruits and vegetables. New York: Dover Publications, Inc. 374 p.

Hickman CP, Roberts LS. Biology of animals, 6th ed. William C. Brown, 1994. Reproduced with permission of the McGraw-Hill Companies, 764 p.

Hooke R. 1665. Micrographia or some physiological descriptions of minute bodies made by magnifying glasses with observations and inquiries thereupon. Facsimile by Dover Publications, Inc. (1961), New York.

Huxley L. 1901. Life and letters of Thomas Henry Huxley, Vol. II. New York: D. Appleton & Company. 541 p.

Jensen MM, Wright DN, Robinson RA. 1997. Microbiology for the health sciences, 4th ed. Upper Saddle River, NJ: Prentice Hall. 530 p.

409

McClintic JR. 1980. Basic anatomy and physiology of the human body, 2nd ed. New York: John Wiley & Sons. 694 p.

Meyers M, Neafie RC, Marty AM, Wear DJ. 2000. Pathology of infectious diseases volume I: helminthiases. Washington, DC: Armed Forces Institute of Pathology, American Registry of Pathology. 530 p.

Moniez R. 1896. Traite de parasitologie: animale et vegetale, applique a la medecine.: Paris: Librairie J.-B. Bailliere et Fils. 680 p.

Morris D. 1962. The biology of art. London: Methuen & Co., Ltd. 176 p.

National Centre for Biotechnology Education, The University of Reading, U.K. 1997. Restriction analysis and electrophoresis of DNA. Burlington, NC: Carolina Biological Supply. 8 p.

Olsen O. 1967. Animal parasites: their biology and life cycles, 2nd ed. Minneapolis, MN: Burgess Publishing. 431 p.

Poirier FE. 1990. Understanding human evolution, 2nd ed. Englewood Cliffs, NJ: Prentice Hall. 356 p.

Rapoza M, Kreuzer H. 1999. Transformations: a teacher's manual. Burlington, NC: Carolina Biological Supply. 21 p.

Sealander JA. 1979. A guide to Arkansas mammals. Conway, AR: River Road Press. 313 p.

Steele JD, Jenks JWP. 1887. Popular zoology. New York: Chautauqua Press. 319 p.

Wallace AR. 1880. The Malay Archipelago: the land of the orang-utan and the bird of paradise, a narrative of travel, with studies of man and nature, 7th ed. London: Macmillan and Co. 653 p.

Wallace RL, Taylor WK. 1997. Invertebrate zoology: a laboratory manual, 5th ed. Upper Saddle River, NJ: Prentice Hall. 336 p.

Credits

10.2 Dennis and Kristen Richardson **10 end** Harter, 1988

CHAPTER 11

11.1-11.2 ©1997, University of Reading, Used by Permission of Carolina Biological Supply Company and the University of Reading
11.4 ©1997, University of Reading, Used by Permission of Carolina Biological Supply Company and the University of Reading
11.5 ©1999, Maria Rapoza and Helen Kreuzer, Carolina Biological Supply Company, Used by permission
11.7-11.8 ©1999, Maria Rapoza and Helen Kreuzer, Carolina Biological Supply Company, Used by permission **11 end** Susan Jessup

CHAPTER 12

12.1a-b Mary Evans Picture Library
12.3a Neg.# 326865, courtesy the Library, American Museum of Natural History
12.3b Neg.# 326866, courtesy the Library, American Museum of Natural History
12.7-12.9 Gunstream, 2001/Mary Dersch, CMI/The Professional Edge **12 end** Wallace, 1880

CHAPTER 13

13.1 Neg.# 334420 courtesy the Library, American Museum of Natural History
13.2 Courtesy Opossum Society of the United States/Leslie Hall
13.3-13.7 Barbara Nitchke
13.8a-c Dennis and Kristen Richardson
13.9 Barbara Nitchke **13 end** Harter, 1988

CHAPTER 14

14.1 Centers for Disease Control/PHIL #1792
14.2 Centers for Disease Control/PHIL # 1803
14.3 Jensen, et al., 1997
14.4 Dennis and Kristen Richardson
14.5 Centers for Disease Control/PHIL #1915
14.6a Centers for Disease Control/PHIL #1957
14.6b Centers for Disease Control/PHIL #4139
14.7 Greenaway, 1880 **14 end** Woodcut from Johannis Geiler de Kaiserberg's *Sermones Argentinen predicatoris fructuosissimi*. Strassburg (1518). Beinecke Rare Book and Manuscript Library, Yale University

CHAPTER 15

15.1 Wallace and Taylor, 1997/Michele Johnson
15.2 Susan Jessup
15.3 Wallace and Taylor, 1997/Michele Johnson
15.4-15.5 BioScience 51: 827-841
15.6 Susan Jessup
15.7 Wallace and Taylor, 1997/Michele Johnson
15.8 Centers for Disease Control/PHIL #613
15.9 Wallace and Taylor, 1997/Michele Johnson
15.10 Courtesy of the TDR Image Library #9100165
15.11 Despommier, et al., 2000
15.12 ©1979, Carolina Biological Supply Company, Used by permission
15.13 ©1996, Carolina Biological Supply Company, Used by permission
15.14-15.15 Gunstream, 1999/Mary Dersch, CMI/The Professional Edge **15 end** Harter, 1988

CHAPTER 16

16.2 Audesirk, et al., 2002
16. Audesirk, et al., 2002/John Durham/Science Photo Library
16.4 Dennis and Kristen Richardson
16.6a-b Harter, 1988
16.8 Dennis and Kristen Richardson **16 end** Harter, 1988

CHAPTER 17

17.3 Audesirk, et al., 2002/Carolina/Phototake NYC
17.4 Dennis and Kristen Richardson
17.5 Neg.#992(1), courtesy the Library, American Museum of Natural History
17.6 Dennis and Kristen Richardson
17.7 Audesirk, et al., 2002/Milton Rand/Tom Stack and Associates **17 end** Harter, 1988

CHAPTER 18

18.4-18.10 Gunstream, 1999/Mary Dersch, CMI/The Professional Edge **18 end** Barbara Nitchke

CHAPTER 19

19.2 Dennis and Kristen Richardson
19.4-19.6 Wallace and Taylor, 1997/Michele Johnson
19.9 Wallace and Taylor, 1997/Michele Johnson

19.10 Courtesy Dr. Thomas J. Nolan
19.11 Susan Jessup
19.12 Dennis and Kristen Richardson
19.13 Dennis and Kristen Richardson
19.14a *Journal of Parasitology* 81: 767-772
19.14b ©T.J. Ulrich/VIREO
19.15, 19.16 Susan Jessup
19.17 Wallace and Taylor, 1997/Michele Johnson
19.19a-b Dennis and Kristen Richardson
19.20 Courtesy Dr. B.W. Payton **19 end** Goodrich, 1859b

CHAPTER 20

20.1-20.2 Wallace and Taylor, 1997/Michele Johnson
20.3a-b Dennis and Kristen Richardson
20.4a Centers for Disease Control/PHIL #1125
20.4b *Journal of the American Medical Association* 188:33-36
20.5 Dennis and Kristen Richardson
20.6 University of Nebraska Department of Entomology/Jim Kalisch
20.7-20.9a-b Susan Jessup
20.10 Dennis and Kristen Richardson
20.11a-b University of Nebraska Department of Entomology/Jim Kalisch
20.12a-b Wallace and Taylor, 1997/Michele Johnson
20.13 Goodrich, 1859b
20.14 Susan Jessup **20 end** Harter, 1979

CHAPTER 21

21.2 Wallace and Taylor, 1997/Michele Johnson
21.4a-d ©Wildlife Conservation Society, Headquartered at the Bronx Zoo
21.5 Hickman, C.P. Jr., & Larry S. Roberts. *Biology of Animals*, 6th ed. William C. Brown, 1994. Reproduced with permission of the McGraw-Hill Companies
21.6 Poirier, 1990
21.7 ©Diane Shapiro **21 end** Goodrich, 1859b

CHAPTER 22

22.2 Gunstream, 1999/Mary Dersch, CMI/The Professional Edge
22.6 Gunstream, 1999/Mary Dersch, CMI/The Professional Edge
22.9 Kodak, Inc. Used by permission **22 end** From Vesalius, *Fabrica* (1543). Graphically remastered image courtesy of Daniel H. Garrison.

CHAPTER 23

23.4 Dennis and Kristen Richardson
23.6a-b Gunstream, 1999/Mary Dersch, CMI/The Professional Edge
23.8 Gunstream, 1999/Mary Dersch, CMI/The Professional Edge
23 end Harter, 1979

CHAPTER 24

24.2 Audesirk, et al., 2002
24.3 Gunstream, 1999/Mary Dersch, CMI/The Professional Edge
24.5 Audesirk, et al., 2002 **24 end** From Fulton, John F. and Leonard G. Wilson. *Selected Readings in the History of Physiology*, 2nd ed. 1966. Courtesy of Charles C. Thomas, Publisher, Ltd., Springfield, IL.

CHAPTER 25

25.1 Centers for Disease Control/PHIL #2611
25.2-25.5 Susan Jessup
25.7 Susan Jessup
25.8 Dennis and Kristen Richardson
25.9b Dennis and Kristen Richardson
25.10 Susan Jessup **25 end** Harter, 1979

CHAPTER 26

26 end Barbara Nitchke

CHAPTER 27

27.1-27.2 Dennis and Kristen Richardson
27.8-27.9 Courtesy John A. Sealander and the Arkansas Natural Heritage Commission **27 end** Goodrich, 1859b

CHAPTER 28

28.2 Goodrich, 1859b
28.3a-h Dennis and Kristen Richardson **28 end** Harter, 1979

CHAPTER 29

29.1 *Journal of Parasitology* 58:169-177
29.2 Ahmadjian, Vernon, and Surindar Paracer, Fig. **8.5** from *An Introduction to Biological Associations* ©1986 by Clark University, reprinted with permission of University Press of New England

29.3 Dennis and Kristen Richardson
29.4-29.5 Drawing by Jon Karapelou, from *Parasitic Diseases*, 4th ed. 2000. Apple Trees Productions, LLC
29.6 Reprinted from *Trends in Parasitology* (formerly *Parasitology Today*) Vol. 4, Flisser, A. *Neurocysticercosis in Mexico*, 131-137. ©1988, with permission from Elsevier Science
29.7 Drawing by Jon Karapelou, from *Parasitic Diseases*, 4th ed. 2000. Apple Trees Productions, LLC
29.8 Courtesy Armed Forces Institute of Pathology/M.M. Wong. From Meyers, et al, 2000
29.9a-b Olsen 1967, Burgess
29.10a Drawing by Jon Karapelou, from *Parasitic Diseases*, 4th ed. 2000. Apple Trees Productions, LLC
29.10b *New England Journal of Medicine* 328:927. ©1993, Massachusetts Medical Society. All Rights Reserved
29.11 Drawing by Jon Karapelou, from *Parasitic Diseases*, 4th ed. 2000. Apple Trees Productions, LLC

29.12a-b Photos by Dr. Nora Labiano-Abello, Courtesy Peter Hotez
29.13a Dennis and Kristen Richardson
29.13b Courtesy Armed Forces Institute of Pathology, Neg. #92-6805, 1373174. From Meyers, et al., 2000
29.14 Dennis and Kristen Richardson
29.15 National Geographic Society #762975/Karen Kasmauski **29 end** Moniez, 1896

CHAPTER 30

30.1 Dennis and Kristen Richardson
30.2 Susan Jessup
30 end Goodrich, 1859b

APPENDIX 3

1-5 Dennis & Kristen Richardson
6 University of Nebraska Department of Entomology/Jim Kalisch
7-10 Dennis & Kristen Richardson
11 Courtesy Laura M. Duclos
12-57 Dennis & Kristen Richardson